MATHEMATICS FOR PLUMBERS AND PIPEFITTERS

Seventh Edition

D0207370

MATHEMATICS FOR PLUMBERS AND PIPEFITTERS

Seventh Edition

Lee Smith

Australia Brazil Canada Mexico Singapore Spain United Kingdom United States

DELMAR
CENGAGE Learning™

Mathematics for Plumbers and Pipefitters, Seventh Edition

Lee Smith

Vice President, Technology and Trades ABU:
David Garza

Director of Learning Solutions:
Sandy Clark

Managing Editor:
Larry Main

Executive Editor:
Steve Helba

Development:
Dawn Daugherty

Marketing Director:
Deborah Yarnell

Marketing Manager:
Kevin Rivenberg

Marketing Coordinator:
Mark Pierro

Director of Production:
Patty Stephan

Production Manager:
Stacy Masucci

Content Project Manager:
Philippa Lehar

Editorial Assistant:
Dawn Daugherty

© 2008 Delmar Cengage Learning

ALL RIGHTS RESERVED. No part of this work covered by the copyright herein may be reproduced, transmitted, stored or used in any form or by any means graphic, electronic, or mechanical, including but not limited to photocopying, recording, scanning, digitizing, taping, Web distribution, information networks, or information storage and retrieval systems, except as permitted under Section 107 or 108 of the 1976 United States Copyright Act, without the prior written permission of the publisher.

For product information and technology assistance, contact us at
Cengage Learning Customer & Sales Support, 1-800-354-9706

For permission to use material from this text or product,
submit all requests online at **cengage.com/permissions**
Further permissions questions can be emailed to
permissionrequest@cengage.com

Library of Congress Control Number: 2007015451

ISBN-13: 978-1-4283-0461-1

ISBN-10: 1-4283-0461-4

Delmar Cengage Learning
5 Maxwell Drive
Clifton Park, NY 12065-2919
USA

Cengage Learning products are represented in Canada by Nelson Education, Ltd.

For your lifelong learning solutions, visit **delmar.cengage.com**

Visit our corporate website at **www.cengage.com**

Notice to the Reader
Publisher does not warrant or guarantee any of the products described herein or perform any independent analysis in connection with any of the product information contained herein. Publisher does not assume, and expressly disclaims, any obligation to obtain and include information other than that provided to it by the manufacturer. The reader is expressly warned to consider and adopt all safety precautions that might be indicated by the activities described herein and to avoid all potential hazards. By following the instructions contained herein, the reader willingly assumes all risks in connection with such instructions. The publisher makes no representations or warranties of any kind, including but not limited to, the warranties of fitness for particular purpose or merchantability, nor are any such representations implied with respect to the material set forth herein, and the publisher takes no responsibility with respect to such material. The publisher shall not be liable for any special, consequential, or exemplary damages resulting, in whole or part, from the readers' use of, or reliance upon, this material.

Printed in the United States of America
6 7 11 10

CONTENTS

Section One REVIEW OF BASIC FUNDAMENTALS

Section Two PIPE LENGTH CALCULATIONS

Section Three SHEET METAL WORK

Section Four VOLUMES, PRESSURES, CAPACITIES

Section Five PHYSICS

Section Six HEATING

Section Seven THE BUILDER'S LEVEL

Appendix A

Appendix B

Mathematics for Plumbers and Pipefitters is a specialized course to fill the need for a trade-related mathematics text in the plumbing and pipefitting fields. This is the very best of its kind and has been rated five stars by Amazon.com.

A variety of currently used fitting materials demonstrate the various problem concepts. Emphasis is placed on providing the student with a problem-solving methodology, which can be universally applied to future materials and job situations, as well as existing ones.

NEW TO THIS EDITION

Section One, Review of Basic Fundamentals, has been extensively updated for clarity and several illustrations have been improved. The example on converting has been rewritten.

In Section Two, Pipe Length Calculations, there are several minor changes throughout.

Section Five, Physics, is a new section with Units on Mechanical Advantage, States of Matter, and Gases Under Pressure.

Section Six, Heating, and Section Seven, The Builder's Level, were previously Section Five and Section Six.

Also new to this edition is an Instructor's E-Resource with ExamView testbank, PowerPoint slides, and the Instructor's Guide.

ABOUT THE AUTHOR

Lee Smith has many years of practical experience in the plumbing and pipefitting trades. He is now retired and was a registered master plumber and vocational instructor of all age groups, including apprentices. He holds certificates and degrees in education and applied science and technology.

In addition, Lee Smith is the original author of the textbook *Plumbing Technology: Design and Installation.*

ACKNOWLEDGMENTS

The author and Delmar Cengage Learning would like to acknowledge and thank the instructors who reviewed the book and provided suggestions and comments. Thanks go to:

John Breece
Red Rocks Community College
Denver, CO

Bruce Roberts
Oklahoma State Technical College
Okmulgee, OK

John Williams
Kirkwood Community College
Cedar Rapids, IA

Jeff Brower
Northwest Technical College
Bemidji, MN

Special thanks and appreciation go to the contributors who worked on the Instructor's E-Resource CD-ROM. Mike Standifird of Angelina College in Lufkin, Texas, prepared the ExamView testbank and Chimborazo prepared the PowerPoint slides. Thank you Mike and Paul Nagin from Chimborazo.

Section One
REVIEW OF BASIC FUNDAMENTALS

Unit 1

Some Basic Rules for Mathematics

Objectives

- Define the kinds of measurements used.
- Review what the mathematical symbols mean.

Introduction

Before going into the solutions to problems that apply to the plumbing and pipe fitting trades, it is appropriate to review some of the mathematical basics. The student is encouraged to make sure that these are completely understood before proceeding with the rest of this book. When the student has difficulty arriving at the correct answers in the later units, the reason for the difficulty will almost certainly be found in these first six units. These units review the basics.

This is not a textbook for teaching basic mathematics. As is true for all texts of this kind, some assumptions have to be made about the student's previous experience in the subject matter. In this case the assumptions are:

- The ability to do the basic mathematical operations of addition, subtraction, multiplication, and division of positive and negative integers, fractions, percentages, and decimal fractions.
- Basic experience with substituting values into a formula and deriving the solution.
- Familiarity with measurements of length, angle, weight, and volume.
- The ability to visualize drawings and sketches and to extract information from charts and graphs.
- An understanding of positive and negative quantities and the number line.

Work out the problems as you go along *even if the answer is given.* Invent similar problems and work them out. Above all, if you feel lost, get help now. It probably won't become more clear later on. No sports star got as good as he/she is without hours and hours of practice. Mathematics also requires practice.

Kinds of Measurements: A Review of Terminology

All mathematics must start with some measurements, or at least numbers which are agreed upon for their accuracy. This textbook will generally refer to the **English style** of measurement, that is, measurements in feet, inches, and fractional parts of an inch, e.g., 3 feet 2 and ¾ inches or 3' 2¾". The **metric style** of measurement will be used occasionally where it is applicable.

Linear Measure

Linear measure is simply **a measurement of a distance.** It is usually obtained using a measuring device called a measuring rule, or tape. It can also be read from architectural drawings (blueprints). The distance from here to there is a linear measure. A **dimension** is a linear measurement. The length of a box is a dimension and also a linear measurement. However, it is well to understand that the distance around a water heater is also a linear measurement. If it can be measured meaningfully with a steel tape, that measurement is linear. Even if we must bend our rule or steel tape to take the measurement, it is still a linear measurement.

In linear measure, 1 foot is equal to 12 inches. This is a **one-dimensional measurement.**

Square or Area Measure

When we need to cover a surface, we may need to know how many square feet or how many square inches it contains. This is area measure. Thus *square measure describes the size of the surface of an object.* If a square measure or a cubic measure (see following) is not a whole number, it should be expressed either as a mixed fraction or in decimal form. For example, if we need six and two-thirds square feet of plastic shower pan material and we express the measure as 6 square feet plus 96 square inches, the vendor who must supply the material will probably not understand our measurement because he/she is not accustomed to handling area measure in this form. We should specify either $6\frac{2}{3}$ or 6.667 square feet.

Square measure is obtained by using math tools on linear measurements. When dealing with area or square measure, you will generally use two measurements at right angles (like the right edge and bottom edge of this page) to determine the answer.

In square measure, 1 square foot is equal to 144 square inches (12 times 12). This is a **measurement in two dimensions.** Notice that two linear twelves, the number of inches in a linear foot, are multiplied together to find the number of square inches in a square foot.

Cubic or Volume Measure

To find out how many gallons a tank will hold, we need to know a cubic measurement. This is determined by taking linear measurements of the tank and using math tools to calculate the cubic measure. *Cubic measure tells us how much a container will hold.*

In cubic measure, 1 cubic foot is equal to 1,728 cubic inches (12 times 12 times 12). This is a **three-dimensional measurement.** Notice that three linear twelves are multiplied together to find the number of cubic inches in a cubic foot. When calculating volume or cubic measure of rectangular tanks, you will use three values multiplied together (like the edges of a cardboard box).

Weight and Miscellaneous Measure

These will be explained as necessary in the lessons to which they apply.

Answer these questions:

1. What kind of measure is used to describe the size of a car's fuel tank?
2. What kind of measure is used to order sheet lead?
3. The distance between cities is expressed with what kind of measure?

Symbols: Mathematics in Algebraic Form

Plumbers and other craftspeople are called upon to use **formulas** in the planning stages of tasks. These are always simple formulas and require no heavy math logic. Consider the grade school problem, $10 = 2 \times 5$. If we needed to use a formula that stated that $A = F \times S$ and were told that F was equal to 2 and S was equal to 5, then the problem is precisely $10 = 2 \times 5$. It's just a simple substitution of real values for the letters. The mathematician who invented the formula $A = F \times S$ used the first letters of some words that he had in mind: ANSWER, FIRST, and SECOND. In other words, "The answer is equal to the first value times the second value." It's that simple.

In algebra, symbols are used to represent values. The symbols, usually letters of the alphabet, stand in place of values. Some of these values are predetermined and are called constants, like the speed of light. Others vary with each problem, like the length of a room. These are called **variables.**

These symbols may be assembled into an **equation.** The equation is a problem solving tool. Scientists and mathematicians use the logic rules of math and algebra to manipulate ideas. Plumbers and pipefitters can use equations and formulas to find answers to complex problems.

A letter of the alphabet, a word, or a short phrase on a printed page: these are symbols. When we see these symbols our brains use them to give access to information stored in our memories. The word-symbol "war" brings certain thoughts and feelings into focus. The word-symbol "kitten" brings other thoughts and feelings to the front of our consciousness.

Symbolism can be very helpful. If we think of the phrase symbol "pipe wrench," a mental picture of the tool may instantly flash into our mind's. That is important, but even more important is that the correct usage of the tool—if we have learned about pipe wrenches—is made available to us. The science called mathematics has a small number of fixed word, phrase, and character symbols that are used to stand in the place of math usages in the same manner as the phrase symbol "pipe wrench" stands for and gives access to the memories we have about pipe wrench usage. If there is no usable mental picture or process produced when the word "exponent" or any of the other mathematical terms are seen or heard, then the math learning process is incomplete in that area and is in need of some repair. When the symbols and their processes are learned, then plumbing and pipefitting math becomes easy.

Most of us are familiar with the math symbols +, –, ×, /, and =. These are the common indicators, or *operators,* for the math processes *add, subtract, multiply,* and *divide;* the equality symbol = is also shown. There are many more mathematics symbols and some of them stand for the same math operations that those above do. These always mean the same math operation wherever they appear.

The symbols above should not be changed because they have universally understood usage and meaning. Symbols can be "made up" as they are needed, so long as the symbol used in this manner is not one of the above or other universally accepted math symbols. For instance, the math statement $\mathbf{R = F \times S}$ can be made. Because we understand the universal symbols, it can be immediately said that this means "the value R is equal to the value of F times the value of S." This is because the mathematical meanings of = and × are known. However, we don't know what **R** and **F** and **S** stand for without being told, because these are **variables.** Those symbols and the values that they represent which are understood and accepted are called **constants.**

If the statement is made that k = 5 and t = 9 and then the problem X = k + t is presented, we can then substitute the values given for the variables k and t and find the answer X.

> **This symbol ⇒ means that the equation to the left of it, after some computation, yields the equation on the right.**

$$X = k + t \Rightarrow X = 5 + 9 \Rightarrow X = 14$$

Sample Problems

In each equation, $a = 4$, $b = 5$, $c = 6$. Solve each for a numerical answer.

1. $a + b - c = x$
   $4 + 5 - 6 = x$
   $3 = x$

2. $2(a + b) = m$
   $2(4 + 5) = m$
   $2 \times 9 = m$
   $18 = m$

3. $\frac{7 + a}{2} = p$
   $\frac{7 + 4}{2} = p$
   $\frac{11}{2} = p$
   $5\frac{1}{2} = p$

4. $\frac{ab}{c} = y$
   $\frac{4 \times 5}{6} = y$
   $3\frac{1}{3} = y$

5. $2a + b = n$
   $2 \times 4 + 5 = n$
   $8 + 5 = n$
   $13 = n$

6. $\frac{7}{2} + a = q$
   $\frac{7}{2} + 4 = q$
   $3.5 + 4 = q$
   $7.5 = q$

Review

Simple addition:

Two years ago we had 20 trucks on the road. We purchased 8 more trucks this year, so now we have 28 trucks.

20 trucks two years ago
+ 8 trucks purchased this year
28 trucks this year

Simple subtraction:

Two years ago we had 20 trucks on the road. Last year 5 of them were permanently disabled, so now we only have 15 trucks.

20 trucks two years ago
– 5 trucks disabled
15 trucks this year

Addition of negative numbers:

Last year we lost 5 trucks to road hazards, and this year we lost 3. That puts us down 8 trucks from two years ago.

(–5) trucks gone last year
+ (–3) more trucks gone this year
–8 trucks lost to our business

> **Note: A number with a negative (–) or positive (+) symbol preceding it is a signed number. These symbols/signs are not the same as mathematics operators like subtract or add.**

Addition of negative and positive numbers:

Last year we lost 5 trucks to road hazards, but this year we bought 3 new ones. That puts us down only 2 trucks from two years ago.

$$\begin{array}{rl} (-5) & \text{trucks lost to road hazards last year} \\ \underline{+\,(+3)} & \text{but we bought three new ones this year} \\ -2 & \text{down from two years ago} \end{array}$$

Subtracting a negative number:

This year we're short 8 trucks from previous *losses*, but perhaps we shouldn't count 5 of those *losses* because they were reimbursed by insurance. So let's subtract those 5 *losses* from the books. *Note: When a number is positive, it requires no sign, and the parentheses merely serve to separate the subtraction operator (–) from the sign of the number.*

$$\begin{array}{rl} (-8) & \text{We previously thought we had a loss of eight trucks.} \\ -\,(-5) & \text{We were reimbursed for some, so let's subtract those losses.} \\ -3 & \text{The rule is to change the sign of the number subtracted,} \\ & \text{whether positive or negative, and then proceed as if it were} \\ & \text{an addition problem.} \end{array}$$

So: $\begin{array}{r} (-8) \\ \underline{-\,(-5)} \\ -3 \end{array}$ becomes $\begin{array}{r} (-8) \\ \underline{+\,(+5)} \\ -3 \end{array}$ becomes $\begin{array}{r} -8 \\ \underline{+\,5} \\ -3 \end{array}$

Subtraction of negative numbers can be tricky, and in stating the previous problem, we hinted at a better way. Suppose we said: So this year we're short 8 trucks from previous losses, but we've been reimbursed for 5. Let's see, we were down 8, so *adding* back 5 would bring us up to being only 3 trucks short [(–8) + (5) = (–3)]. It helps to visualize the famous number line from grade school:

THE NUMBER LINE

Numbers less than 1 | Numbers more than 1

–9 –8 –7 –6 –5 –4 –3 –2 –1 0 1 2 3 4 5 6 7 8 9

Using the number line, we can say we lost 2 trucks last year (move your pencil point from 0 to –2) and then 3 more this year (move your pencil point 3 numbers going left, subtracting, to –5). But we were reimbursed for 4 trucks, so we can hardly count them as a loss. (Count to the right now, because we're adding.) Move 4 numbers to the right, and you end up on –1. So actually our losses amount to only one truck.

The number line is very useful when doing measurement math on the job. You can use it to add and subtract measurements accurately, avoid borrowing inches, and be automatically correct for unlike fractions without using anything but two folding rules or tapes. (Rules are better because they have no tendency to abruptly roll up inside the case.) Lay the rules side by side in opposite directions. Let's say you have an overall measurement of 52¼ inches and you want to allow (subtract) for a fitting that takes up a space of 6 9/16 inches. Measure from 52¼ inches on one rule in the minus direction with your other rule, and read the answer directly on the first rule. No math, yet accurate. What was your answer?

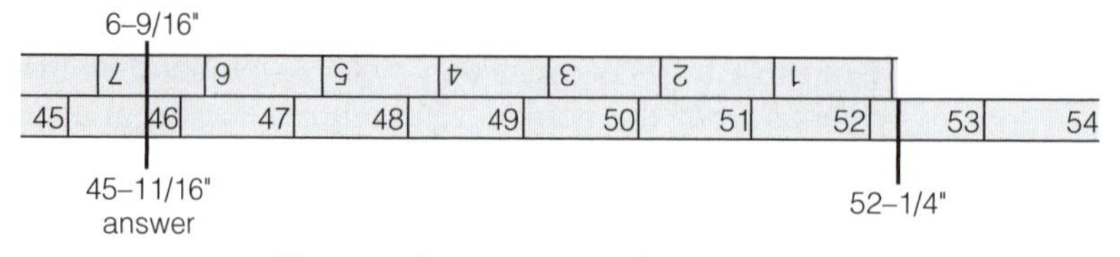

Use two rules as a number-line calculator

Multiplication:

When we multiply two numbers together, we are adding the first number to itself as many times as the second number or adding the second number to itself as many times as the first number. It's the same regardless of which number we use as the multiplier.

$$(3 \times 4 = 4 + 4 + 4) \text{ or } (4 \times 3 = 3 + 3 + 3 + 3)$$

With multiplication, the order of doing the work makes no difference in the answer.

$$2 \times 3 \times 5 \times 8 = 240$$
$$3 \times 8 \times 2 \times 5 = 240$$
$$5 \times 8 \times 3 \times 2 = 240$$

Signed numbers *do* make a difference in the answer.

$$(-2) \times 3 \times 5 \times 8 = -240$$
$$(-2) \times (-3) \times 5 \times 8 = 240$$
$$(-2) \times (-3) \times (-5) \times 8 = -240$$
$$(-2) \times (-3) \times (-5) \times (-8) = 240$$

What do you notice about the signs of the problem parts and the signs of the answer parts that will help you predict the sign of the answer *before* you multiply? Can you just look at the signs, multiply all the numbers together without signs, and then put the correct sign to the answer from now on?

Division:

Division can be expressed like this: $\frac{104}{9}$

How many times can I add the bottom of the fraction to itself without having a larger answer than the top of the fraction? Or, in the parlance of mathematicians: How many times can I add the divisor to itself before the quotient is larger than the dividend? Eleven times. Here is a series of division problems:

$$104 \div 9 = 11.556$$
$$-104 \div 9 = -11.556$$
$$104 \div -9 = -11.556$$
$$-104 \div -9 = 11.556$$

Looking at this series of division problems, what can you predict about the answer's sign (positive or negative) in the following question?

Question: $-104 \div -2 \div 5 \div 6 = (?1.733)$.

$$2 \div 3 \div 4 \div 5 = 0.0333$$
$$3 \div 4 \div 5 \div 2 = 0.075$$
$$4 \div 5 \div 2 \div 3 = 0.133$$

Look again at this series of division problems. What would you say about the order of the divisors in a division problem affecting the answer?

The student must be aware that the order of doing the work can make a difference. This will be addressed in a later unit.

Angles:

An angle is a direction from a reference starting line. In the pipe trades, we define angles in a number of ways: degrees of a circle, fractional divisions of a circle, and by fitting name.

Degrees of a circle: There are 360 degrees in a circle. If we stand and fix our eyes forward on a small object and turn our body around without advancing in any direction until we are once again seeing our starting point straight on, we have turned 360 degrees. The line from our eyes to the small object is our reference line in this case. If we had turned halfway around, we would have turned 180 degrees ($360 \times \frac{1}{2}$) and would be looking in the opposite direction from that at which we started. If we had turned a quarter of the way around, we would have turned 90 degrees ($360 \times \frac{1}{4}$). If we had turned an eighth of the way around, we would have turned 45 degrees ($360 \times \frac{1}{8}$). We could also use a compass to measure the stopping points in our turn because the compass is divided into 360 degrees and the marked points of it are arranged at angles useful to the craftsperson.

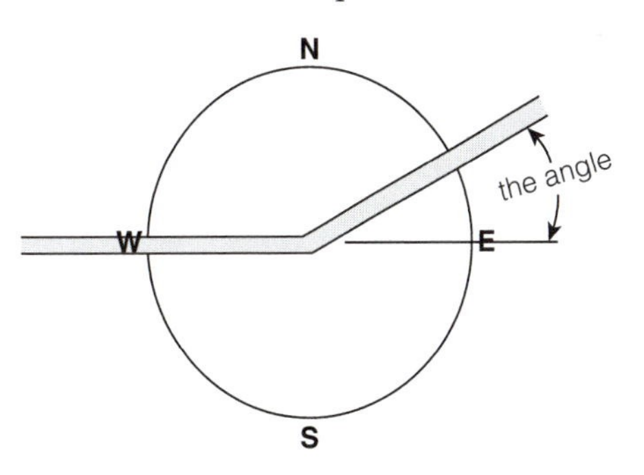

Fractional divisions of a circle: The quarter bend, the eighth bend, the sixteenth bend, and the thirty-second bend are fittings that turn the direction of our pipeline in the portion of a circle that their names indicate. Other bends are also described in this fashion.

Fitting name: A sweep turns the direction of the pipe 90 degrees. A return bend turns the direction of the pipe back on itself, 180 degrees. This is useful for laying out heating pipe.

By far, the most frequent angles that the plumber/pipefitter uses are the 90° angle (90° elbow, sweep, quarter bend, rectangle, rectangular prism) and the 45° angle (45° elbow, eighth bend, triangle, square).

Pipework geometry:

The importance of having the ability to visualize the basic geometric shapes and to use them not only in cutting and fitting pipe, but also in fixture and appliance layout, heating design, pipe and duct routing, trenching, and many other applications can hardly be overestimated.

Walls, floors, and ceilings are rectangles. Buildings are rectangular prisms, as are the vast majority of interior spaces. Pipe and duct layouts often form geometric shapes. Drainage piping has fall; one end must be higher than the other to function correctly. From the side, this forms a right triangle. Pipe offsets form right triangles also. The sun beating on a solar collector forms an imaginary right triangle with a tangent to the circle of the earth. Wind against a wall produces a force vector that is a triangle also. Right triangles are half of a rectangle. Any triangle is half of a parallelogram. Pipe bends are portions of a circle. Tanks may have circular ends. Pipe ends are circular. Belt pulleys are circles. Geometry, geometry—it's all based on geometry.

Exercise

The formula for the distance around a circle (circumference) is $\pi \times D$ where $\pi = 3.14$ and D is the diameter (greatest distance across) of the circle.

1. A circle with a D of 5 must have a circumference of ____________________ .
2. A circle with a D of 3 must have a circumference of ____________________ .
3. A circle with a D of 12.19 must have a circumference of ________________ .

The formula for the area of a rectangle is Side A × Side B = Area where Side A and Side B are sides joined at one end.

4. A rectangular yard whose sides are 34 feet and 50 feet and 34 feet and 50 feet would need ________________ square feet of grass sod to cover it completely.
5. A cellar floor whose sides are 122 feet and 70 feet and 122 feet and 70 feet would need enough concrete paint to cover ________________ square feet.

Add:

6. $\begin{array}{r} -4 \\ \underline{-2} \end{array}$ 7. $\begin{array}{r} 8 \\ \underline{-11} \end{array}$ 8. (−1) + 3 + 9 + (−2) = ________________

Subtract:

9. $\begin{array}{r} -4 \\ \underline{-2} \end{array}$ 10. $\begin{array}{r} 8 \\ \underline{-11} \end{array}$ 11. $\begin{array}{r} \mathbf{-1} \\ \underline{\mathbf{-1}} \end{array}$

Divide:

12. 12 / 3 = ________________ 13. 12 / 12 / 3 = ________________

Unit 2

Formulas

Objectives

- Define the use of symbols in mathematics.
- Review math procedure and math precedence.
- Relate geometry to piping mathematics.

Symbols in Formulas: Applying the Symbols

Whoever makes up a formula has the right to assign whatever variable symbols to quantities that he or she wishes, so long as they are not any of the universally accepted math symbols. However, the maker of the formula must supply this information with the formula. An important rule is that a symbol can have only one meaning in any one problem, i.e., if there is more than one **K** in the formula, then **K** always represents the same value in that formula. When an **equation** (formulas are in the form of equations) is confusing, examine all of the accompanying information for the meaning or values of these variables.

A mathematical equation states an equality. This means that the mathematical quantity on the left side of the = sign is equal to the mathematical quantity on the right side. In the case of the equation $\mathbf{R = F \times S}$, this means that **R** is equal to whatever the answer is when we multiply **F** and **S** together. The value on the left side of the equal symbol is the same as the value to the right of the equal symbol. Can you see that if the assumption is made that the variables in that equation are positive whole numbers (1, 2, 3 . . . etc.) then neither **F** nor **S** can be larger than **R**?

Math Precedence: Working Out Formulas

There is an order in which mathematical problems must be solved to arrive at the correct answer consistently.

Multiply $4 \times 6 \times 8$. Now multiply $8 \times 6 \times 4$. Are the answers the same?

Add $4 + 6 + 8$. Now add $8 + 6 + 4$. Are the answers the same?

Subtract $4 - 6 - 8$. Now subtract $8 - 6 - 4$. Are the answers the same?

Divide $4 / 6 / 8$. Now divide $8 / 6 / 4$. Are the answers the same?

As you can see, it can make a difference depending on what the math operator is and the order in which numbers are acted upon. **Math operators:** +, ×, –, /, etc.

When there is no math operator between two values, this calls for multiplication (ab = 41). A dot between two values is also considered a multiplication operator ($\pi \cdot D = 9.3126$).

We will refer to parentheses, (); brackets, []; and curly braces, {}, simply as "brackets," for simplicity.

Powers and Roots

A number that is "raised to a power" has a superscript value appended to its right indicating to which power it must be raised (4^2, r^2). The value of 5^4 (five raised to the fourth power) is 625. If we suddenly had an urge to know the fourth root of 625 we would ask ourselves: What number when multiplied by itself 4 times ($N \times N \times N \times N$), would be equal to 625? By trial and error, we would find our answer, 5.

The most frequent use of powers and roots is with values taken to their second powers or second roots. These are given the familiar names of squares and square roots (0.783^2 and $\sqrt{0.783}$). The process for deriving square roots manually is fully explained in Unit 4. Cubed numbers, cube roots (1.414^3 and $\sqrt[3]{1.414}$), and all higher powers and roots have a lesser use in vocational math.

Solving Formulas

What is 1.414 squared? Extra credit: What would be the length of the hypotenuse of a right triangle whose other sides are both equal to 1?

Rule 1

Do all work in brackets *as if these were separate math problems* and then replace the bracketed quantity with the answer to the mini-problem it represents.

Example

$(3 + 2) \times 9 \Rightarrow 5 \times 9 \Rightarrow 45$

Rule 2

Always work from the innermost set of brackets outward.

Example

$[(3 + 2) \times 4] \times 9 \Rightarrow [5 \times 4] \times 9 \Rightarrow 20 \times 9 \Rightarrow 180$

Rule 3

Work from left to right, but do division and multiplication before addition and subtraction. (Also check rules 4 and 5, which follow.)

Do this problem: $3 + 2 \times 9$. Did you get 21? That's correct! Because multiplication takes precedence over addition, you must multiply 2×9 before adding 3.

Rule 4

Powers of numbers and roots of numbers take precedence over other operations, including multiplication and division. Calculate powers and roots within brackets first, then in the rest of the equation before proceeding with other operations.

Example

$[(3 + 2) \times 4 - 3^2] \times 9$

a. Do the (3 + 2) first because this is the innermost bracket set.

Now we have $[5 \times 4 - 3^2] \times 9$

b. Do the powers work in the last bracket set next.

Now we have $[5 \times 4 - 9] \times 9$

c. Complete the work in the last bracket set, multiplication first.

Now we have 11×9

d. Complete the calculation. $11 \times 9 = 99$

Rule 5

A long division bar implies problem grouping in the same manner as brackets do.

Do the work above and below the bar separately and then do the division by dividing the remaining value above the bar by the value below the bar.

Example

$$\frac{124 \times 5}{5^4} \Rightarrow \frac{625}{5^4} \Rightarrow \frac{625}{625} \Rightarrow \frac{1}{1} \Rightarrow 1$$

Practice Exercises

1. $3 + 3 \times 4 =$
2. $(3 + 3) \times 4 =$
3. $3 + (3 \times 4) =$
4. $3 + 3 \times 4^2 =$
5. $3 + (5 + 4)^2 =$
6. $(3 + 5) + 4^2 =$
7. $3 + (5 + 4^2) =$
8. $3 + 4 \times 3 - 6 =$
9. $3 - 6 + 4 \times 3 =$
10. $(3 - 6 + 4) \times 3 =$
11. $3 - (6 + 4 \times 3) =$
12. $3 - [(6 + 4) \times 3]$
13. $\frac{11 \times 6^2}{3^2}$

Geometry in Plumbing: Visualize the Geometric Shape

Geometry has a large and absolutely essential role to play in all of the construction trades. The ability to visualize problems in geometric shapes and forms is vital in decision making. Piping length calculations always come down to one or another common plane figure: a square, a rectangle, a triangle, a circle, or even a simple line. Once a problem has been resolved to one or more of these, a corresponding formula may be easily applied and the solution is close at hand.

The problem may require the application of a number of geometric shapes before the solution can be deduced. When calculating heat loss from buildings, the areas and volumes of walls and rooms are often found by breaking the overall shape of the room into its geometric parts and then adding together the areas and volumes of these individual parts. The same breakdowns must be made when calculating the volumes of tanks and excavations.

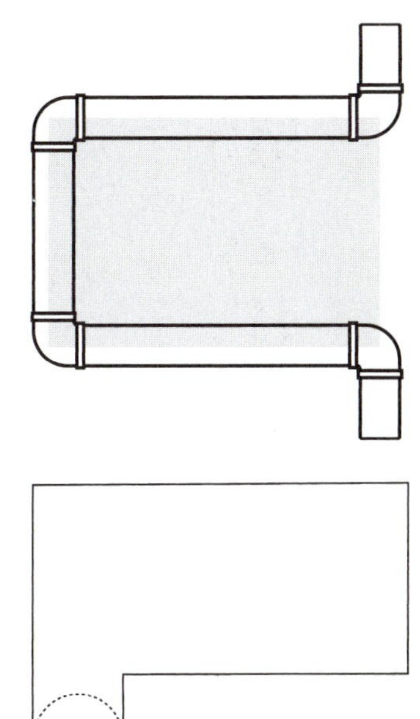

To calculate the area of the ceiling of the room at bottom right, how many geometric shapes would have to be calculated?

If you said three, you were correct! Two rectangles and a circle are in the shape. Of course, we would need to divide the circle's area by 2, because half of the circle is included within one of the rectangles. Can you see that?

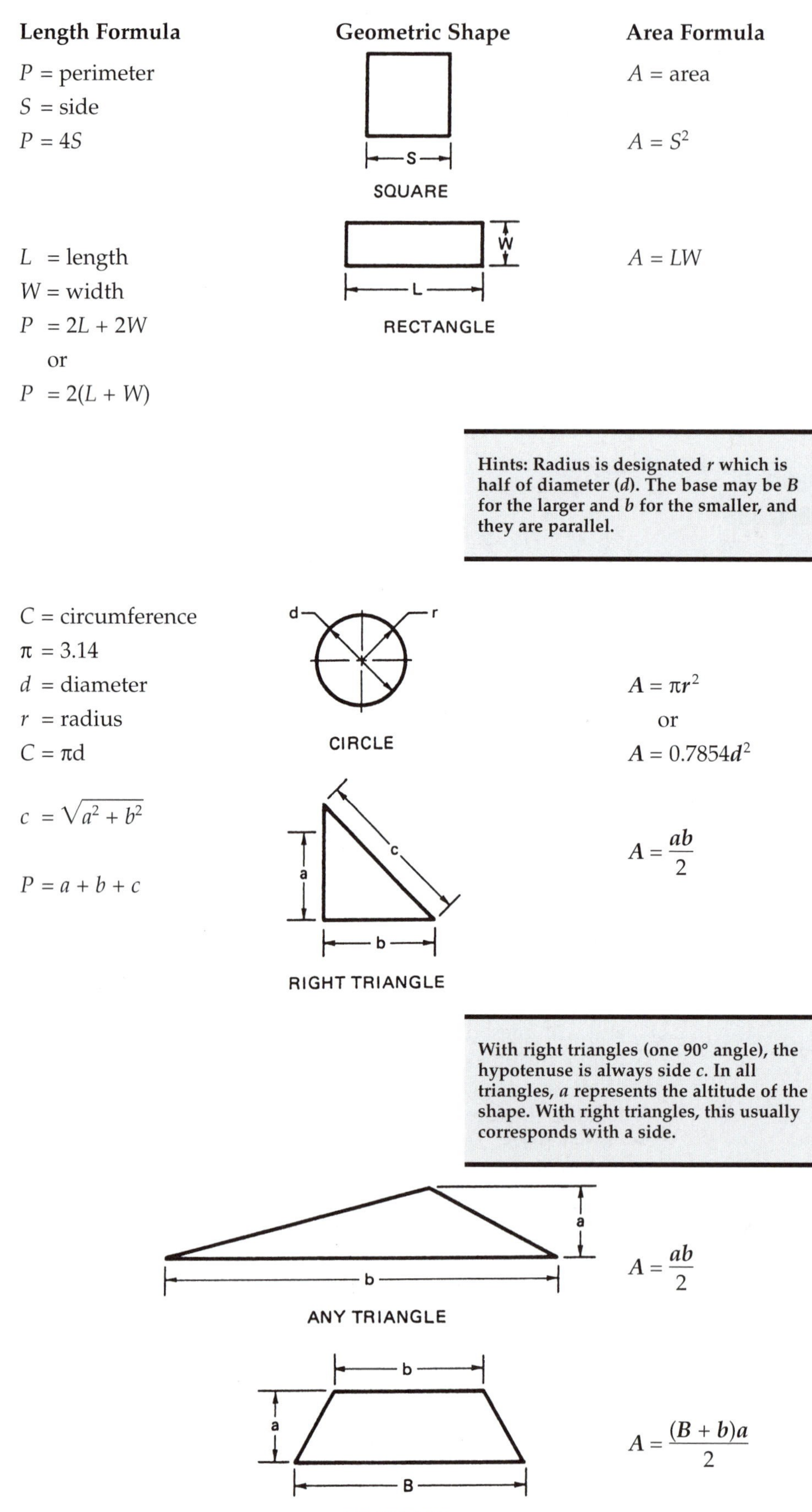
Length Formula
Geometric Shape
Area Formula
P = perimeter
S = side
$P = 4S$
S
SQUARE
A = area
$A = S^2$
L = length
W = width
$P = 2L + 2W$
or
$P = 2(L + W)$
W
L
RECTANGLE
$A = LW$
Hints: Radius is designated r which is half of diameter (d). The base may be B for the larger and b for the smaller, and they are parallel.
C = circumference
$\pi = 3.14$
d = diameter
r = radius
$C = \pi d$
d
r
CIRCLE
$A = \pi r^2$
or
$A = 0.7854d^2$
$c = \sqrt{a^2 + b^2}$
$P = a + b + c$
a
b
c
RIGHT TRIANGLE
$A = \frac{ab}{2}$
With right triangles (one 90° angle), the hypotenuse is always side c. In all triangles, a represents the altitude of the shape. With right triangles, this usually corresponds with a side.
a
b
ANY TRIANGLE
$A = \frac{ab}{2}$
b
a
B
TRAPEZOID
$A = \frac{(B + b)a}{2}$

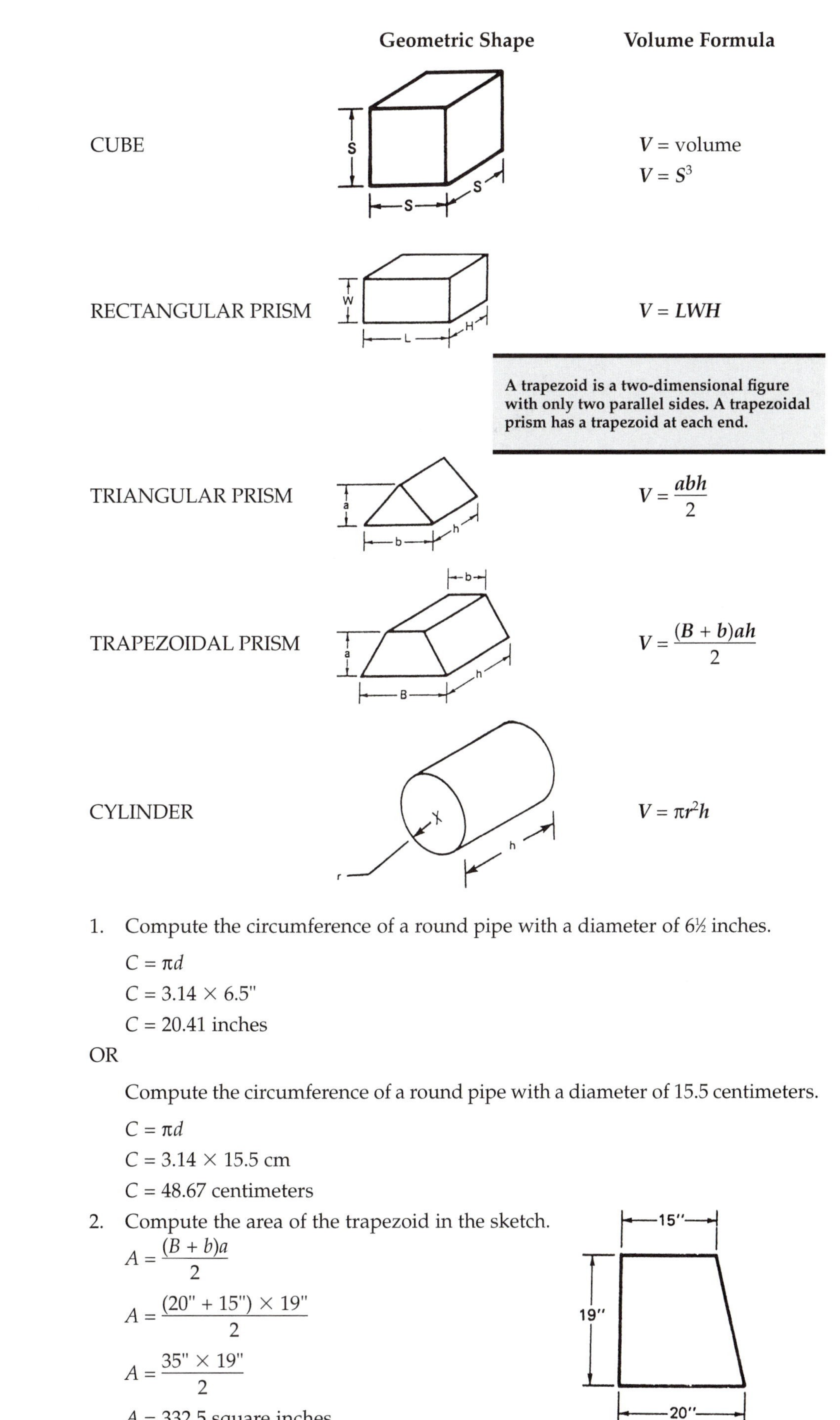

	Geometric Shape	Volume Formula
CUBE		V = volume $V = S^3$
RECTANGULAR PRISM		$V = LWH$
TRIANGULAR PRISM		$V = \frac{abh}{2}$
TRAPEZOIDAL PRISM		$V = \frac{(B + b)ah}{2}$
CYLINDER		$V = \pi r^2 h$

A trapezoid is a two-dimensional figure with only two parallel sides. A trapezoidal prism has a trapezoid at each end.

Sample Problems

1. Compute the circumference of a round pipe with a diameter of 6½ inches.

   $C = \pi d$

   $C = 3.14 \times 6.5''$

   $C = 20.41$ inches

OR

Compute the circumference of a round pipe with a diameter of 15.5 centimeters.

$C = \pi d$

$C = 3.14 \times 15.5$ cm

$C = 48.67$ centimeters

2. Compute the area of the trapezoid in the sketch.

   $A = \frac{(B + b)a}{2}$

   $A = \frac{(20'' + 15'') \times 19''}{2}$

   $A = \frac{35'' \times 19''}{2}$

   $A = 332.5$ square inches

OR

Compute the area of the shape in the sketch.

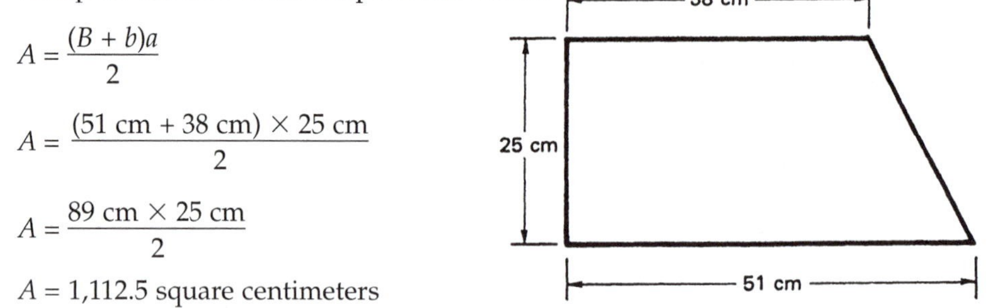

$A = \frac{(B + b)a}{2}$

$A = \frac{(51 \text{ cm} + 38 \text{ cm}) \times 25 \text{ cm}}{2}$

$A = \frac{89 \text{ cm} \times 25 \text{ cm}}{2}$

$A = 1{,}112.5$ square centimeters

3. Compute the volume of a cylinder that is 14 inches in diameter and 20 inches long.

   $V = \pi r^2 h$

   $V = 3.14 \times 7'' \times 7'' \times 20''$

   $V = 3{,}077.20$ cubic inches

OR

Compute the volume of a cylinder that is 36 centimeters in diameter and 51 centimeters long.

$V = \pi r^2 h$

$V = 3.14 \times 18 \text{ cm} \times 18 \text{ cm} \times 51 \text{ cm}$

$V = 51{,}885.36$ cubic centimeters

Exercise

A. Solve for each length.

1. Perimeter of a room 12' 6" × 12' 6"
2. Perimeter of a room 10.5 meters × 10.5 meters
3. Perimeter of a room 18' 9" × 15' 0"
4. Perimeter of a room 6.75 meters × 3.0 meters
5. Circumference of a pipe 12¾" inches in diameter
6. Circumference of a pipe 32.3 centimeters in diameter
7. 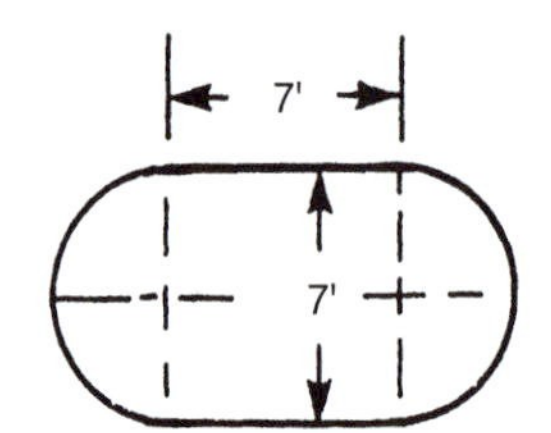

   Solve for the perimeter in feet for the shape shown by the sketch.

B. Solve for each area.

1. The square inches of copper in a sheet 14½" × 14½"
2. The square centimeters of copper in a sheet 35 centimeters × 35 centimeters
3. The square feet of lead in a sheet 3' 6" × 4' 6"

4. The square meters of a floor 3.5 meters × 4.5 meters
5. 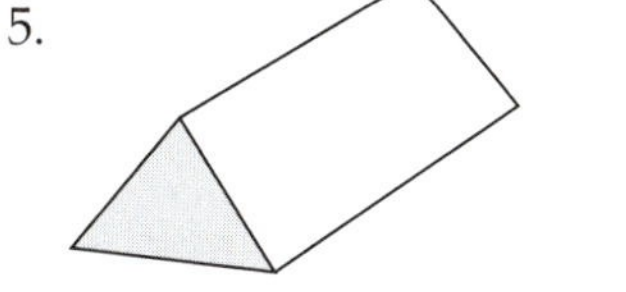  The square feet in a house gable 26' 0" × 9' 0"
6. The square inches of opening in a trapezoid-shaped hole with parallel lines 14 inches and 22 inches that are 12 inches apart
7. The square centimeters of opening in a trapezoid-shaped hole with parallel lines 28 centimeters and 44 centimeters that are 24 centimeters apart
8. Area of a circle 18 inches in diameter
9. Area of a circle 52 centimeters in diameter

10. Compute the area in square centimeters for the shape shown by the sketch.

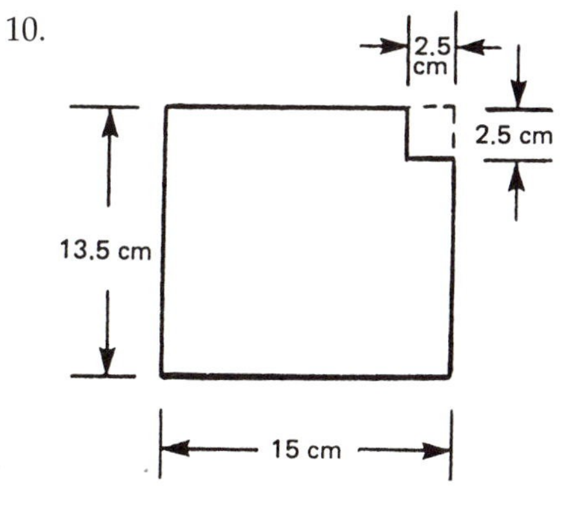

C. Solve for each volume.

1. The cubic inches in a cube 12 inches on a side
2. The cubic centimeters in a cube 10 centimeters on a side
3. The cubic yards for a cellar hole 30' 0" × 36' 0" × 4' 6" deep
4. The cubic meters for a cellar hole 10 meters × 11 meters and 1.4 meters deep
5. The cubic feet of an attic 8' 0" high for a 24' 0" × 34' 0" house with a gable roof
6. The cubic centimeters of a triangular steel wedge 5.8 centimeters × 5.8 centimeters and 18 centimeters from head to cutting edge
7. The cubic inches for a cylindrical pail 9 inches in diameter and 11 inches high
8. The cubic centimeters for a cylindrical pail 23 centimeters in diameter and 28 centimeters high
9. Compute the volume of the semicircular trough in cubic inches.

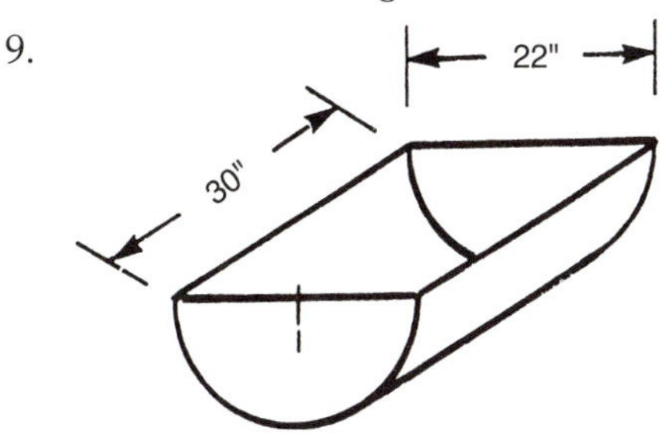

10.

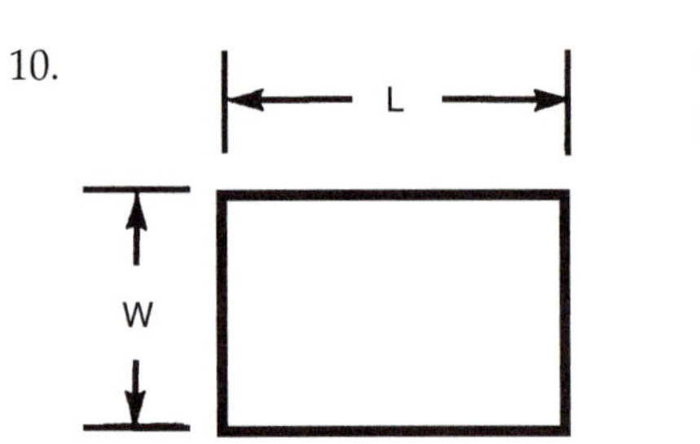

Write, in math shorthand, the way to solve for the area (A) of the rectangle shown.

A =

11.

Write, in math shorthand, the way to solve for the volume (V) for the tank shown in the sketch.

V =

Unit 3

Solving Formulas/Equations

Objectives

- Define formulas/equations.
- Review formula solving.
- Demonstrate the Number Template Technique.

Formulas and Equations: What is a Formula?

The words "equation" and "formula" are sometimes used interchangeably and that is not strictly accurate. The **equation** is a math tool that assists us when we are thinking about a complex math problem. It has letter symbols and values on both sides of an equal sign, $2(y + z) = 2y + 2z$. In some cases there may simply be a 0 (zero) on one side, $2k - t = 0$. We can use an equation to develop a useful formula.

> **Basic Rule:**
>
> **If we increase or decrease the complete value on both sides of the equality symbol by the same amount, the equation is still true.**

e.g., $2(y + z) = 2y + 2z$ or $2k - t = 0$.

The **formula** is an equation in which only one unknown symbol is on one side and a series of known values is on the other, as in $A = \pi r^2$. We need the value on the left side of the formula and have at our disposal the values on the right side of the formula. The formula can be thought of as a recipe. Add the given amounts to one side, stir well, do the indicated math, and the unknown value appears!

Formula Solving: The Formal Method

Problem

$C = \pi D$, where π is equal to 3.14 and D is equal to 3.

Step 1: Substitute into the formula the values that are known.

$C = 3.14 \times 3$

Step 2: Do the math.

$C = 9.42$

> **A common format for simple formulas is $A = B \times C$. That is, two terms multiplied together equal a third term: the answer.**

Often, the value that is normally the answer is known and one of the values from the other side is not known. It is essential to be able to rearrange a formula to obtain the value of any unknown.

Remember: the value on the left side of the equality symbol (=) is the same as the value on the right side.

> **Another way to indicate multiplication is to use no symbol at all. XY^2 means "multiply X by the square of Y."**

Example

$C = P \times D$, where P is equal to 3 and D is equal to 5.

Since the value of the right side is equal to 15, i.e., 3×5, then the value of the left side must also be 15. So when the symbol C is seen representing the left side, it can be stated that "C is equal to the value 15."

Rearranging Formulas

Sometimes values are not conveniently set off by themselves on one side of the equation as C is. It is not unusual to have a formula like $C = P \times D$ and know what C and P are equal to and need to determine the value of D. Because D is not located by itself on one side of the equation, we must do some rearranging. With a little practice, it is easy to do.

Things that are known:

1. The value of everything on the right side of the equality symbol is the same as the value of everything on the left side.
2. It can be said that C is equal to P times D. It can also be said that P times D is equal to C.

What may not be so obvious is that *so long as we increase or decrease the value of one side of the equality symbol (=) by the same amount as the other side, the equation is still correct.* This fact can be used to solve the equation for the unknown variable.

Problem

$C = P \times D$, where C is equal to 6 and D is equal to 2. What is the value of P?

In this case the value of P is needed and in order to find that, we need to have P by itself on one side of the equals symbol (P = ? or ? = P, same thing). Then, when the other side is worked out, the answer will appear.

To solve this problem we must first ask, *what is the opposite of multiplication?* If you said division you were correct! To get P to appear by itself on one side of the equation we need to remove the symbol D from that side. Since D is multiplied into the right side (C = P [multiplied by] D), it can be removed by dividing the right side by D, whatever that value may be. If this is done, D will disappear from the right side.

WHY: D divided by D is equal to 1;
Any number divided by itself is equal to 1;
Any number multiplied by 1 remains unchanged.

SO:

a. $C = P \times D$;
b. $C = P \times D / D$
c. $C = P \times 1$;
d. $C = P$ (*Incorrect at this point!* To stay in balance we must now change the left side in the same manner.)
e. $C / D = P$ (Now the equation is in balance.)

Now the equation is balanced and true again, and we now know that P is equal to 3. Remember that when the problem was first stated, C was given the value of 6 and D was given the value of 2; 6 divided by 2 is equal to 3.

The object was to get P on one side of the equation by itself. We essentially moved the D from the right side to the left by dividing both sides of the equation by D. What triggered that was the fact that D was originally multiplied on the right side. And so the "opposite" math operation, division, was used to remove it.

Another set of opposites are "the square of" and "the square root of." If S^2 (S squared) is equal to 64, then the value of S can be found by "taking the square root of" 64.

If $S^2 = 64$, then $S = \sqrt{64}$ and therefore $S = 8$.

$S^2 = 64$ became $S = 8$ because the square root of both sides of the equation was taken.

When we take the square root of S^2, the answer is simply S.

To change $S = \sqrt{64}$ back into $S^2 = 64$, what must be done to both sides of the equation?

A Number Substitution Solution Method: For Simple Formulas

Two true statements are $2 \times 3 = 6$ and $6 = 2 \times 3$.

Let us say that $Z = 20$ and $B = 4$ and the problem is $Z = J \times B$. To solve it using number substitution, set it up like this:

$Z = J \times B$	J is the value that we are not given. What is it?
$6 = 2 \times 3$	An easy number template to help visualize

The value after the equal symbol is the unknown value. Using the two values in $6 = 2 \times 3$ corresponding to the known values in $Z = J \times B$ (which are Z and B), ask yourself: what must be done between the 6 and the 3 in the lower equation to arrive at 2? If you thought, divide 6 by 3, you were right. Now do the same corresponding math operations in the problem equation.

$Z / B = J$ and $20 / 4 = 5$. Therefore, J must be equal to 5.

Note that the "opposite" math operation will always be used. The opposite of multiplication is division. The opposite of addition is subtraction.

The number of real-life formula problems that can be solved using this simple technique is amazing. Think about the familiar circle area formula $A = \pi r^2$. Can the number substitution technique be used here?

Let's say that A is 114 (we know that π is equal to 3.14159) and we want to solve for the value of R^2. Here we'll treat the term R^2 as we did J in the previous problem. Remember that the • symbol means multiplication.

$A = \pi \bullet R^2 \Leftarrow$ The unknown value is in the last position.

$6 = 3 \times 2 \Leftarrow$ The value 2 occupies the last position here.

If 6, first term position, is divided by 3, second term position, we obtain the value 2, so perform the same operation on the upper equation, yielding $A / \pi = R^2$. Since the equality symbol (=) indicates that one side of the equation has the same value as the other, why can't we exchange them? We can. Therefore, $R^2 = A / \pi$.

Substituting our numbers in the formula, we have $R^2 = 114 / 3.14159$, or $R^2 = 36.29$.

Checking Our Rearrangement for Validity

The assertion has been made that $R^2 = A / \pi$ based on our rearrangement of the proven valid formula $A = \pi \bullet R^2$ formula. We do this by working out a simple problem using the formula that we are sure of and then plugging those same numbers into our rearranged formula.

The value of π is 3.14159, so let's make up an easy value for R. Let's use the value 5.

$A = \pi \bullet R^2$ yields $A = 3.14159 \times 5^2$. We know that 5^2 means 5×5 or "five squared." So $A = 3.14159 \times 5 \times 5$. Working the problem out, the result is $A = 78.54$.

And now let's test our rearranged formula: $R^2 = A / \pi$, so 25 should be equal to 78.54 divided by 3.14159, i.e., $25 = 78.54 / 3.14159$. When the right side is worked out, we arrive at $25 = 25$, and now we know that our rearrangment is valid.

Practice Problems

Formula: $A = L \times W$: this is the formula for the area of a rectangle. Area is equal to length multiplied by width.

1. Given that the length of a rectangle is equal to 12 and the width is equal to 5, what is the area?
2. Given that the length of a rectangle is equal to 4.5 and the width is equal to 2, what is the area?
3. Given that the width of a rectangle is 5 and the area is 100, what is the length?
4. Given that the width of a rectangle is 3 and the area is 36, what is the length?
5. Given that the length of a rectangle is 15 and the area is 255, what is the width?
6. Given that the length of a rectangle is 13 and the area is 58.5, what is the width?

Exercise

Complete the following problems; solve for the unknown quantity in each.

1. A square room has an area of 225 square feet. How long is one side?
2. A square room has an area of 16 square meters. How long is one side?
3. A square plot has a perimeter of 66 feet. How long is one side?
4. A square plot has a perimeter of 84 meters. How long is one side?
5. A circle has a circumference of 22 inches. Compute the diameter.
6. A circle has a circumference of 56 centimeters. Compute the diameter.
7. How wide is a piece of sheet lead that is 30 inches long if the area is 144 square inches?
8. How wide is a piece of sheet lead which is 50 centimeters long if the area is 200 square centimeters?
9. A circle has an area of 50.24 square inches. How long is the radius? How long is the diameter?
10. A circle has an area of 113.04 square centimeters. How long is the radius? How long is the diameter?

Unit 4

Square Root

Objectives

- Review "taking" the square root.
- Using square root to solve triangles.

Finding Square Root: "Taking" the Square Root of a Number

Taking the square root of a number means finding a number that when multiplied by itself equals the number being considered. If I am considering the number 25, for instance, I would ask myself, "What number when multiplied by itself is equal to 25?" The answer, of course, is the whole number 5. Square roots may also be decimal numbers.

By far the easiest way to discover the square root of a number is to use a calculator which has a square root key, usually marked with the square root radical $\sqrt{\ }$. In practice, the user enters in the calculator display the number whose square root is required and then presses the square root key. What is then displayed is the square root of the original number.

Solving Right Triangles with Square Root: Finding All of the Sides

You have learned that the hypotenuse of a right triangle may be deduced with the formula $C = \sqrt{A^2 + B^2}$. This is derived by rearrangement of the formula called the Pythagorean Theorem: $C^2 = A^2 + B^2$.

We may use this formula to find any of the sides of a **right triangle,** so long as the lengths of the other two sides are known. The hypotenuse is always represented by the letter *C* and is always the longest side. *A* and *B* can be used for either of the other two sides.

Remember that we used $6 = 2 \times 3$ as a template for rearranging simple formulas which use multiplication back in Unit 2. Another template for *addition* may be constructed to help in the arrangement of $C^2 = A^2 + B^2$. This could be $6 = 4 + 2$. Actually any correct equation of different numbers would do.

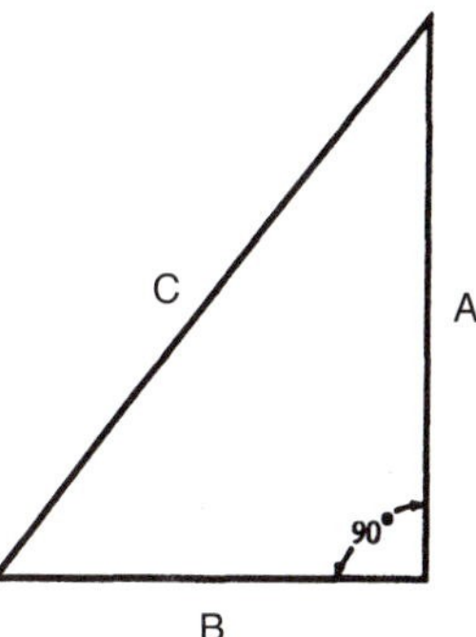

To obtain the length of side A we would ask, "What would I do between 6 and 2 to obtain 4?" And the answer, of course, is subtract 2 from 6. Now just do the same things to the terms in the same positions in the equation $C^2 = A^2 + B^2$ and thus obtain $C^2 - B^2 = A^2$. A^2 is now on one side of the equation by itself. It is a simple matter to calculate $C^2 - B^2$, $(C \times C) - (B \times B)$. Now find the square root of the result, A^2, and this will be the answer.

Problem: If $C = 20$ and $B = 16$, what would be the length of side *A*? *C* squared is 400. *B* squared is 256. Therefore, the square of *A* would be 144, i.e., 400 – 256. *A*, then, is 12. As can be seen, if any two sides of a right triangle are known, the remaining side can always be found by rearranging this famous formula.

Notice that the sides of the right triangle being described at the left are multiples of 3, 4, and 5. When this is the case, the triangle is always a right triangle.

Sample Problem

Solve for the length of the hypotenuse *c*.

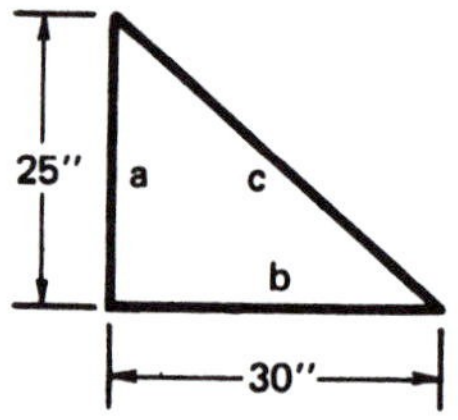

The Pythagorean theorem

$c^2 = a^2 + b^2$

Take the square root of both sides of the equation:

$c = \sqrt{a^2 + b^2}$

$c = \sqrt{625 \text{ sq in} + 900 \text{ sq in}}$

$c = \sqrt{1{,}525 \text{ sq in}}$

$c = 39.05"$

$c = 39\frac{1}{16}$

Square Root Solutions with a Pencil

1. Locate the decimal point.

$\sqrt{1{,}525.}$

2. Pair off both ways from the decimal point.

$\sqrt{15\ 25.00\ 00}$

3. Examine the first pair under the radical sign (15) and determine the approximate square root. In this case, the square root is more than 3 but less than 4—between $3 \times 3 = 9$ and $4 \times 4 = 16$.

$\sqrt{15}$

4. Place the 3 above the radical sign over the 15. Multiply the 3 by itself and place the product (9) under the 15.

$\overset{3}{\sqrt{15}}$

5. Subtract as in a division problem to get the remainder.

$\frac{9}{6}$

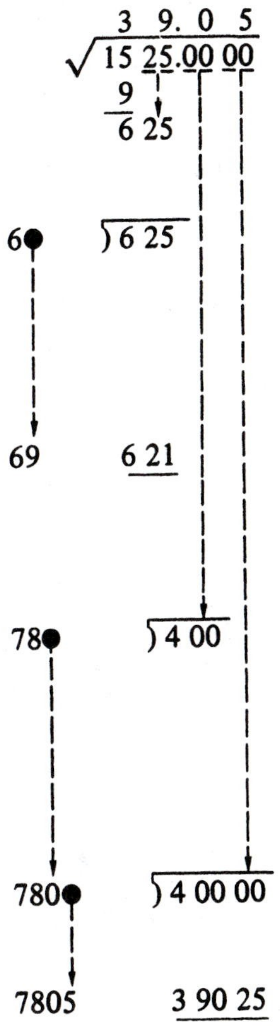

6. Bring down the next pair of numbers (25).

```
  3  9. 0  5
√15 25.00 00
  9
  6 25
```

7. Double the answer number, 3, to use as a trial divisor and write the 6 outside the division frame in the tens column.

```
6●   )6 25
```

8. Determine how many times the "sixty something" may be divided into 625. The number in the unit column must be the same as that used as a multiplier. In this case, try 9. Place it in the answer and after the 6 in the divisor. $69 \times 9 = 621$. Place this under the 625.

```
69    6 21
```

9. Subtract and bring down the next pair of numbers (00).
10. Again, double the present answer (39) as a trial divisor for 400. Place the result (78) in the hundreds and tens columns.

```
78●   )4 00
```

11. How many times will "seven hundred eighty-something" go into 400? Obviously, none, as it is already larger. So, place a zero in your answer above the two zeros after the decimal point. Place a zero after the 78 in the trial divisor to become 780 and bring down the next pair of numbers (00).

```
780●   )4 00 00
```

12. How many times will "seven thousand eight hundred something" go into 40,000?

```
7805    3 90 25
```

$5 \times 7{,}805 = 39{,}025$

The $\sqrt{1525} = 39.05$

Note: *Steps 1, 2, 3, and 4 are used once in each problem.*
Steps 5, 6, 7, and 8 show how each number in the answer is obtained—after the first one.
Step 11 shows the use of zero in the answer.

Factoring Method

The preceding step-by-step procedure for extracting the square root of a number can sometimes be simplified or bypassed. Such a shortcut is possible if the number under the radical sign can be broken down into numbers whose square root is already known or can be found more easily. This method is called *factoring.*

Sample Problem

Find $\sqrt{400}$

1. Break the number into numbers whose square roots are known.

   $\sqrt{4} \times \sqrt{100}$

2. Extract the square root of each factor.

   2×10

3. Multiply the factors.

   $2 \times 10 = 20$

Use of Square Root Tables

Square root tables for numbers 1.000 to 99.90 are included in the Appendix (see pages 239–242). These tables can be used for numbers greater than 100.

Sample Problem

Find $\sqrt{1{,}525}$

1. Factor the number. $\sqrt{1{,}525} = \sqrt{100} \times \sqrt{15.25} = 10 \times \sqrt{15.25}$
2. Find $\sqrt{15.25}$ from the table.

Excerpt of Table C Appendix

	0	1	2	3
14.	3.742	3.755	3.768	3.782
15.	3.873	3.886	3.899	3.912
16.	4.000	4.012	4.025	4.037
17.	4.123	4.135	4.147	4.159
18.	4.243	4.254	4.266	4.278

a. Read 15.2 and 15.3 on Table C.

b. The square root of 15.25 may be estimated as the average of the two answers:

$\sqrt{15.2} = 3.899$

$\sqrt{15.3} = 3.912$

$3.899 + 3.912 = 7.811$

$7.811 \div 2 = 3.905$

3. $\sqrt{1{,}525} = \sqrt{100 \times 15.25} = 10 \times 3.905 = 39.05$

Exercise

Solve for the hypotenuse (c) in each of the given right triangles.

	a	*b*	*c*
1.	6"	8"	
2.	8"	15"	
3.	20"	25"	
4.	27"	32"	
5.	44"	19"	
6.	12"	12"	
7.	1"	1"	
8.	10 cm	11.25 cm	
9.	16 cm	9.5 cm	
10.	17.5 cm	23.25 cm	

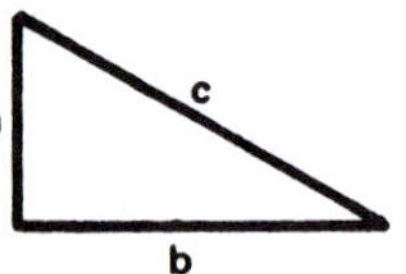

Solve for the altitude (a) in each of the given right triangles.

Note: *Rearrange the formula* $c^2 = a^2 + b^2$

	a	*b*	*c*
11.		12"	15"
12.		2"	3"
13.		13"	18"
14.		24 cm	25 cm
15.		6 cm	40 cm

The use of a calculator is shown in Appendix B. The steps to find a square root are reduced to the use of one key. The learning experience is that the use of a calculating device best follows an understanding of the mathematics.

Unit 5

Review of Angle Measure

Objectives

- Learn what an angle is.
- Use a protractor to measure angles.
- Describe the relationship of angles formed by intersecting lines.

Length, height, and width are measurements of linear distance. An angle, on the other hand, is a *measurement of direction.* Angles are formed when two straight lines meet or intersect. Let's examine some statements about angles.

1. "The surface sloped upward at a 20-degree angle from that point."
2. "Gramby Street and Church Avenue run at a 90° angle to each other."
3. "The river runs generally north and south at Sharpesburg."
4. "The cold chisel edge should be ground at a 60° angle."

Notice that an angle always has a reference line. Sometimes that reference line is only implied.

In statement 1, the reference line is the average angle of the earth's surface, i.e., zero degrees. If we use the term *flat* to mean the average slope of the earth's surface, then it can be said that the surface changes from flat to an upward angle of 20 degrees.

Note: It can be maintained that the earth's average surface is curved because the earth itself is a sphere. But for the short distance that we require for our reference line, the reference line of the earth's surface can be considered straight.

In statement 2, each of the two streets serves as a reference line for the other.

In statement 3, the map of the earth's surface with lines of longitude running between the poles of the earth becomes the reference line.

In statement 4, the two reference lines meet at a *vertex* and serve as reference lines for each other (see the illustration). The vertex is the point at which the two lines which define an angle meet.

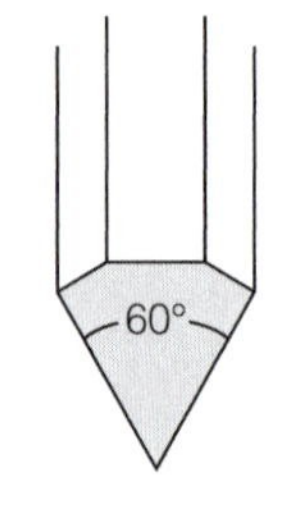

A Cold Chisel Point

Ancient Egyptian astronomers devised angle measure. They thought of themselves as at the center of a circle with the sun rising and setting on different points of the circumference. The vertex of an angle is the center of a circle and the angle is formed by radius lines.

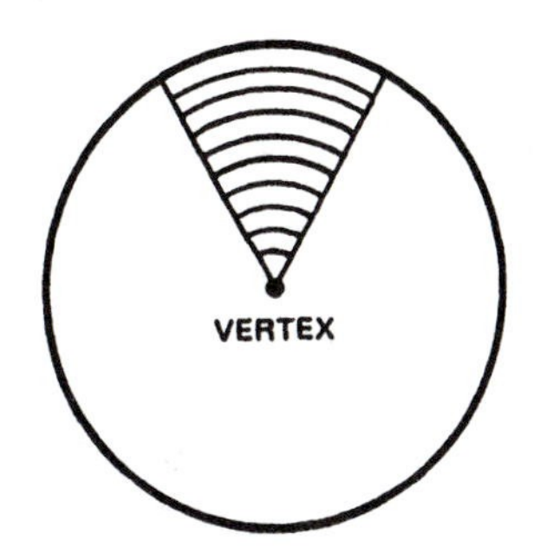

The Degrees of a Circle

The Egyptians set the number of degrees for a circle at 360 as their year had 360 work days and five feast days. The 360° circle has continued in use. A degree is $\frac{1}{360}$ of a full circle.

The Protractor

The protractor is used to measure and construct angles. The center point of the protractor must be at the vertex of the angle. One of the radius lines is located along the base line of the protractor. The second radius line is located at a number of degrees from the first. The protractor has two rows of numbers so that angles can be measured clockwise (outer numbers) or counterclockwise (inner numbers). The lengths of the radius lines do not affect the angle measure between them.

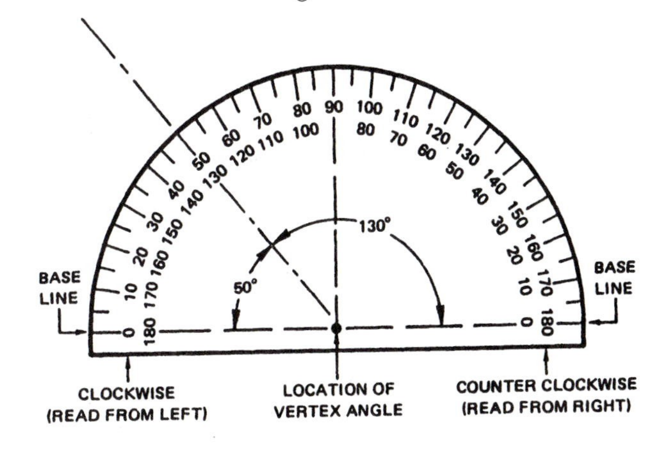

The Straight Angle

Two lines that meet to form a straight line are at an angle of 180°, or half a circle.

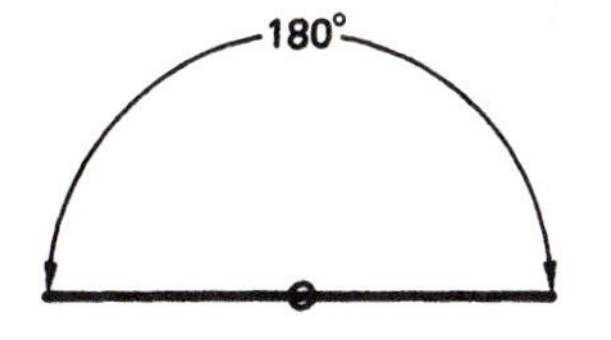

The Right Angle

The right angle is formed when two perpendicular lines intersect. A right angle and a 90° angle are the same. The corners of this textbook are formed by perpendicular lines and that makes the corners right angles. Walls and floors should be at right angles to each other. In building construction, workers often construct a perfect right angle to "square up" walls with each other by nailing three pieces of lumber together to form a 3' by 4' by 5' triangle. These measurements always result in a right angle.

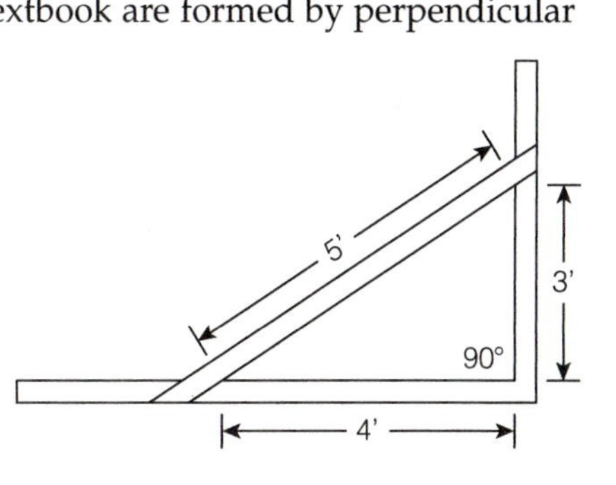

The Sum of Angles in a Triangle

The sum of the three angles in any triangle is 180°. Angle A plus angle B plus angle C equal 180°.

Opposite Angles

When two lines intersect (cross), the opposite angles are equal.

Angle A = Angle B

Angle C = Angle D

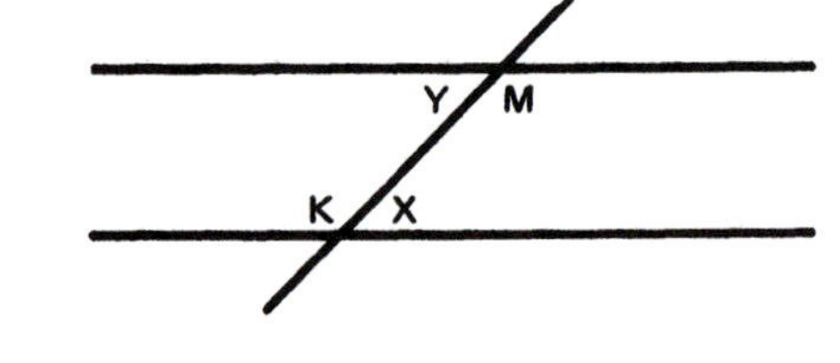

Alternate and Corresponding Angles

When parallel lines are intersected (crossed) by a third line, the alternate interior angles are equal.

Angle X = Angle Y

Angle K = Angle M

When parallel lines are intersected by a third line, the corresponding angles are equal.

Angle Z = Angle M

Angle W = Angle Y

Angle X = Angle N

Angle K = Angle L

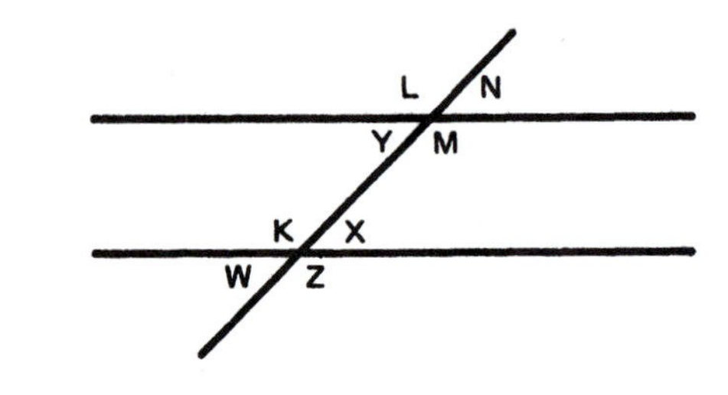

Sample Problems

Solve for the number of degrees in each lettered angle. State the basis for the method used.

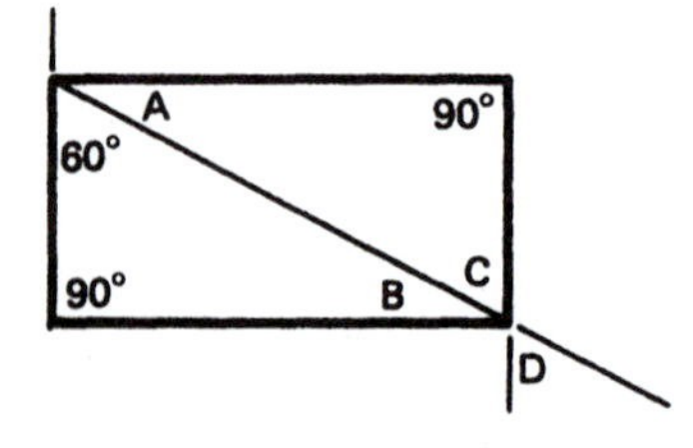

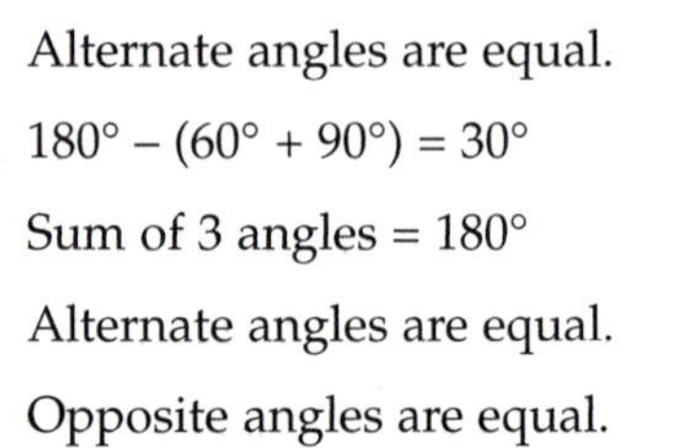

1. Angle C = 60° — Alternate angles are equal.
2. Angle A = 30° — 180° – (60° + 90°) = 30°
   Sum of 3 angles = 180°
3. Angle B = Angle A = 30° — Alternate angles are equal.
4. Angle D = Angle C = 60° — Opposite angles are equal.

Exercise

1. a. Use a protractor to make a figure similar to the sample problem, using a 48° angle in place of the 60° angle.
   b. Measure each angle with a protractor and write down your readings.
   c. Solve for each angle and compare with angle measurements.
2. Read the values of the angles indicated by each letter:

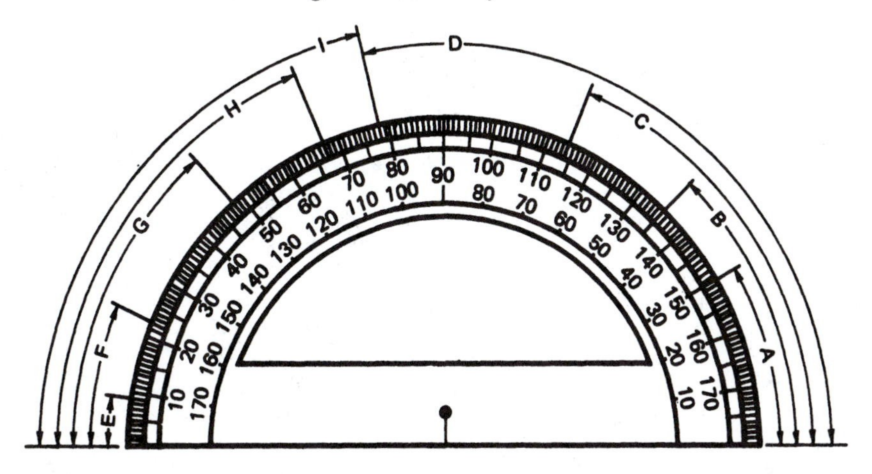

3. Measure each of the illustrated angles with a protractor:

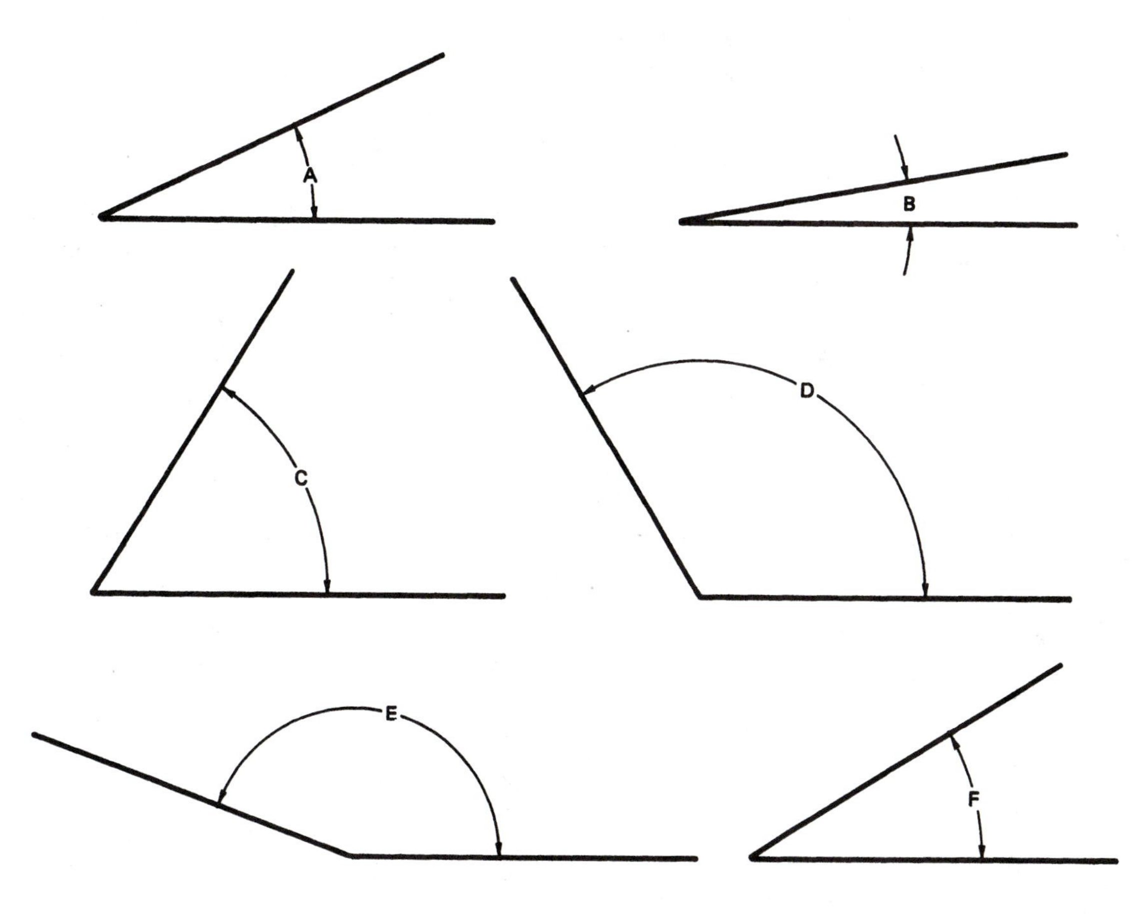

Unit 6

Conversion of Length Measure

Objectives

- Use the proper unit of measure for each task.
- Convert between different dimensional standards.

Math tasks in the building trades will often involve the application of different measuring systems, and the student should be familiar with metric standards and protocols. The United States largely uses English units of measure. Therefore, most of this book will refer to and use feet and inches, gallons, pounds, etc.

Most other countries use the metric system of measurement. The United States of America uses the English system of measurement.

One of the customs of mathematics is that the answer to a problem should be expressed in the same units and style that were originally given in the problem. In other words, if the area of a rectangle that is 5 feet wide and 6 feet long is asked for, the answer should be in square feet, and not in square yards or square inches.

Occasionally, formulas require the use of constant values, which are usually expressed in decimal form. The trigonometric function that we will later use for calculating lengths that lie along a 45-degree angle has a decimal value of 1.414, for instance. This value is the cosecant (a trigonometric function) of 45°.

Conversion of Measurements: Feet to Inches, Meters to Feet, etc.

Conversions are required between measurements in the same standard, e.g., feet to inches, and between measurements from different standards, e.g., meters to feet.

Some lineal measurements in descending lengths are:

English System	*Metric System*
miles	kilometers
yards	meters
feet	decimeters
inches	centimeters
fractions of inches	millimeters

Thinking About Converting

A conversion problem might present itself in this way: "How many inches are there in seven feet?" At an early point in the problem solving process, we will say to ourselves, "One foot is equal to twelve inches," or "There are twelve inches in a foot." Set this and all conversion problems up in the following way: Say to yourself, "The number of inches is equal to twelve times the number of feet." It is important to use the phrase "the number of" in this re-statement because if you said that "a foot is equal to twelve times an inch" it might lead you to state the (incorrect!) equation $F = 12 \times I$. A correct statement of the sentence in equation form would be $NI = 12 \times NF$, the number of inches is equal to 12 times the number of feet. It is not necessary to use NI and NF as variables; any symbols that make sense would be acceptable.

Now to solve our originally stated problem, "How many inches are there in seven feet?" just "plug in" the number 7 to our equation and easily deduce the answer:

$$NI = 12 \times 7, \text{ or } NI = 84$$

Notice that $NI = 12 \times NF$ is now in the form of our previously discussed $6 = 2 \times 3$ formula rearranger template. This can be used to obtain inches in a problem that requires inches to be derived from feet.

If the problem was "How many inches are there in seven feet, five inches?" We would simply subtract the 5 inches at the beginning and add them in with the 84 inches obtained from the 7 feet to obtain a final answer.

Note: It may be tempting to enter N16s as the Number of Sixteenths variable. Avoid putting numbers in a variable, as they will make the mathematics confusing.

Example

A measurement of 2.375" needs to be converted to inches and fractions of an inch with an accuracy to 1/16 of an inch. When we visualize the answer before beginning (a good practice, by the way), we see that it will be 2 inches plus an unknown number of sixteenths.

Hint: When using English measuring tools, answers are obtained in sixteenths, eighths, quarters, and halves of an inch. The terms of these fractions may be reduced by dividing both the top and the bottom by 2 as many times as necessary until the top number is odd.

Knowing that the decimal part of our problem (.375) could also be treated as a percentage, it is a simple matter to just multiply that by the number of sixteenths there are in an inch. Since sixteenths are the smallest divisions on our rule, we can derive any usable fraction we might need from them. We wish to come up with what 37.5 percent of 16 might be. Because $16 \times .375 = 6'$, we know that the answer is 6/16, or properly reduced that would be 3/8. Now we would also know that the answer to our original problem—what would 2.375" be on our steel tape measure—is precisely 2 3/8".

Suppose the decimal-inch measurement was a little more obscure like 2.711". Find 71.1 percent of 16: $16 \times .711 = 11.37$ sixteenths. Make a decision: Will it be 11 or 12 sixteenths? Look at digit to the right of the decimal point. Is it less than 5? Yes. We do not increase the 11 to 12. The answer to the nearest sixteenth inch is 2 11/16".

Changing Fractions to Their Decimal Equivalents

To change $\frac{11}{16}$ to a decimal form, simply divide the top of the fraction by the bottom. Then $\frac{11}{16}$ becomes 0.6875. It is a good practice to place a zero before the decimal point, as we have done here, to avoid confusion if the decimal point is erased or obscured.

If the fractional part of the question were actually a whole number along with a fraction, the mathematical progression might look like the following:

$$9\tfrac{3}{4} = 9 + \tfrac{3}{4} = 9 + 0.75 = 9.75$$

Some Measurement Relationships

There are	1,760	yards	in a	mile
	5,280	feet		mile
	3	feet		yard
	12	inches		foot
	1,000	meters		kilometer
	10	decimeters		meter
	100	centimeters		meter
	1,000	millimeters		meter

Dimensions are better tools when they are in the form that best suits the particular problem. In some cases, dimensions in feet and inches may be easiest to work with; in others, inches alone may be best. Calculations are often best figured in decimals, but need to be converted to feet and inches for measuring purposes. The recommended dimension form for various problems is used in each unit of this book. This unit covers methods of converting from one form to the other.

There are four standard methods of graduating measuring tools: (1) inches and fractional parts of an inch; (2) inches and decimal parts of an inch; (3) feet, inches, and fractions; and (4) feet and decimal parts of a foot.

The plumber's rule reads in inches and sixteenth-inches. The steel tape reads in inches and eighth-inches. Rules are made in inches and decimal parts, although decimal inches are generally confined to calculations in plumbing. Surveyors and architects often use feet and decimal parts of a foot for land measure.

Note that in English measure, the dash is sometimes used as a separator so that dimensions are not confused, i.e., 11'-3" does not indicate a subtraction of 3 inches from 11 feet.

Sample Problems

1. Fractional inches to decimal inches

   Change $\frac{11}{16}$" to decimal inches.

   $\frac{11}{16}$ is 11 divided by 16

   $\frac{11}{16}$" = 0.6875"

2. Decimal inches to nearest sixteenth inch

   Change 0.725" to nearest 16th inch.

   0.725" × 16 = 11.6 ⇒ 12/16 ⇒ 3/4"

3. Feet and inches to inches

   Change 8' 5" to inches.

   8' 5" = 96" + 5" = 101"

4. Feet to inches

   Change 0.69 foot to inches.

   0.69 × 12 = 8.28

   0.69 foot = 8.28 inches

5. Inches to feet and inches

   Change 110 inches to feet and inches.

   110 ÷ 12 = 9 and 2 remainder

   110" = 9' 2"

6. Inches to decimal foot

   Change 11½ inches to feet.

   11.5 ÷ 12 = 0.958

   11½" = 0.958'

7. Feet to nearest sixteenth inch

   Change 1.78 feet so it can be measured on a rule.

   1.78 × 12 = 21.36

   1.78' = 21.36"

   0.36" × 16 = 5.76 ⇒ 6/16 ⇒ 3/8"

   1.78' = 21 3/8"

8. Meters to centimeters

   Change 1.75 meters to centimeters.

   1.75 × 100 = 175

   1.75 meters = 175 centimeters

9. Centimeters to meters

   Change 239 centimeters to meters.

   239 ÷ 100 = 2.39

   239 centimeters = 2.39 meters

Exercise

Convert each length into the form indicated by the column heading.

	Inches and Fractional Parts of Inch	Inches and Decimal Parts of Inch	Feet, Inches, and Fractional Parts of Inch	Feet and Decimal Parts of Foot
1.	19 11/16"			
2.	42 3/8"			
3.		26.35"		
4.		69.94"		
5.			8' 4 1/2"	
6.			2' 9 3/16"	
7.				5.57'
8.				15.09'
9.		103.72"		
10.	79 3/4"			

11. How many centimeters is 0.85 meter?
12. How many meters is 872 centimeters?

Section Two
PIPE LENGTH CALCULATIONS

Unit 7

Standard Weight Pipe

Objectives

- Learn the pipe weights.
- Use a pipe data sheet.
- Calculate pipe clearances.

Pipe is made in many "weights." By far the most common are *standard, extra-heavy,* and *double extra-heavy*. These correspond directly to the more modern pipe weight terms: *schedule #40, schedule #80, and schedule #120.*

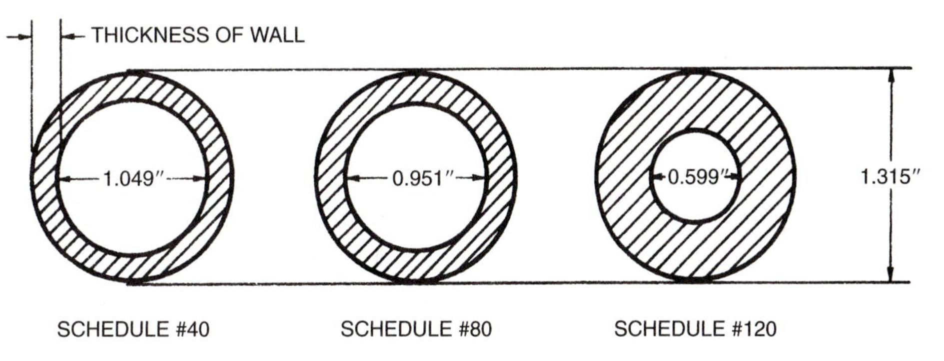

The cross sections show the inside and outside diameters for the three weights in one-inch nominal size pipe. The outside diameters are alike so that the same threading dies will fit all three weights of pipe. The internal pressure determines the weight of pipe needed. Most plumbing is done with standard weight pipe.

Nominal Size Pipe

The term "nominal" size means that the size of the pipe in question is determined by the measurement across the opening of the pipe. It means, in other words, the inside diameter of the pipe. On occasion the inside diameter is far from the "name pipe size" (nominal). In the case of ⅛-inch pipe the inside diameter is greater than ¼ of an inch. But generally, in the larger sizes, the pipe size can be determined by measuring the inside diameter.

The American National Pipe thread has a slight taper so that tightening a pipe into a fitting causes the fit between pipe and fitting to become tighter and tighter. A properly threaded pipe will turn into the fitting 2½ to 3 turns with the hand alone.

Understanding a Pipe Data Sheet

See below and the Data #2 sheet in the Appendix.

Diameters:

O.D. The outside diameter of the pipe.

I.D. The inside diameter of the pipe.

Wood auger for pipe Another term for this is "clearance drill size." This is the size of a hole that must be drilled for the pipe to pass through. The hole must be somewhat greater than the outside diameter of the pipe.

Tap drill The size of the hole that must be made to "tap" or make an internal thread for this size of pipe.

Threads:

Number per inch The number of threads per inch. Each complete turn of the fitting on the pipe moves the fitting the distance consumed by one thread up the pipe. If the pipe size required 16 threads per inch, then one complete turn of the fitting would move it 1⁄16 of an inch along the pipe.

Length on pipe How much, in inches, of the pipe will be covered by threads.

Threads on pipe How many times the threader will turn before the thread is complete.

Thread-in or thread engagement How much of the pipe end will be inside the fitting.

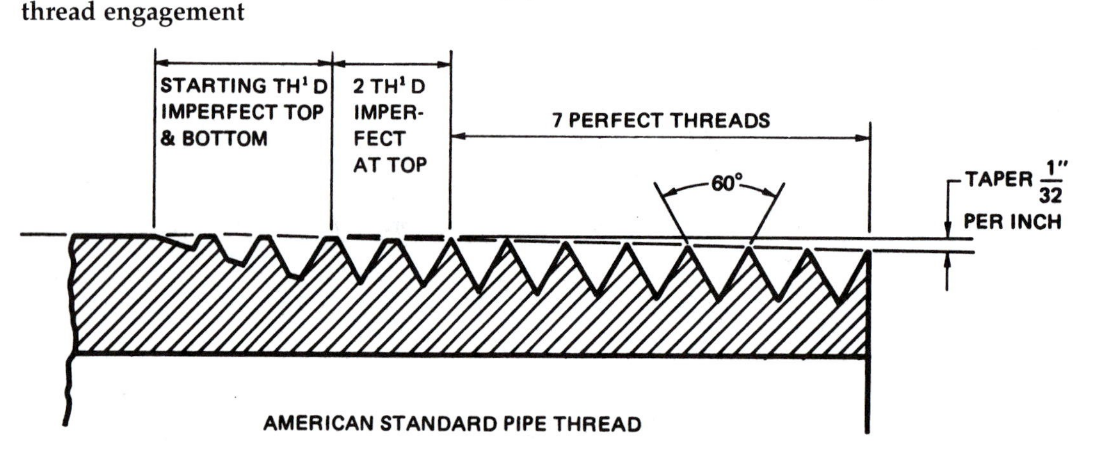

Sample Problem 1

Compute the maximum clearance between a ½" pipe and the wood auger hole as shown in Data #2, Appendix.

Solution

Step 1: Determine diameter of wood auger hole required for ½" pipe = 15⁄16"

Step 2: Determine O.D. of ½" pipe = 0.840"

Step 3: Convert decimal to fraction (16ths) 0.840" × 16⁄16 = 13.44⁄16" or 13⁄16"

Step 4: Subtract O.D. of ½" pipe from diameter of auger hole 15⁄16" − 13⁄16" = 2⁄16" or ⅛" clearance

Sample Problem 2

How far will one full turn advance a ¼" pipe into the fitting?

Solution

Data #2 shows 18 threads per inch for ¼" size pipe. Each full turn advances the pipe into the fitting 1/18" (0.0555" or about 1/16").

Exercise

Answer the following questions based on Data #2, Appendix.

1. Compute the maximum clearance between a ¾" pipe and the wood auger hole.
2. Compute the maximum clearance between wood auger holes and 1", 2½", and 6" pipe.
3. What is a tap drill (Data #2)?

   Why is it smaller than the O.D. of the pipe?
4. How far will a 2½" pipe advance into a fitting with one full turn?
5. How far will a 1" pipe advance into a fitting with one full turn?

Unit 8

Allowance for Threaded Fittings

Objectives

- Learn types of measurements.
- "Take off" for fittings.
- State generic rules for fitting allowance.

The process for fitting allowance is the same for every kind of fitting:

1. The measurements of the space in which the pipe is to be installed are obtained either by taking dimensions from a drawing or by actually measuring the space in which the pipe is to be installed.
2. The center-to-center measurements, described below, are obtained by the use of geometry and simple mathematics.
3. The end-to-end measurements, described below, are obtained by *subtracting* the fitting allowance from the center-to-center measurements.

Pipe is indicated on the architectural blueprint by displaying only the centerline of the pipe. The fittings are described by hash marks across the indicated pipe. See Figure 8–1.

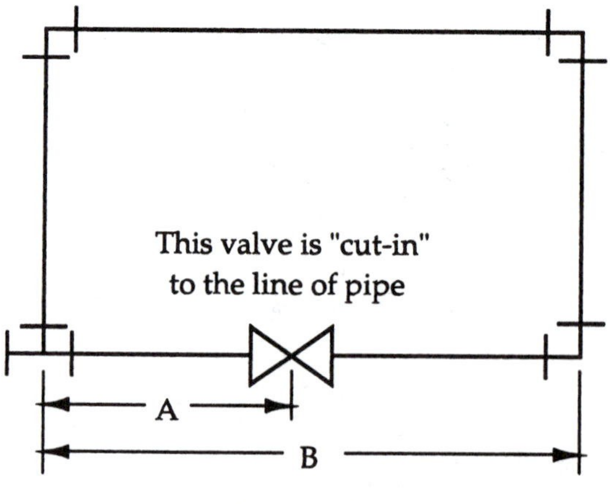

Figure 8–1

Fittings have one fitting allowance for each pipe opening. The pipe mechanic usually will be concerned with only one allowance at a time. However, when a fitting is being "cut-in" to a line as is the valve in Figure 8–1, the allowances for both sides of the valve must be taken into account when dealing with the center-to-center measurement "B."

With some exceptions, the fitting allowances on multiple-opening fittings will all be the same. In other words, the fitting allowances for a ½" CPVC tee might be 7/16" and 7/16" and 7/16". Fittings designed for drainage work occasionally violate this rule.

The allowance for a given fitting is the distance between the end of the pipe, *which may be and probably is inside the fitting itself,* and the center of the fitting. See the illustration below.

Can you see three possible fitting allowances inside the tee fitting illustrated below?

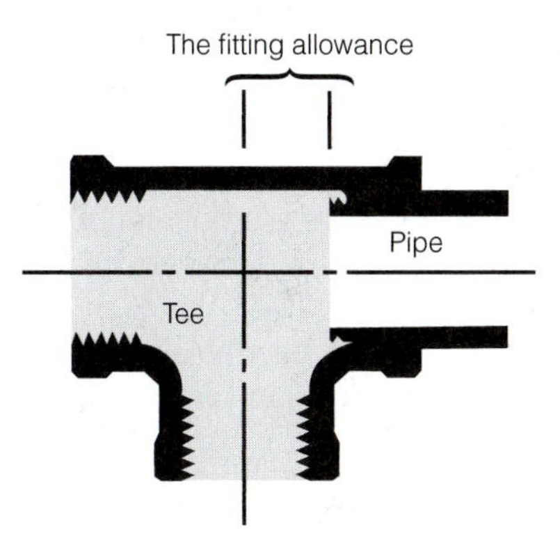

Pipe length is measured along the centerlines. When two centerlines cross there is a center point. These center points are located in a fitting. Center-to-center and end-to-center measurements are often made on the job. The pipe is cut to an end-to-end length. The end-to-end is always shorter than the center-to-center because the pipe does not thread into a fitting as far as the center point. The illustration shows various ways of measuring pipe.

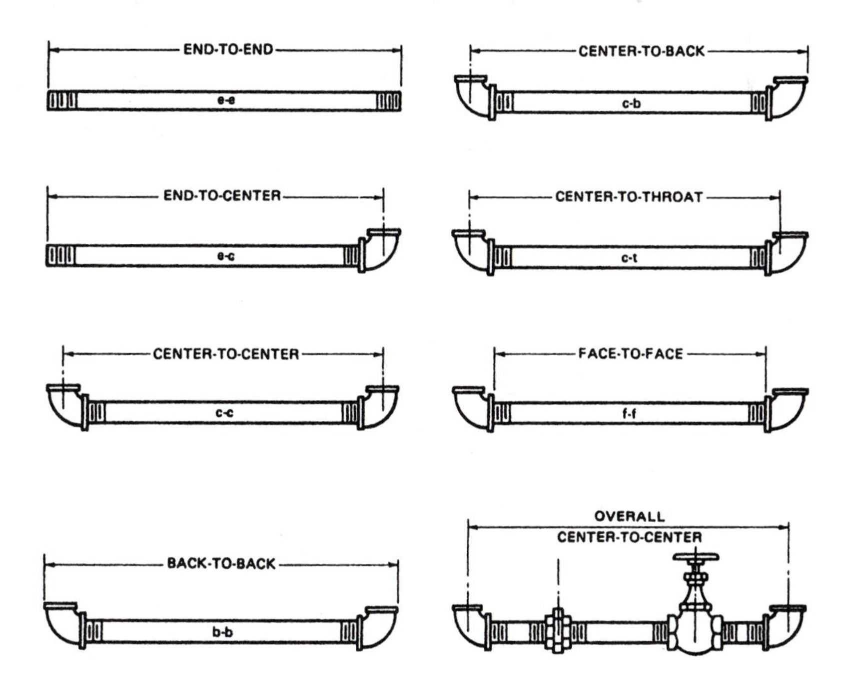

The length to cut threaded pipe cannot be measured directly. It is necessary to make an allowance for fittings. This allowance provides space for the fitting and for the pipe thread into the fitting to make the connection.

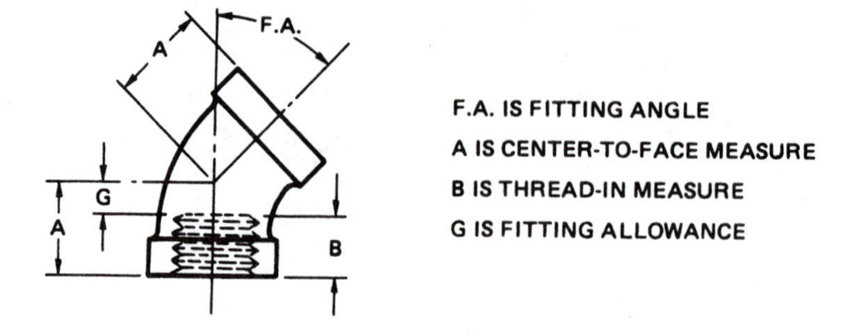

Threaded fittings are not standardized except in thread dimensions. The center-to-face measurement varies with manufacturers. Also, there are long and short patterns of fittings. Data sheets, in the Appendix, show measurements for the fitting allowance to be used with the problems of this book. These tables were made by measuring fittings in the shop. On the job, measure the fitting!

Note: *The metric measurement of pipe should not be a problem to the plumber or pipefitter. When the entire industry, architect, fitting and pipe manufacturer, fixture manufacturer, and other trades use metric, the pipe trades can turn to metric measuring tools and tables without difficulty. Until these changes are made, the use of feet and inches will be common. A mixing of both measuring systems would be difficult. This text uses foot and inch calculations with an introduction to metric in the basic problems.*

Sample Problem Solve for e-e length of pipe K.

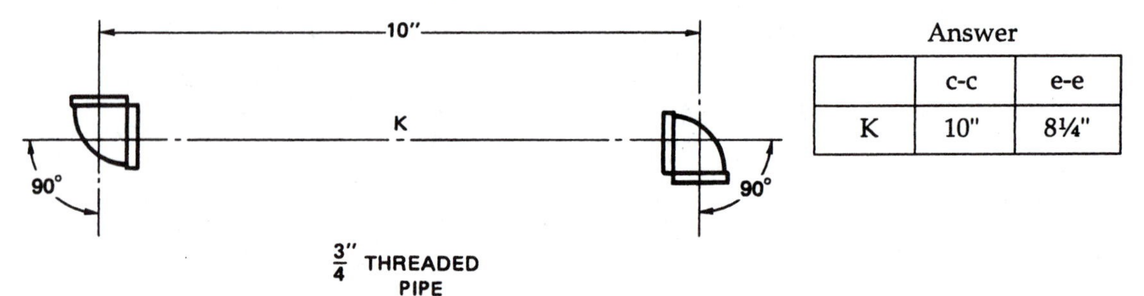

Answer

	c-c	e-e
K	10"	8¼"

Solution

Solve for pipe K: c-c = 10″

Step 1: Find the center-to-face measure and the thread-in measure using Data #5, Appendix.

¾", threaded, 90° ell—A = 1⅜"

B = ½"

Step 2: Express the total fitting allowance for 2 ells.

Total fitting allowance = Two center-to-face measures minus two thread-in measures.

2(1⅜") – 2(½")

Step 3: Find the end-to-end measure

e-e = c-c – (total fitting allowance)

e-e = 10″ – [2(1⅜") – 2(½")]

e-e = 10″ – (2¾" – 1")

e-e = 10″ – 1¾"

e-e = 8¼"

Exercise

Solve Problems 1–5 for e-e length of threaded pipe required. Use the table of fitting sizes, Data #5 and #6 in the Appendix for standard fittings.

	c-c	Threaded Pipe	Ells	e-e
1.	18"	¾"	90° (Data #5)	
2.	2' 1½"	½"	90° (Data #5)	
3.	3' 4"	1"	90° and 45° (Data #5)	
4.	40"	1¼"	45° (Data #5)	
5.	5' 3"	1½"	60° (Data #6)	

Unit 9

Copper Tubing

Objectives

- Describe the domestic uses for copper tubing.
- Understand the uses for the various copper tubing weights.
- Calculate tubing wall thicknesses.

Copper tubing has good performance and long life for water (pressure) lines and for waste line plumbing. For copper plumbing, the tubing is designed for pressure use or for drainage, waste, and vent use. The lightweight tube is less costly than that designed to hold greater pressures. The copper tube is connected to the fitting by a soldering operation.

Copper tubing is made in four weights called types. The types are K, L, M, and DWV, as shown by the following diagrams and table of information.

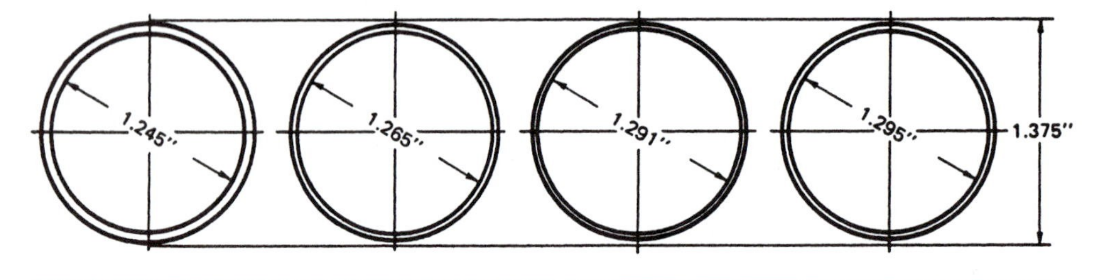

Type	K	L	M	DWV
Recommended for	Underground and interior service	Interior service (underground)	Hot water heating, drainage waste, or vent (interior)	Drainage, waste, and vent service
Available in	Hard or soft temper	Hard or soft temper	Hard temper	Hard temper

The cross sections show the inside and outside diameters of 1¼" nominal size copper tubing. The four types have the same outside diameter. Hard temper tubing is available in 20' straight lengths while soft temper tubing is in coils as long as 100' for the smaller sizes.

Tubing is available in nominal sizes similar to the nominal sizes of threaded pipe. The nominal size of tubing is an approximation of the inside diameter. The outside diameter is greater than the nominal size.

Sample Problem

Compute the wall thickness for K-type tube of 1¼" diameter. Refer to Data #4 in the Appendix.

Solution

Step 1:	Determine O.D.	1.375"
Step 2:	Determine I.D.	1.245"
Step 3:	Subtract I.D. from O.D.	0.130"
Step 4:	Divide by 2	0.065"

Exercise

Answer the following questions.

1. Compute the wall thickness for 3" DWV-type copper tube.
2. Compute the wall thickness for 3" L-type copper tube.
3. Which two types of copper tubing would be recommended for water supply inside a house?
4. What is the advantage of soft temper K-type tube over hard temper K-type tube for a water supply installed in a ditch?
5. Would K-type tube give satisfactory service on DWV use? Give a reason for your answer.
6. Would M-type tube be expected to be satisfactory for interior water lines?

Unit 10

Allowances for Copper Fittings

Objectives

- Explain the difference between pressure and DWV fittings.
- Measure a fitting for copper tubing.

Fittings for copper tubing have a solder cup to make the connection to the tube. For pressure plumbing, such as water lines, the solder cup is deeper than for DWV fittings. The DWV fittings are lightweight because they are used with DWV copper pipe in nonpressure plumbing applications. This is an economy in cost and in use of copper. The method of making allowance for fittings is the same for all copper-to-copper fittings.

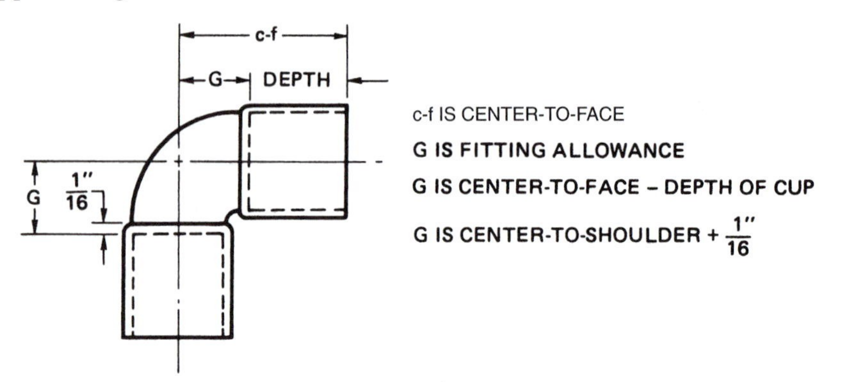

The allowance for a copper fitting may be made by one of two methods. In the first, the fitting is considered as similar to the threaded fitting. Thus, the allowance for fitting equals a center-to-face measurement minus a depth of solder cup. Both of these measurements can be made from the fitting. The measurement is made from the centerline to 1/16" beyond the shoulder. The tables in the Appendix show the allowance for fittings as dimension "G" in the diagram. They are taken from manufacturers' catalogs and are to be used for the problems in this text. In real life, measure the fitting on the job.

Sample Problem

Solve for e-e length of pipe K.

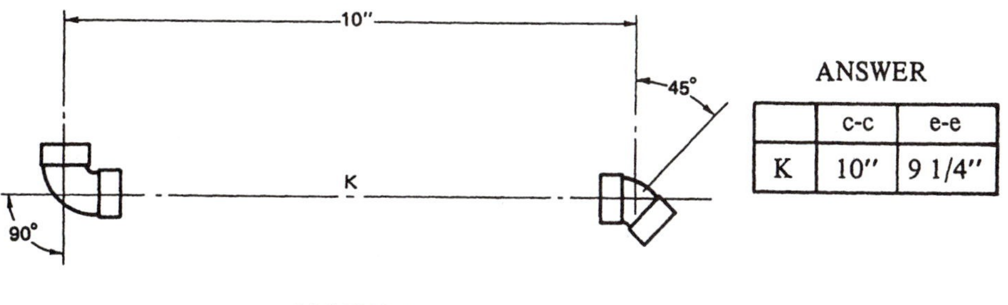

	c-c	e-e
K	10"	9 1/4"

Solution

Solve for pipe K: c-c = 10"

Step 1: Find the fitting allowance for each ell using Data #5, Appendix.

¾", copper, 90° ell—½"

¾", copper, 45° ell—¼"

Step 2: Express the total fitting allowance for 2 ells.

½" + ¼"

Step 3: Find the end-to-end measure.

e-e = c-c – (total fitting allowance)

e-e = 10" – (½" + ¼")

e-e = 10" – ¾"

e-e = 9¼"

Exercise

Solve problems 1–5 for e-e length of copper tube required. Use the allowances for fittings in the Appendix.

	c-c	Copper Tube	Ells	e-e
1.	18"	¾"	90° (Data #5)	
2.	7'-3"	3"	90° (Data #5)	
3.	31½"	1¼"	90° and 45° (Data #5)	
4.	43"	2"	90° and 22½° (Data #5 and #6)	
5.	6'-9"	1½"	90° and 60° (Data #5 and #6)	

Note: *Copper tube and fittings (from Data #5 and #6, Appendix) may be used for many of the problems in the following units of this text. This will not affect c-c answers. Spaces are provided for e-e answers for both threaded and copper, or cast iron and copper, where copper can be used as an alternate method. The use of two materials in a problem allows a comparison of e-e answers. Also, problems may be solved for one e-e answer at the direction of the instructor.*

Unit 11 Plastic Pipe

Objectives

- Present the styles of and weights of plastic pipe.
- Learn the chemical composition of plastic pipe material.

The plastics industry supplies a variety of compositions of plastic tubing (in coils) and rigid plastic pipe (in lengths) and also the fittings that are used for a wide range of applications in the pipe trades. Because of differences in composition and weights of plastic tubing and pipe, the plumber or pipefitter should follow specifications for the installation on which the work is done.

Rigid polyvinyl chloride pipe (PVC) is used for pressure, drainage, waste, and vent installations. PVC pipe is made in three weights for different pressure installations.

The strength of plastic pipe decreases as temperature increases. A rigid chlorinated polyvinyl chloride pipe (CPVC) is available for hot water use. It is industrial thermoplastic piping with an upper working temperature of 215 degrees Fahrenheit, about 60 degrees above that of other rigid vinyl materials. Fittings should match the specifications of the pipe being used.

The outside diameters of PVC pipe are the same as for threaded pipe (Data #2 in the Appendix). The same dimensions are not used for other plastic pipe materials.

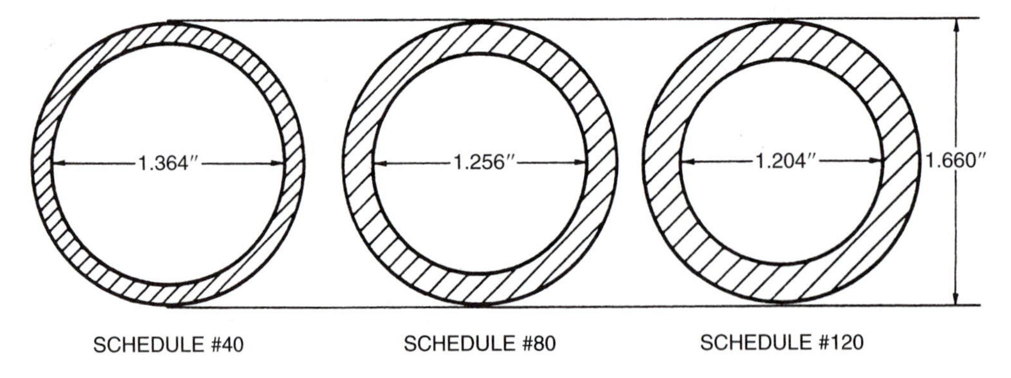

Type	Schedule #40 PVC	Schedule #80 PVC	Schedule #120 PVC
Recommended for	Water Pressure to 160 psi Water to 73.4°F 23°C	Water Pressure to 200 psi Water to 72.4°F 23°C	Water Pressure to 315 psi Water to 73.4°F 23°C
Note: CPVC is to be considered for temperatures of 120°F and higher.			

The illustrated cross sections show the inside and outside diameters of 1¼" nominal size PVC pipe. The three types have the same outside diameter. Standard lengths are 10 feet and 20 feet.

Sample Problem

PVC pipe is rated at 62 percent of strength at 100°F compared to its strength at 73.4°F. What water pressure may be used at 100°F for Schedule 40 PVC pipe?

160 psi (at 73.4°F) × 0.62 = 99.2 psi (at 100°F)

Exercise

Multiple Choice: Select the best answer to complete each statement.

1. The wall thickness for 1¼" schedule 40 PVC pipe is

   a. 0.296" c. 0.148"

   b. 0.202" d. 0.228"

2. Schedule 40 PVC pipe has a pressure rating factor of 22 percent at 140°F. The recommended psi for 140°F is

   a. 195.2 c. 44

   b. 35.2 d. 69.3

3. The recommended PVC pipe for drainage, waste, and vent service is

   a. Schedule #40 c. Schedule #120

   b. Schedule #80 d. CPVC (schedule #40)

4. The recommended PVC pipe for 50 psi cold water service is

   a. Schedule #40 c. Schedule #120

   b. Schedule #80 d. CPVC (schedule #40)

5. The recommended PVC pipe for 50 psi hot water service is

   a. Schedule #40 c. Schedule #120

   b. Schedule #80 d. CPVC (schedule #40)

6. The O.D. of 2" steel pipe is

   a. 2³⁄₁₆" c. 2⅜"

   b. 2" d. 2¹⁄₁₆"

7. The O.D. of 2" PVC pipe is

   a. 2³⁄₁₆" c. 2⅜"

   b. 2" d. 2¹⁄₁₆"

Unit 12

Allowance for Plastic Fittings

Objectives

- Compare plastic fittings to copper fittings.
- Allow for the plastic hub type fitting.

PVC fittings have a close tolerance cup that fits over the end of the joining pipe. It is similar to copper fittings in this way. The joint is made by use of a combined solvent-cement that softens the surfaces and fills the space between the cup and the pipe. Pressure fittings are made in three weights to match the PVC pipe. There also are CPVC fittings for higher temperature use.

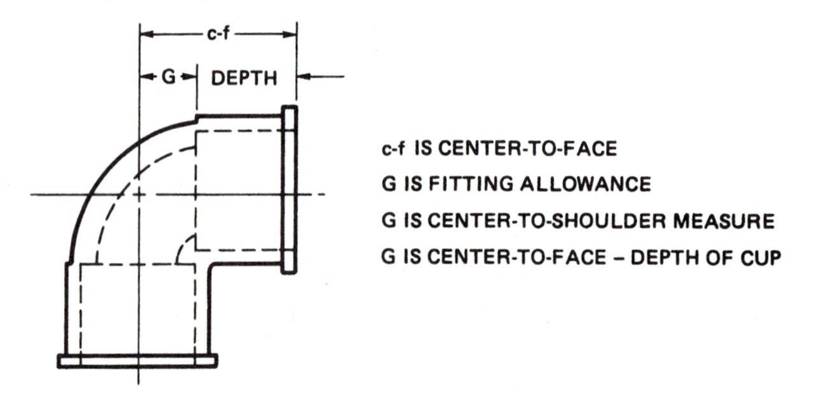

The allowance for PVC fittings is similar to copper fittings. It is possible to measure the center to face and subtract the measurement for cup depth to get the fitting allowance. The fittings are more often measured as illustrated from center to the outside line of the cup. The tables in the text are from manufacturers' catalogs and from measured fittings. Measure the fittings on the job! Use the tables of fitting allowances for the problems in this text.

Sample Problem

Solve for e-e length of pipe K.

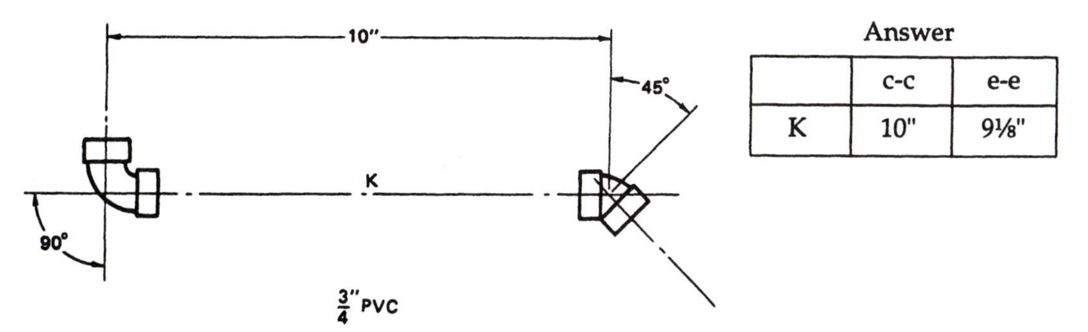

Answer

	c-c	e-e
K	10"	9⅛"

Solution

Solve for pipe K: c-c = 10"

Step 1: Find the fitting allowance for each ell using Data #5, Appendix.

¾", PVC, 90° ell—9/16"

¾", PVC, 45° ell—5/16"

Step 2: Express the total fitting allowance for 2 ells.

9/16" + 5/16"

Step 3: Find the end-to-end measure.

e-e = c-c (total fitting allowance)

e-e = 10" – (9/16" + 5/16")

e-e = 10" – ⅞"

e-e = 9⅛"

Exercise

Solve problems 1–5 for e-e length of PVC pipe required. Use the allowances for fittings in the Appendix.

	c-c	PVC Pipe	Ells	e-e
1.	7½"	¾"	90° (Data #5)	
2.	6'–2"	3"	90° (Data #5)	
3.	37"	1¼"	90° and 45° (Data #5)	
4.	42¼"	2"	90° and 22½° (Data #5 and #6)	
5.	7'–8"	1½"	90° and 60° (Data #5 and #6)	

Note: *The center-to-center length in any pipe diagram is the same no matter which pipe and fittings material is used. The fittings allowance is different with different materials. However the plastic and copper allowances are made by the same steps, while threaded fittings require extra calculation steps to allow for thread-in.*

The end-to-end computations can be done for more than one pipe material from the same center-to-center calculation; student interest and needs should determine which e-e calculations to use in the problems. Experience has been that copper and plastic e-e answers are easy for the student after the e-e for threaded pipe is mastered.

PVC pipe in schedule #80 and #120 weights may be threaded with standard pipe thread dies. There are also threaded PVC fittings.

PVC pipe can be welded; a hot gas method is recommended with a torch temperature of 600° F and a welding temperature of 500°F. Successful welding of PVC requires training.

Unit 13 Welded Steel Pipe

Objectives

- Define the terms used in welded piping.
- Show the kinds of materials used.
- Describe the considerations for fitting allowance.

In pipe fitting, the term "butt-welded" is often used, incorrectly, to describe all kinds of pipe joining by the arc or gas welding process. To clarify this point, there are two major kinds of welds: the butt weld and the fillet weld (see figure at right).

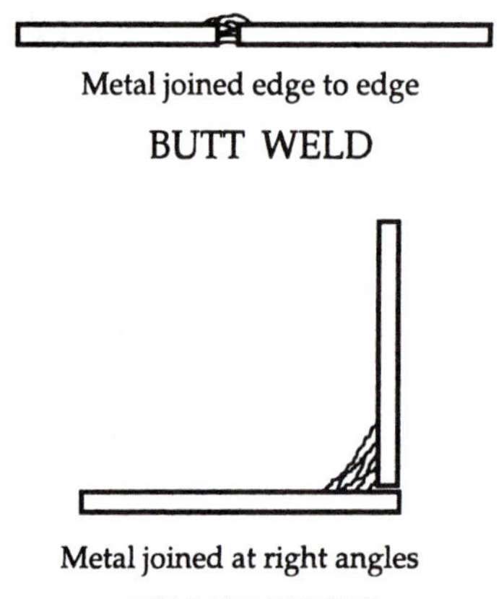

The terms "lap weld," "socket weld," and others are all derived from these two basic welds.

Welded joints are used when exceptionally rugged service is expected of the pipe assembly in question. This may be for pipe which is expected to carry higher than ordinary internal pressures, as in high-pressure steam and hydraulic piping. Welded pipe joints should also be used in high vibration applications or when the pipe is subject to severe shock.

The pipe used is steel pipe in any weight in any of the steel pipe manufacturing methods, butt-weld, lap-weld, or seamless.

Fittings for Welded Pipe

Fittings for welded applications are available in weights of up to 2000 PSI and greater.

Flanges

While socket-weld unions are available, the most common weld-type fitting used to obtain portable joints is the flange. Flanges may be "slip-on," "weld-neck," or "socket-weld."

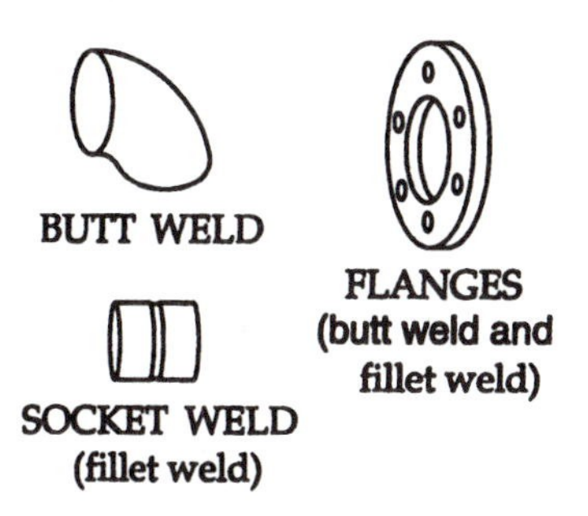

The slip-on flange simply slides over the outside diameter of the pipe; the pipefitter makes two fillet welds and the job is complete. The weld-neck flange has a projection which matches the inside and the outside diameters of the pipe; a simple butt weld joint is made.

Allowances for flanges must take a number of factors into account:

Slip-on

1. The distance that the pipe is withdrawn inside the flange. This is the wall thickness of the pipe to which the flange will be joined.
2. The thickness of the gasket to be used between the two flange facts.

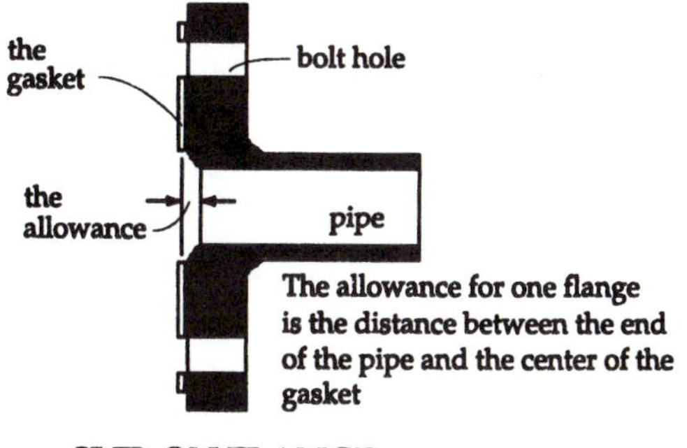

SLIP-ON FLANGE

Weld-neck

1. The distance from the face of the flange to the butt-weld end of the flange.
2. The thickness of the gasket to be used between the two flange faces.

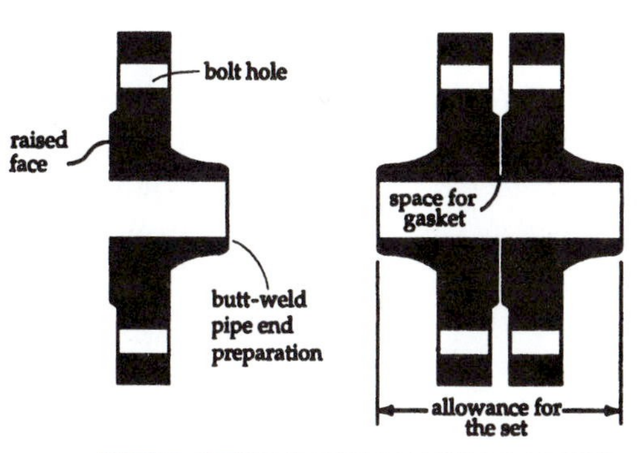

WELD-NECK, RAISED FACE, FLANGES

Bolts or cap screws

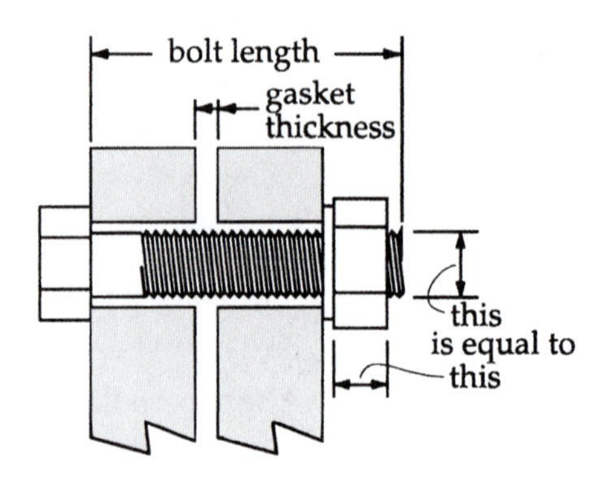

1. The outside diameter of the bolt is equal to the diameter of the bolt-hole in the flange minus one-eighth of an inch.
2. The length of the flange bolt (the distance between the underside of the bolt head and the end of the threads) is equal to the thickness of both flanges added together plus the thickness of the gasket plus the thickness of the washer plus the length of bolt which must protrude through the other side to accept a nut. This last distance is equal to the bolt diameter. (See figure above.)

Bolt lengths come in increments of one quarter of an inch. For example, a ⅝" bolt might come in sizes of 2", 2¼", 2½", 2¾", etc. It is important that the end of the bolt does not end up below the surface of the nut. Always use the next longer bolt length if this is going to happen.

Butt-Weld Fittings

The butt-weld fitting is unusual in that the pipe does not go inside the fitting as is the case with hubbed fittings. The measurement of the fitting from center to face, plus a small amount for what is called weld penetration, is the allowance for the particular fitting.

1/8 inch

A gap is left between the fitting and the pipe for welding penetration

So-called weld elbows come in two major varieties for the 90° fitting pattern, short radius, and long radius. (See Figures 13–1 and 13–2.) This means that, while the degree of the turn is the same in both cases, the long-radius fitting turns the direction of the flow more gradually than the short-radius fitting.

The butt-weld tee has the same fitting allowance as the short radius elbow. The cut-in allowance for a tee, i.e., the amount that would have to be cut out of a straight line of pipe to accept the fitting, is twice the fitting allowance for a short-radius elbow plus ¼" for weld penetration on both sides of the tee. A 1½-inch butt-weld tee would therefore take up 3¼ inches in a line of pipe.

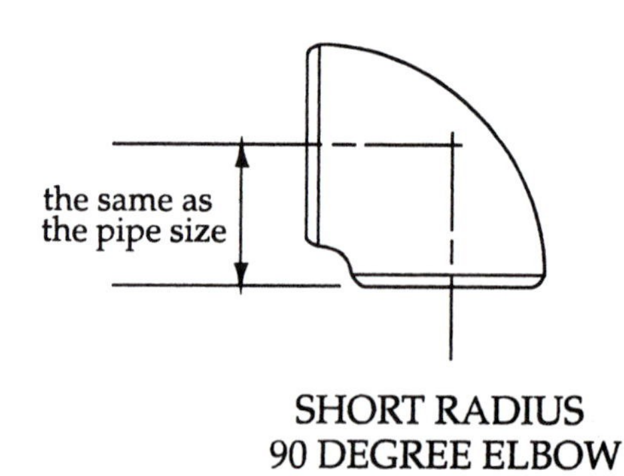

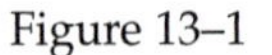
Figure 13–1

The 45° butt-weld elbow has a fitting allowance that is ⅝ times the pipe size plus approximately ⅛ inch for weld penetration. If the pipe size in question was 2 inches then multiply 2 × 5 and then divide the answer by 8.

2 × 5 = 10 and 10/8 = 1¼ inches.

The allowance for a 2 inch 45° butt-weld elbow then is 1¼ inches. Adding ⅛ inch for weld penetration yields working allowance, for the 2-inch 45° butt-weld elbow, of 1⅜ inches.

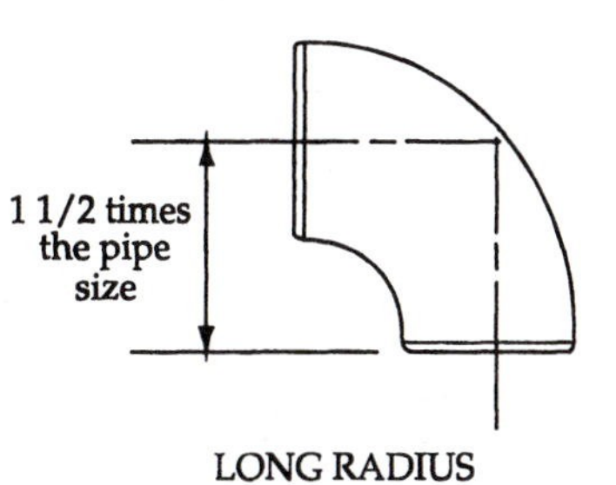

Figure 13–2

Socket-Weld Fittings

The socket-weld fitting is handled like plastic or copper hubbed fittings. The steel pipe in this case slides rather than threads into a steel fitting. Then a fillet weld is made between the face of the fitting and the pipe.

SOCKET-WELD FITTINGS

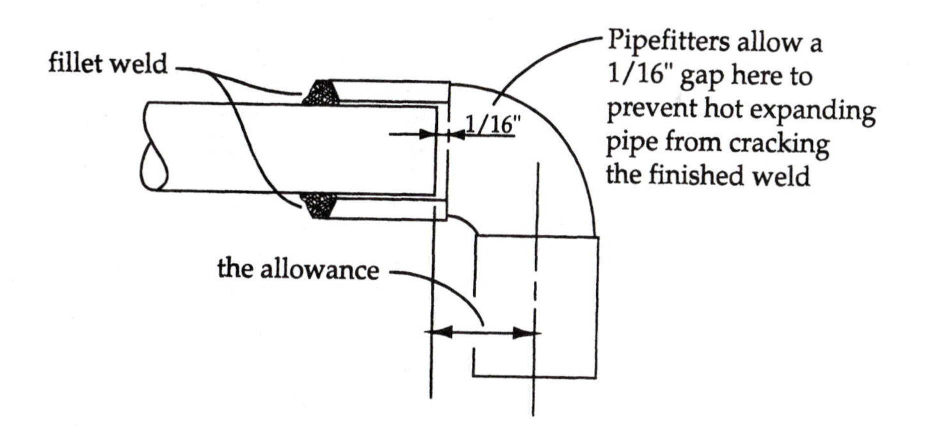

When making an allowance for socket-weld fittings, the pipefitter withdraws the pipe from the full depth of the socket by 1/16 inch. This allows for the pipe's expansion as welding heat is applied, and keeps the expanding pipe from forcing the newly formed weld apart. This 1/16 inch must be added to the total fitting allowance.

REMEMBER: The allowance for any socket-type fitting is equal to the distance from the end of the pipe within that fitting to the center line of the other outlet.

Exercise

Select the BEST answer to the following questions.

1. There are two different kinds of welds. They are the
   a. butt weld and the fillet weld.
   b. socket weld and butt weld.
   c. flange weld and socket weld.
   d. lap weld and butt weld.
2. To provide for proper weld penetration in a butt joint
   a. the pipe and fitting must be in close contact.
   b. the fitting is welded from the other side of a partition.
   c. a small gap between the pipe and the fitting must be permitted.
   d. the fitting must be assembled in reverse order.
3. A ______ fitting allows the pipe subassembly to be portable.
   a. tee
   b. flange
   c. elbow
   d. welded
4. Welded pipe is used in ______ applications.
   a. drainage
   b. low stress
   c. domestic water supply
   d. high vibration
5. Which kind of flange is butt-welded to the pipe?
   a. slip-on
   b. companion
   c. raised-face
   d. weld-neck
6. The kind of flange which has the largest fitting allowance is the
   a. tee flange.
   b. weld-neck.
   c. slip-on.
   d. raised-face.
7. When a slip-on flange is used the total fitting allowance amounts to the ______ plus one-half of the gasket thickness.
   a. wall thickness of the pipe
   b. height of the welding neck
   c. length of the bolt used
   d. 9/16 inch
8. The fitting allowance of a long radius 90° elbow is equal to
   a. 1 times the pipe size.
   b. 5/8 times the pipe size.
   c. 1½ times the pipe size.
   d. 1/8 inch.
9. What would be the total fitting allowance, allowing for weld penetration, for a 3-inch 45° butt-weld elbow?
   a. 4 5/8 inches
   b. 3 1/8 inches
   c. 5/8 inch
   d. 2 inches
10. To calculate the allowance for a standard butt-weld tee just
    a. use the same fitting allowance as the long-radius 90° elbow.
    b. add one-half of the gasket thickness.
    c. use the pipe size.
    d. subtract 3 inches.

Unit 14 Allowing for Welded Fittings

Objectives

- Learn about weld flange allowances.
- Practice allowing for weld-type fittings.

The student may need to refer to the previous unit, *Welded Steel Pipe*, for the theory of allowing for welded fittings when working out the practical problems in this unit.

Butt-Weld Fittings

Allowances for the major butt-weld fittings, specifically the elbows and the tees, can be calculated as follows:

Long-radius 90° butt-weld elbow	1½ times the pipe size
Short-radius 90° butt-weld elbow	1 times the pipe size
45° butt-weld elbow	⅝ times the pipe size
Butt-weld tee	1 times the pipe size

It is important to remember that the fitting must be slightly separated from the pipe to which it is going to be welded. An allowance of 3/32 inch is standard for the smaller pipe sizes. One-eighth inch is a suitable allowance for most cases, and that is what we will use to solve these practice problems. The end-to-end measurement of the pipe will have to be reduced by an additional ⅛ inch for each butt-weld fitting.

Slip-On Flanges

Slip-on flanges are assembled by sliding the flange over the pipe. However, the pipe is not allowed to protrude from the face of the flange or even to come even (flush) with the face of the flange. To provide for two welding surfaces, the end of the pipe is withdrawn from the face of the flange by a distance equal to the pipe's wall thickness (see accompanying figure).

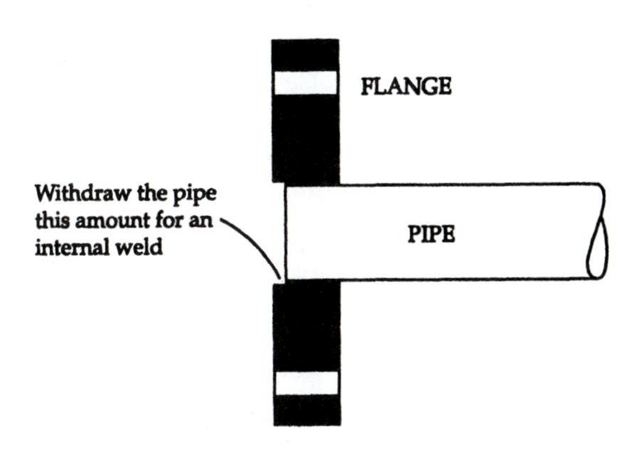

The allowance for this flange now amounts to this withdrawn amount. The total allowance would be this amount *plus the thickness of the flange gasket.*

Table 14-1. Flat-Faced, Slip-On Flanges

(Use this chart for the sample and practice problems.)

Pipe size in inches	1½	2	3	4	5
Hub diameter	2⅞	3⁵⁄₁₆	4⅝	5¾	6¾
Diameter of flange	6	6½	8¼	10	11
Thickness of flange	1¹³⁄₁₆	⅞	1⅛	1¼	1⅜
Pipe wall thickness	³⁄₁₆	³⁄₁₆	¼	¼	¼
Diameter of bolt circle	4½	5	6⅝	7⅞	9¼
Number of bolts	4	4	8	8	8
Diameter of bolt holes	¾	¾	⅞	⅞	⅞
Gasket thickness	⅛	⅛	⅛	⅛	⅛

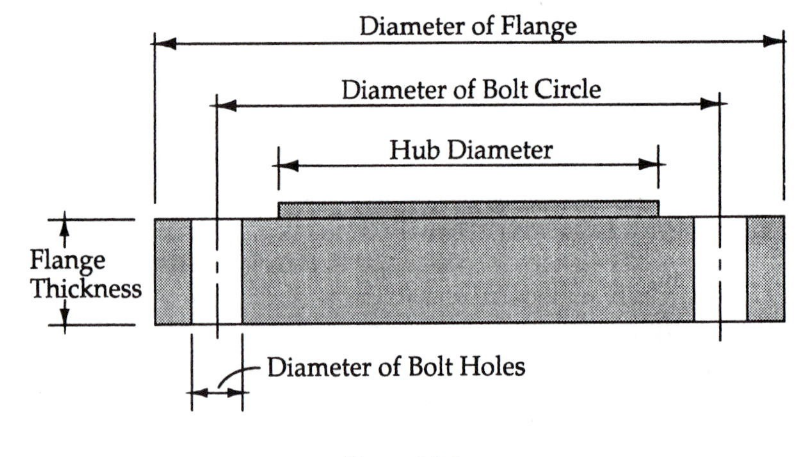

Figure 14–1

Can you see that if we were given a measurement to the center of a pair of flanges bolted together, the measurement would fall in the center of the gasket? In this case the total allowance for either one of the flanges would include the pipe drawback plus one-half the thickness of the gasket. If the starting measurement began at the surface of one flange, the total companion flange's allowance would include the pipe drawback plus the *whole* gasket thickness. It all depends on where the stated measurements begin. In one case the measurement was to the center of the two flanges. In the other case the measurement was to the surface of one of the two flanges.

Table 14–1 gives the measurements for a specific kind of flange and a limited number of pipe sizes. Figure 14–1 shows a detail drawing that explains the measurements displayed in the table. In the case of slip-on flanges in which the allowance is determined by the amount of pipe drawback and the thickness of the gasket, the dimensions are useful mainly for determining the number of bolts required for the job, the size and the length of the bolts, and the size of the gasket needed.

Sample Problem Solve the following problem for the end-to-end lengths. Use the drawing below for the working dimensions and a pipe size of 1½ inches.

Step 1: Calculate the center-to-center dimensions for the pipe cuts to be made. (A): Subtract 12 from 32. Then subtract another 1/16 inch to get to the center of the left gasket because the given dimension is from the face of the flange on the tank. (B): The center-to-center measurement for this one is 12 inches, the same as the given measurement. (C): We can obtain this measurement by subtracting the sum of 9 and 32 from 72 (31 inches). In this case we find an unknown horizontal measurement by subtracting the known horizontal measurements from a known overall measurement. What is left over is the unknown horizontal measurement. (D): This is given on the drawing as 24 inches. (E): This is given on the drawing as 9 inches.

Step 2: Calculate the allowances that are going to be used. One 1½-inch slip-on flange: One pipe wall thickness (3/16") plus one-half of a gasket thickness (½ × ⅛) = 1/16 inch. Total ¼". This may seem odd, but remember: The fitting allowance is the distance from the end of the pipe within the flange to the face of the flange.

One 1½-inch butt-weld tee: 1 times the pipe size plus ⅛ inch for weld penetration. Total 1⅝".

One 1½-inch long-radius 90° butt-weld elbow: 1½ times the pipe size plus ⅛ inch for weld penetration. Total 2⅜".

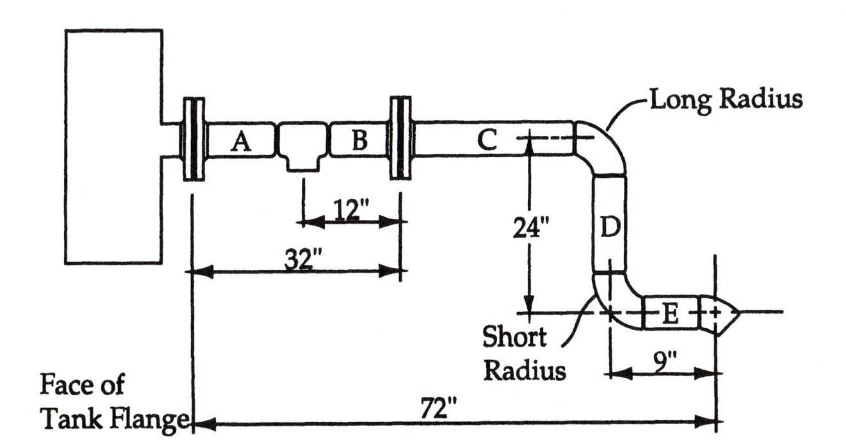

One 1½-inch short-radius 90° butt-weld elbow: 1 times the pipe size plus ⅛ inch for weld penetration. Total 1⅝".

One 1½-inch 45° butt-weld elbow: ⅝ times the pipe size plus ⅛ inch for weld penetration. Total 1 1/16".

Eight ⅝ × 2½-inch bolts, eight ⅝-inch nuts, eight ⅝-inch washers: There are two sets of flanges with four bolt holes in each. This means quantities of eight. The bolt size is ⅛ inch less than the bolt hole diameter. The length of the bolt is equal to 2 times the flange thickness plus 1 gasket thickness, plus 1 washer thickness, plus the thickness of one ⅝" nut (⅝ inch).

Step 3: Subtract the allowances from Step 2 from the center-to-center measurements from Step 1.

Solutions

Length	c-to-c	Total Fitting Allowances		e-to-e
A	19 15/16"	– flange (1/4")	– tee (1 5/8")	= 18 1/16"
B	12"	– tee (1 3/8")	– flange (1/4")	= 10 1/8"
C	31"	– flange (1/4")	– LR elbow (2 3/8")	= 28 3/8"
D	24"	– SR elbow (1 5/8")	– LR elbow (2 3/8")	= 20"
E	9"	– SR elbow (1 5/8")	– 45° elbow (1 1/16")	= 5 9/16"
Bolts	8 ea 5/8" × 2 1/2"			
Nuts	8 ea 5/8"			
Washers	8 ea 5/8"			
Gaskets	2 ea 1/8" thick			

Practice Problems

Do the following problems. A chart is provided below for your answers. It is suggested that you work out the problems using the same method and step-wise problem solving as is demonstrated in the previous sample problem.

Size inches		c-to-c	– Fitting one	Allowance	– Fitting two	Allowance	e-to-e
2	A						
	B						
	C						
	D						
	E						
3	A						
	B						
	C						
	D						
	E						
4	A						
	B						
	C						
	D						
	E						
5	A						
	B						
	C						
	D						
	E						

Unit 15 Equal Spacing

Objectives

- Calculate equal center-to-center measurements.
- Convert from fractions to decimals.
- Make an equal spacing pipe cut list.

In pipe fitting, equal spaces are equal c-c lengths. Often the e-e lengths are also equal, but not always. Different e-e lengths would result if the allowance for fitting were not the same for each pipe.

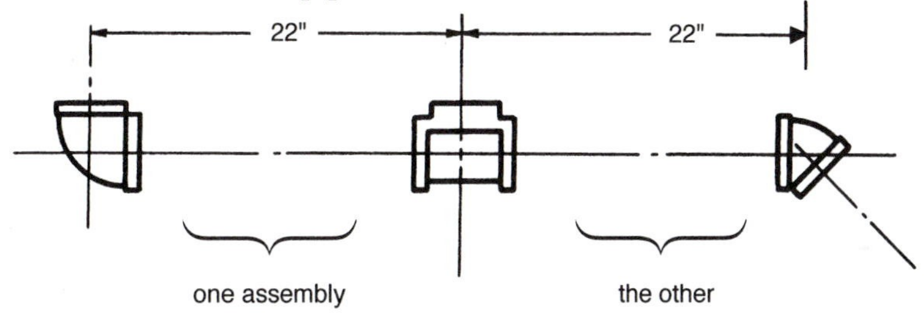

If one pipe assembly used a 90° ell and the other a 45° ell, then the c-c lengths would be alike. The e-e lengths, however, would differ because of the difference in fitting allowance between a 90° ell and a 45° ell.

Sample Problem

Distance c-c is $20\frac{5}{16}$". Divide into three equal spaces.

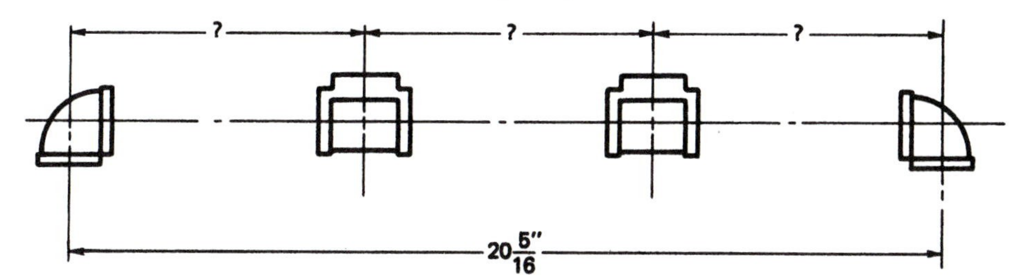

Solution

Step 1: Convert to whole and decimal inches.

$20\frac{5}{16}$" = 20.312" (See Data #1, Appendix.)

Step 2. Divide by 3.

20.312" ÷ 3 = 6.770"

Step 3. Convert 0.77" to nearest 16th inch.

0.77" = ¾"; 6.77" becomes 6¾"

Exercise

In each of the following equal-spaced pipe problems, solve for c-c and e-e lengths for each lettered pipe. (Solve e-e length for threaded, copper, and PVC pipes.)

1.

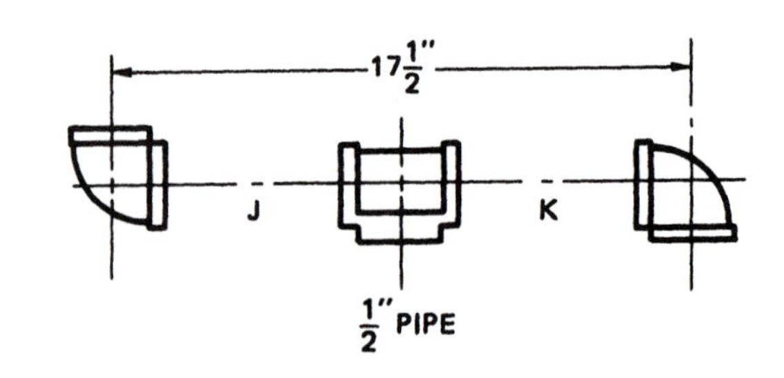

	c-c	Threaded e-e	Copper e-e	PVC e-e
J				
K				

2.

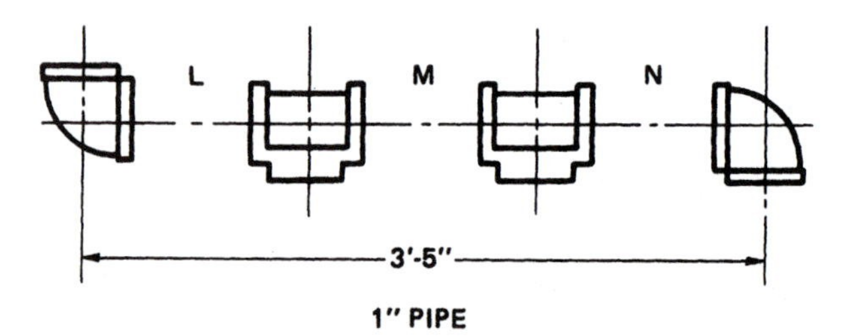

	c-c	Threaded e-e	Copper e-e	PVC e-e
L				
M				
N				

3.

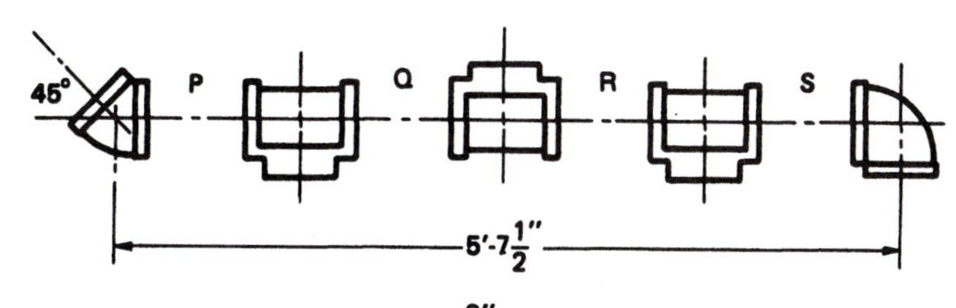

	c-c	Threaded e-e	Copper e-e	PVC e-e
P				
Q				
R				
S				

4.

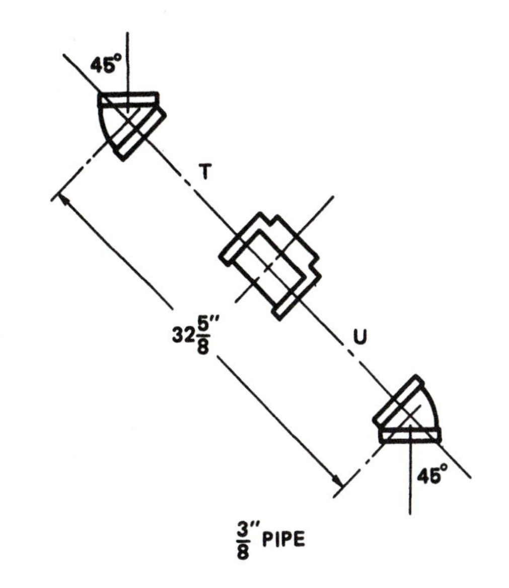

	c-c	Threaded e-e	Copper e-e	PVC e-e
T				
U				

Unit 16 Angles in Plumbing

Objectives

- Describe how a fitting angle is measured.
- Compare the bend or fitting angle to a circle.
- Define the complementary angle.

The smaller sizes of pipe, under 1¼", have fittings available in 90° and 45° fitting angles. The larger sizes have fittings of 60°, 22½°, and 11¼°, as well as the 90° and 45° fittings. In cast iron pipe, there is a 72° fitting, but no 11¼° fitting.

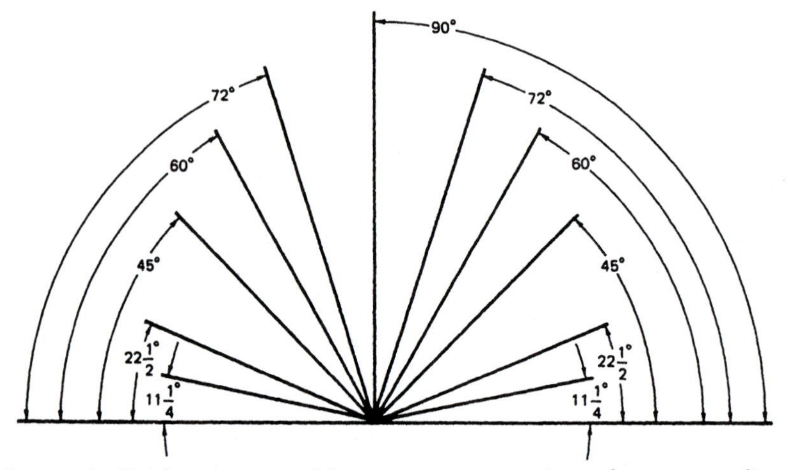

The *fitting angle* (F.A.) is measured between an extension of one centerline to the other centerline. The larger the fitting angle, the greater the angle measure.

The *complementary angle* (C.A.) is often mistaken for the fitting angle. The complementary angle is found by subtracting the fitting angle from 90°. The larger the fitting angle, the smaller the complementary angle.

C.A. = 90° – F.A.

C.A. = 90° – 60°

C.A. = 30°

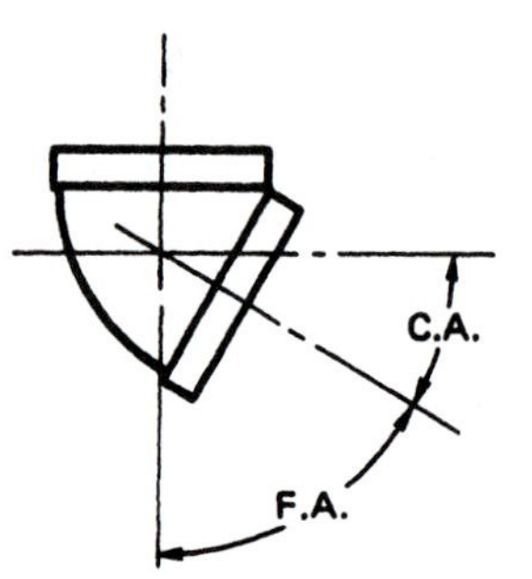

A bend is a fraction of a circle. The bend equals the fitting angle divided by 360°.

$$\text{Bend} = \frac{\text{F.A.}}{360°}$$

$$\text{Bend} = \frac{60°}{360°}$$

$$\text{Bend} = \tfrac{1}{6}$$

A ⅙ bend is a cast iron elbow of 60° fitting angle. A 60° ell, or elbow, is a threaded-style fitting. The fitting angle equals the bend times 360°.

$$\text{F.A.} = \text{bend} \times 360°$$

$$\text{F.A.} = \tfrac{1}{6} \times 360°$$

$$\text{F.A.} = 60°$$

Exercise

1. Complete the missing items in the following table:

Fitting Angle	Complementary Angle	Bend
90°		
72°		
60°	30°	⅙
45°		
22½°		
11¼°		

2. If there were a 50° cast iron fitting, what bend would it be?
3. If there were a ⅕ bend, what fitting angle would it have? What complementary angle?
4 Write the plumbing trade name for each of the angles shown by letters in the following illustration.

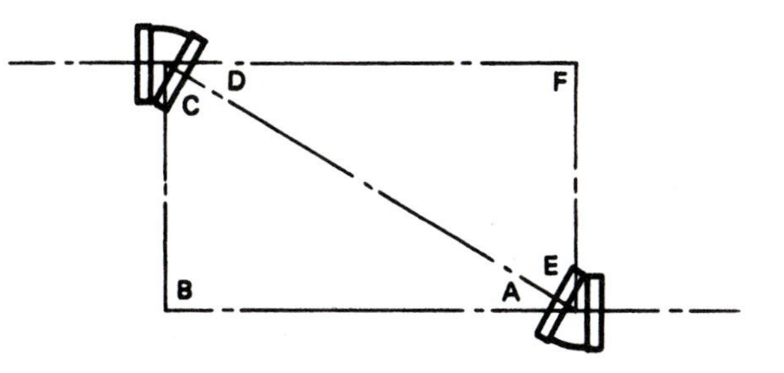

Unit 17 Offset, Diagonal, Rise, and Run

Objectives

- Define the meaning of *offset.*
- Understand rise and run.
- Use a protractor to visualize the offset rectangle.

Offset

When two pipes are parallel to each other, they are an offset distance apart. They may both be horizontal or they may both be vertical. Pipe A is parallel to pipe B and pipe C is parallel to pipe D. The distance between the centerlines of two parallel pipes is called *offset.* The offset is the center-to-center length if the parallel pipes are connected using 90° fittings.

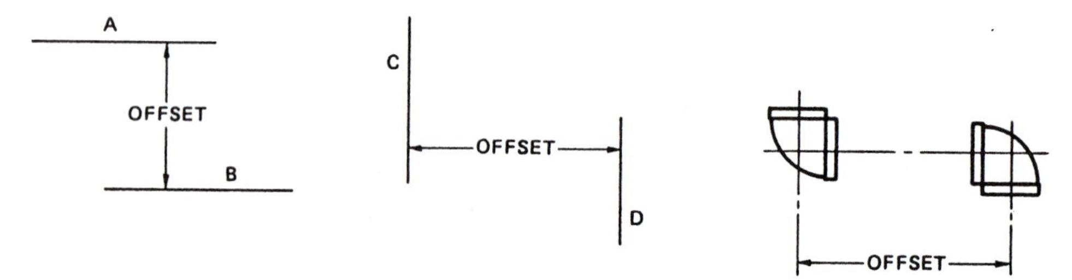

Diagonal

If two parallel pipes are connected by fittings other than 90°, then the center-to-center length of the connecting pipe is a *diagonal.* The name comes from the diagonal of the square or rectangle as shown. The diagonal construction is preferred to the 90° fittings because there is less restriction to flow.

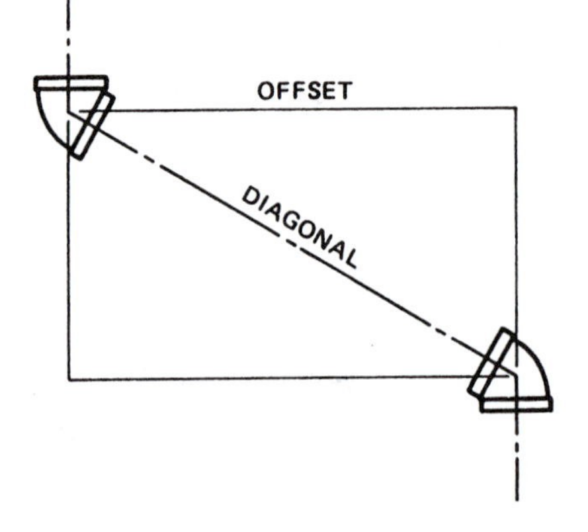

Rise or Run

Rise and run are the same distance. The figures show how the name changes. For parallel vertical pipes, there is a *rise*. But for parallel horizontal pipes, there is a *run*. The term *set* is a shortened version of the word *offset*.

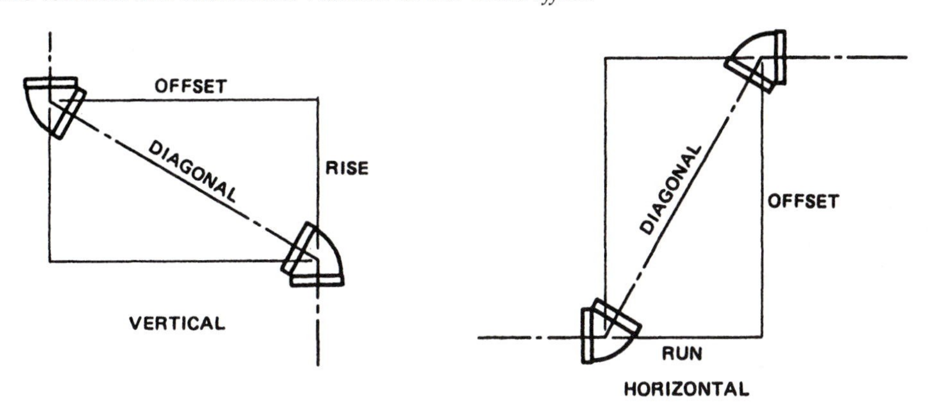

Sample Problem

Make two pipe diagrams, one with pipes horizontal and the other with pipes vertical, using 60° fitting angles. Draw a rectangle around each pipe diagram and label each line with its plumbing trade name. Use a protractor to measure the fitting angles.

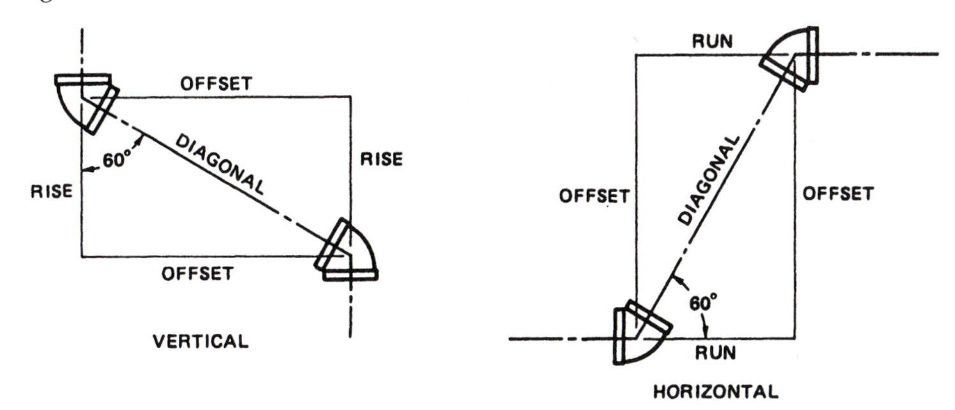

Exercise

For each of the given fitting angles, construct two pipe diagrams, one with the diagonal connecting vertical pipes, the other with the diagonal connecting horizontal pipes. Enclose each diagonal with a square or rectangle and label each line.

1. 45° fitting angle
2. 22½° fitting angle
3. 11¼° fitting angle
4. 72° fitting angle

Unit 18 Solving with 45° Constants

Objectives

- Demonstrate how angle ratio constants are developed.
- Use the Pythagorean theorem in a practical situation.
- Solve offset, run, and diagonal problems.

The diagonal and offset are proportional. In other words, as illustrated by the figure below, the offset and diagonal are dependent upon each other. If one is changed, it produces a relative change in the other.

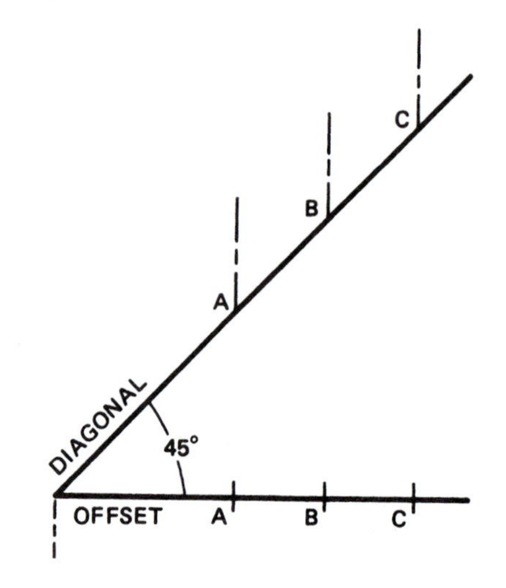

A ratio number called a constant represents this relative change. The diagonal multiplied by a certain constant equals the offset. The offset multiplied by a certain other constant equals the diagonal. For any given angle, these constants do not change.

It is quite simple to look up these constants on a data sheet. Data #20, in the Appendix, lists the constants for 45° fittings. However, understanding how these constants were derived will help you understand how and why they are used. Only mathematics of a general nature is used but there are several steps.

Sample Problem

Solve for the constants for a 45° diagonal, using an offset of 6".

Solution

1. Determine the run. The triangle formed by the run, offset, and diagonal has two equal angles. The sides of this triangle must be equal. Or, the sides of the square formed are equal. Thus, the run equals the offset, or 6".

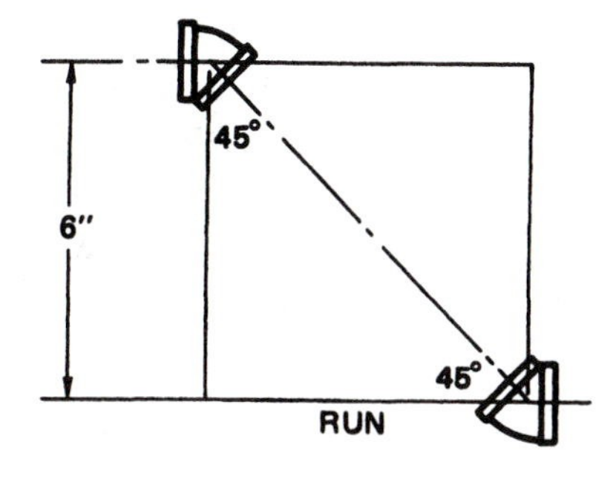

   Remember the Pythagorean theorem from Unit 4?

   $c^2 = a^2 + b^2 \Rightarrow c = \sqrt{a^2 + b^2}$

   In this case the run and the offset of the triangle will replace a^2 and b^2 in the Pythagorean equation.

2. Solve for the diagonal. Diagonal $= \sqrt{\text{run}^2 + \text{offset}^2}$

   $= \sqrt{36 + 36}$ or $\sqrt{72}$

   $= 8.485"$

3. Divide the diagonal by the offset. 8.485" ÷ 6" = 1.414

   The diagonal, then, is 1.414 times as long as the offset. The offset multiplied by 1.414 equals the diagonal.

4. Divide the offset by the diagonal. 6" ÷ 8.485" = 0.707

   The offset is 0.707 as long as the diagonal. The diagonal times 0.707 equals the offset.

Exercise

1. Solve for the two constants as in the sample problem, using an offset of 9". (Show all work.)

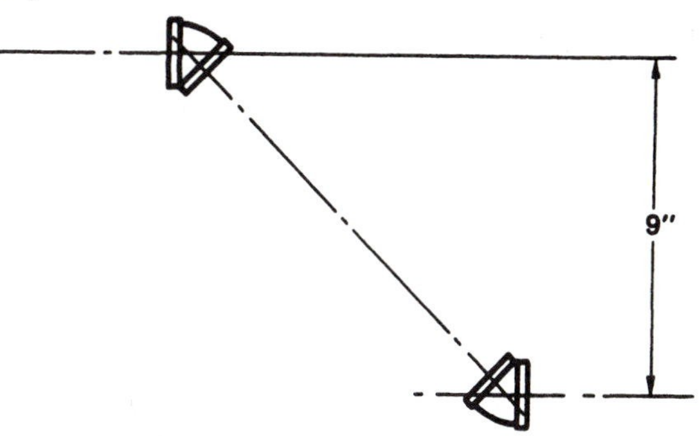

2. Assuming a 45° angle in each of the following, compute the missing information.

	Offset	Run	Diagonal
a.	10"		
b.		1' 4"	
c.			2' 6"
d.	3' 5"		
e.			3' 5"
f.	1' 2"		
g.	20 cm		
h.			150 cm

Unit 19 45° Diagonal and Pipe Length

Objectives

- Calculate pipe diagonals.
- Use a constant to derive pipe lengths.

A sequence of steps enables the plumber to determine the e-e length to cut the pipe to make up the diagonal.

- The offset is measured from the job, but for problem purposes, it is given as a measured distance.
- The diagonal is found by multiplying the offset by a constant; in this case, 1.414. This is the c-c or diagonal length.
- The pipe length is found by making an allowance for fittings. This procedure can be used for any size or any kind of pipe.

Sample Problem

Solve for c-c and e-e lengths for threaded pipe K.

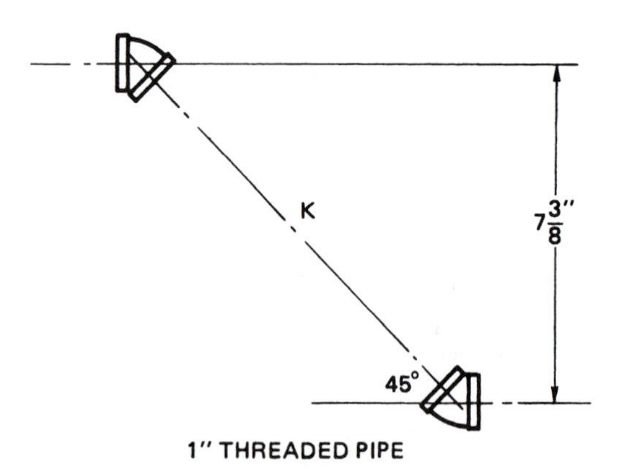

Answers

	c-c	e-e
K	10 7/16"	9 3/16"

Solution

1. Convert offset to decimal inches.

   7⅜" = 7.375" (Data #1)

2. Find c-c length for pipe K. (Data #20)

   diagonal = 1.414 × offset

   c-c = 1.414 × 7.375"

   c-c = 10.428 250" or 10 7/16" (Data #1)

3. Find the total fitting allowance. (Data #5)

   1", threaded pipe, 45° ell – A = 1⅛"

   B = ½"

   2 (1⅛") – 2(½")

4. Find e-e length for pipe K.

   e-e = c-c – (total fitting allowance)

   e-e = 10 7/16" – [2 (1⅛") – 2 (½")]

   e-e = 10 7/16" – 1¼"

   e-e = 9 3/16"

Exercise

For each offset and pipe size determine c-c and e-e for pipe K.

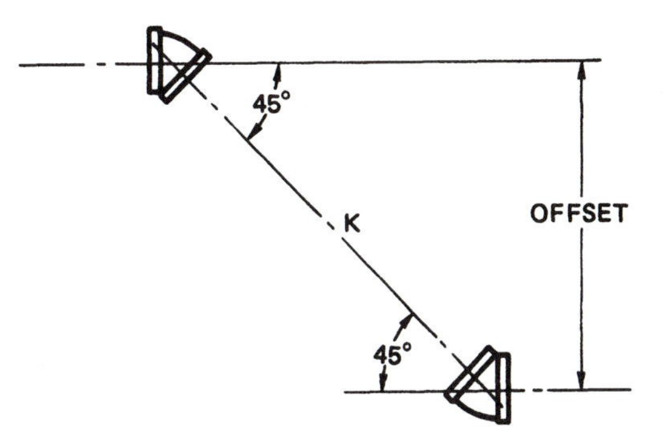

	Offset	Pipe Size	c-c	Threaded e-e	Copper e-e	PVC e-e
1.	8¼"	¾"				
2.	9"	1"				
3.	11½"	½"				
4.	18"	1¼"				
5.	22¾"	1¼"				
6.	24"	1½"				
7.	2'-3½"	1½"				
8.	3'-9"	2"				
9.	5'-6½"	2"				

Unit 20 Three-Pipe Diagrams with a 45° Offset

Objective

- Solve practical 45° offset problems.

The three-pipe diagram is often a part of a more complete pipe assembly. The methods of solving for diagonal length and of making the fittings allowance have been explained and practiced in previous units.

In this unit the rise is needed as a part of the solution for the third pipe. On the job, the plumber would measure the overall, the offset, and the c-c length for the first pipe. These dimensions are given in the problems.

Sample Problem Solve for c-c and e-e dimensions of threaded pipes J, K, and L as shown in the three-pipe diagram, using 45° offset.

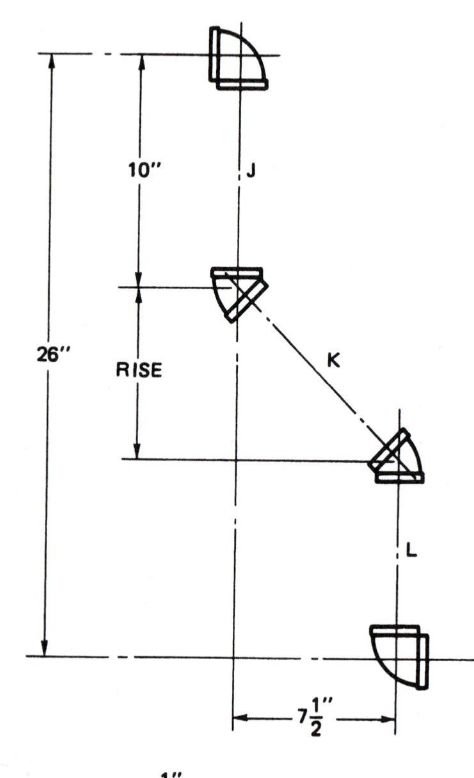

Answers

	c-c	e-e
J	10"	9⅛"
K	10⅝"	10⅛"
L	8½"	7⅝"

Solution

Solve for pipe J: c-c = 10"

Find the fitting allowance. (Data #5)

90° ell – A = 1⅛", B = ½"

45° ell – A = ¾", B = ½"

1⅛" + ¾" – 2(½")

Find e-e for pipe J.

e-e = 10" – [(1⅛" + ¾") – 2(½")]

e-e = 10" – (1⅞" – 1")

e-e = 10" – ⅞"

e-e = 9⅛"

Solve for pipe K: Offset = 7½"

Find c-c for pipe K.

diagonal = 1.414 × offset (Data #20)

c-c = 1.414 × 7.5"

c-c = 10.605" or 10⅝" (Data #1)

Find the fitting allowance. (Data #5)

45° ell – A = ¾", B = ½"

2(¾") – 2(½")

Find e-e for pipe K.

e-e = 10⅝" – [2(¾") – 2(½")]

e-e = 10⅝" – (1½" – 1")

e-e = 10⅝" – ½"

e-e = 10⅛"

Solve for pipe L.

Find c-c for pipe L.

rise for pipe K = offset (Data #20)

rise = 7½"

c-c = 26" – (10" + 7½")

c-c = 26" – 17½"

c-c = 8½"

Find the fitting allowance (same as pipe J).

1⅛" + ¾" – 2(½")

Find e-e for pipe L.

e-e = 8½" – [(1⅛" + ¾") – 2(½")]

e-e = 8½" – (1⅞" – 1")

e-e = 8½" – ⅞"

e-e = 7⅝"

Exercise

Solve for c-c and e-e dimensions for pipes J, K, and L of the three-pipe diagram shown. Use dimensions, pipe sizes given, and 45° offset for each problem.

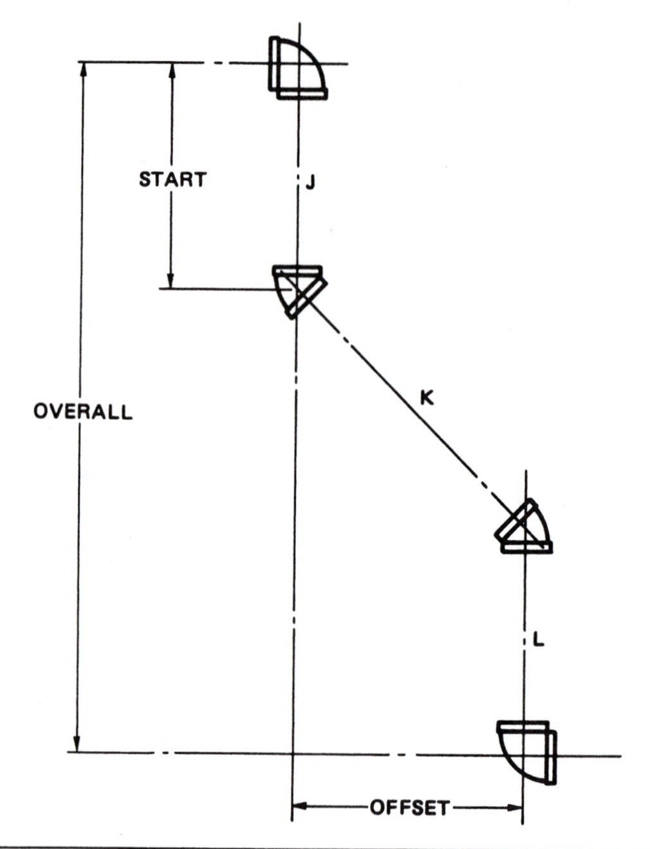

	Overall	Start	Offset	Pipe
1.	30"	9"	11"	1" threaded
2.	42½"	12"	13½"	½" copper
3.	53¾"	15"	16"	¾" threaded
4.	60"	14"	32"	1¼" PVC
5.	70"	Make J and L alike	34"	1½" PVC
6.	6' 0"	Make J and L alike	16¾"	1½" threaded

Solve the problems for c-c and e-e for pipes M, N, and O. Use dimensions, pipe sizes given, and 45° offset for each problem.

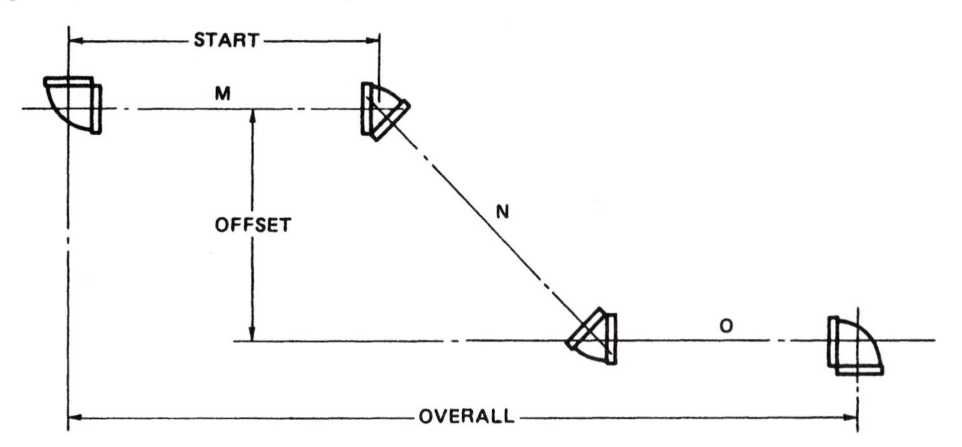

	Overall	Start	Offset	Pipe
7.	17"	5"	5½"	⅜" threaded
8.	33"	7¾"	8"	½" PVC
9.	3' 3"	9½"	9½"	¾" threaded
10.	3' 9¾"	12½"	15½"	1" threaded
11.	4' 0"	Make M and O alike	12"	1" copper
12.	4' 7"	Make M and O alike	21"	1¼" threaded

Unit 21

Right Angle with a 45° Diagonal

Objectives

- Use two 45° elbows to form 90°.
- Use two 45° elbows to avoid an obstruction.

Because two 45° fittings make 90°, it is common practice to use a 45° diagonal to make a right angle. This can be done only with two 45° fittings; therefore, it is a special application. This 45° diagonal allows an easier flow around the turn than would a 90° ell. It also provides an alternative to avoid obstacles. Three on-the-job measurements are needed: the vertical overall, the horizontal overall, and the start to the offset.

Sample Problem

Solve for c-c and e-e dimensions for threaded pipes J, K, and L.

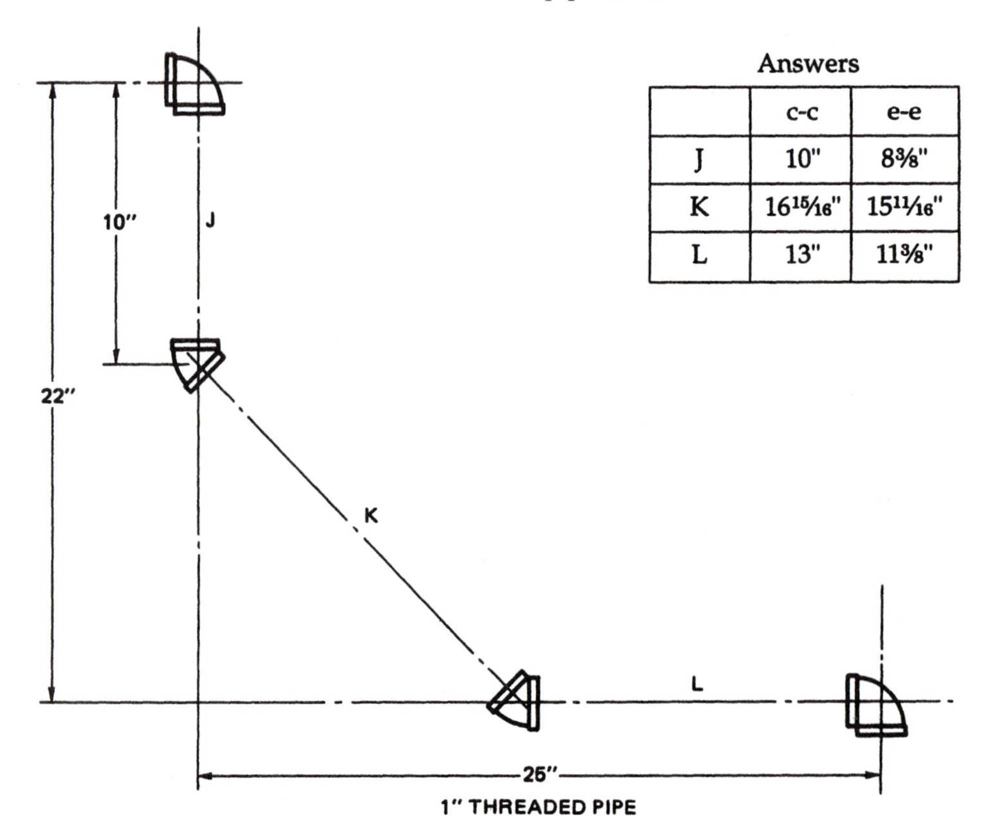

Answers

	c-c	e-e
J	10"	$8\frac{3}{8}$"
K	$16\frac{15}{16}$"	$15\frac{11}{16}$"
L	13"	$11\frac{3}{8}$"

Solution

Solve for pipe J: c-c = 10"

Find the fitting allowance. (Data #5)

90° ell A = 1½", B = ½"

45° ell A = 1⅛", B = ½"

1½" + 1⅛" – 2(½")

Find e-e for pipe J.

e-e = 10" – [(1½" + 1⅛") – 2(½")]

e-e = 10" – (2⅝" – 1")

e-e = 10" – 1⅝"

e-e = 8⅜"

Solve for pipe K: vertical overall = 22"

Find c-c for pipe K.

rise = 22" – 10" or 12"

diagonal = 1.414 × rise (Data #20)

c-c = 1.414 × 12"

c-c = 16.968" or 16¹⁵⁄₁₆" (Data #1)

Find the fitting allowance. (Data #5)

45° ell – A = 1⅛", B = ½"

2(1⅛") – 2(½")

Find e-e for pipe K.

e-e = 16¹⁵⁄₁₆" – [2(1⅛") – 2(½")]

e-e = 16¹⁵⁄₁₆" – (2¼" – 1")

e-e = 16¹⁵⁄₁₆" – 1¼"

e-e = 15¹¹⁄₁₆"

Solve for pipe L: overall horizontal = 25"

Find c-c for pipe L.

offset = rise (Data #20)

offset = 12"

c-c = 25" – 12"

c-c = 13"

Find the fitting allowance (same as pipe J).

1½" + 1⅛" – 2(½")

Find the e-e for pipe L.

e-e = 13" – [(1½" + 1⅛") – 2(½")]

e-e = 13" – (2⅝" – 1")

e-e = 13" – 1⅝"

e-e = 11⅜"

Exercise

Solve for c-c and e-e dimensions of pipes J, K, and L using the dimensions listed in each numbered problem.

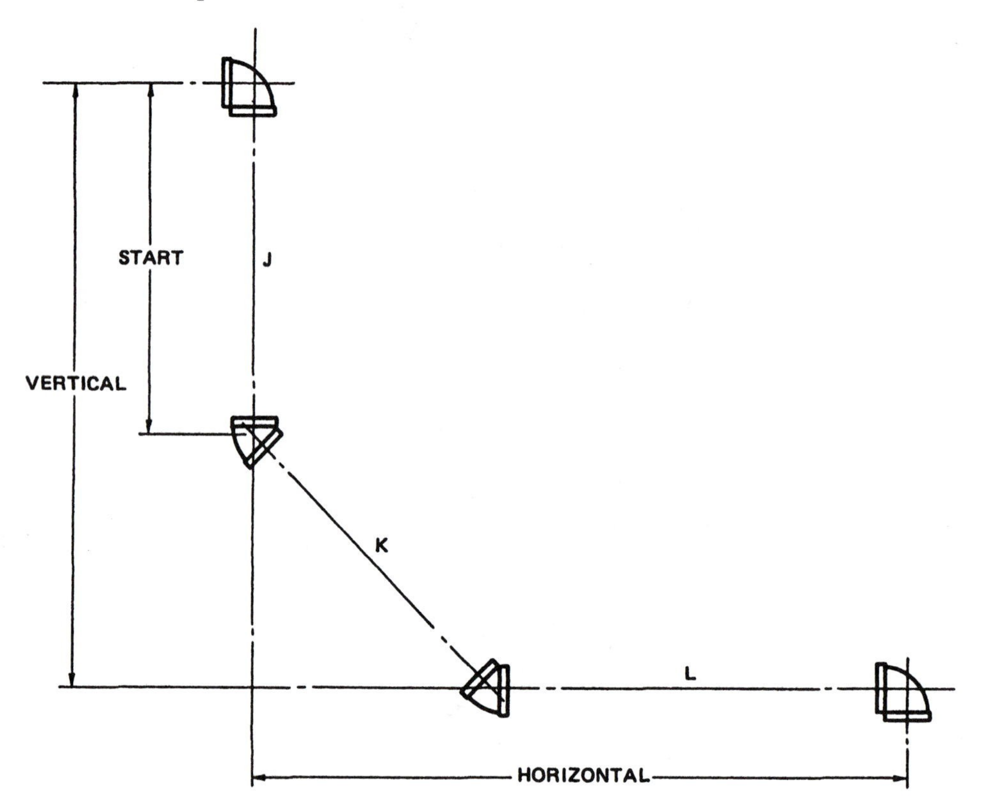

	Vertical	Horizontal	Start	Offset	Pipe
1.	30"	36"	16"		⅜" PVC
2.	43"	38"	21"		2½" threaded
3.	4'	5'	1' 7½"		2" threaded
4.	5' 3"	4' 10"	2' 2½"		1¼" copper
5.	6' 4"	5'-11"		3' 4¼"	¾" copper

Unit 22

45° Offsets with a Wye Fitting

Objective

- Use a wye to add a branch or a cleanout to the offset.

Cast iron drainage fittings are used in hazardous usage areas like warehouses, factories, seagoing vessels, high-rise buildings, and anywhere heavy vibration and moving machinery are likely to be present.

The 45° wye fitting could be used in a pipe diagram similar to the one in the previous unit. The plugged opening provides a way to clean out a pipe. The wye is also used to add a branch so that two fittings can waste into a single stack.

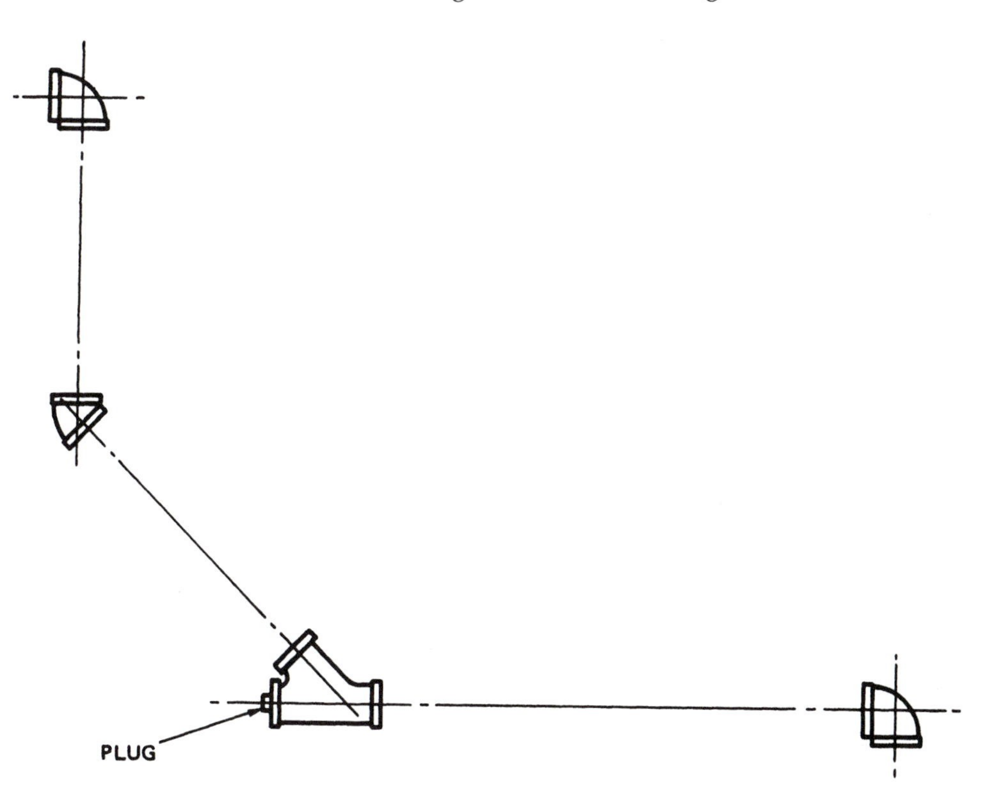

A study of standard fitting in Data #8 in the Appendix and the following illustration shows that the centerlines meet in such a way as to have two long centers and one short center for each wye. In the figure, *C* represents the long center; *D*, the short center. When all three openings are for the same size pipe, both long centers are of equal length. Therefore, in Data #8 only two dimensions are given. In Data #9 for wyes in which the branch is a smaller pipe size than the run, three dimensions are given; the branch is called *E*.

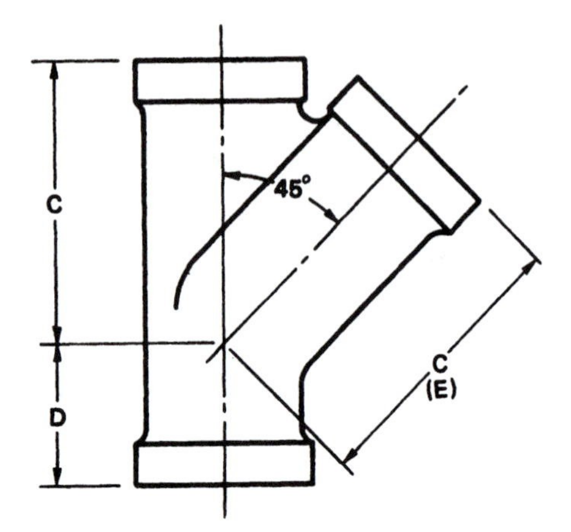

The wye fittings are different from ells that have equal centers. Dimension *A* in the figure shows that the centers in elbows are equal.

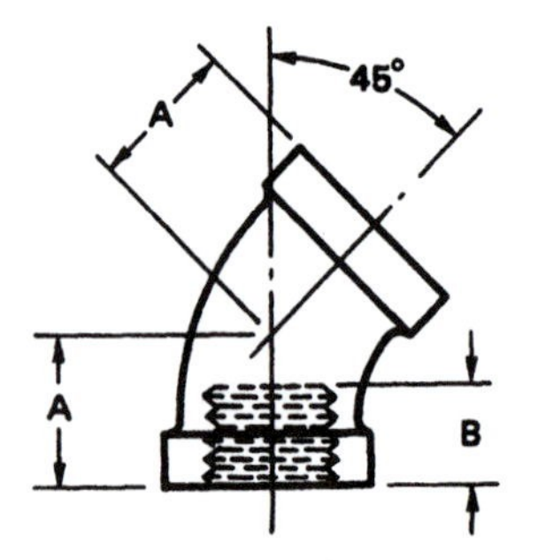

Sample Problem 1

Solve for c-c and e-e lengths for threaded pipes J, K, L, and M as shown in the illustration. (Use Data #5 and Data #8 for fitting allowance.)

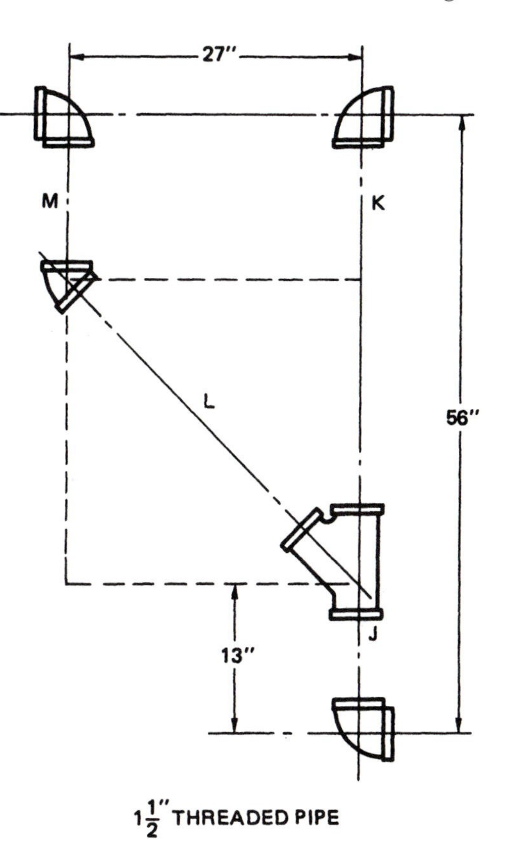

Answers

	c-c	e-e
J	13"	10⅛"
K	43"	38½"
L	38$\frac{9}{16}$"	34$\frac{9}{16}$"
M	16"	13⅝"

Solution

Solve for pipe J: c-c = 13"

(Data #5) 90°, threaded ell – A = 1$\frac{15}{16}$", B = ½"

(Data #8) 45°, threaded wye – D = 1$\frac{15}{16}$", B = ½"

e-e = 13" – [(1$\frac{15}{16}$" + 1$\frac{15}{16}$") – 2 (½")]

e-e = 13" – 2⅞"

e-e = 10⅛"

Solve for pipe K: vertical overall = 56"

c-c = 56" – 13"

c-c = 43"

(Data #8) 45°, threaded wye – C = 3$\frac{9}{16}$", B = ½"

(Data #5) 90°, threaded ell – A = 1$\frac{15}{16}$", B = ½"

e-e = 43" – [(3$\frac{9}{16}$" + 1$\frac{15}{16}$") – 2(½")]

e-e = 43" – 4½"

e-e = 38½"

Solve for pipe L: offset = 27"

(Data #20) diagonal = 1.414 × offset

c-c = 1.414 × 27"

c-c = 38.178" or 38$\frac{3}{16}$" (Data #1)

(Data #8) 45°, threaded wye – C = 3$\frac{9}{16}$", B = $\frac{1}{2}$"

(Data #5) 45°, threaded ell – A = 1$\frac{7}{16}$", B = $\frac{1}{2}$"

e-e = 38$\frac{3}{16}$" – [(3$\frac{9}{16}$" + 1$\frac{7}{16}$") – 2($\frac{1}{2}$")]

e-e = 38$\frac{3}{16}$" – 4"

e-e = 34$\frac{3}{16}$"

Solve for pipe M: vertical overall = 56"

(Data #20) rise = offset

rise = 27"

c-c = 56" – (13" + 27")

c-c = 16"

(Data #5) 45°, threaded ell – A = 1$\frac{7}{16}$", B = $\frac{1}{2}$"

90°, threaded ell – A = 1$\frac{15}{16}$", B = $\frac{1}{2}$"

e-e = 16" – [(1$\frac{7}{16}$" + 1$\frac{15}{16}$") – 2($\frac{1}{2}$")]

e-e = 16" – 2$\frac{3}{8}$"

e-e = 13$\frac{5}{8}$"

Sample Problem 2

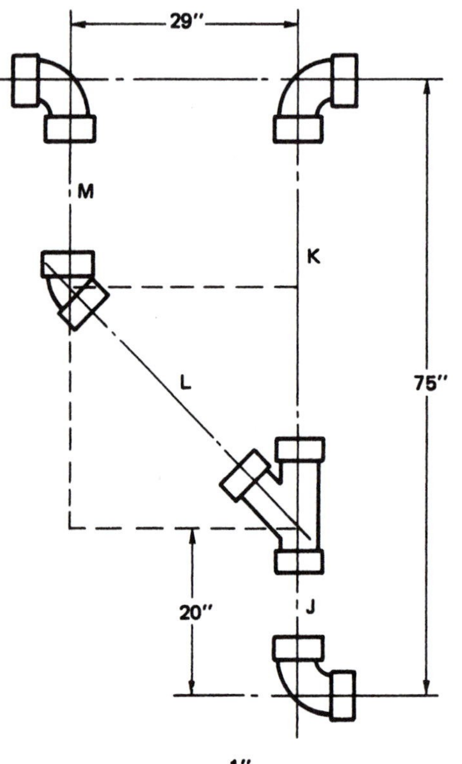

PVC 2" WITH 1$\frac{1}{2}$" BRANCH

Answers

	c-c	e-e
J	20"	16$\frac{5}{8}$"
K	55"	49$\frac{3}{8}$"
L	41"	36$\frac{7}{16}$"
M	26"	23$\frac{1}{8}$"

Solution

Note: J and K are 2" pipe; L and M are 1½" pipe.

Solve for pipe J: c-c = 20"

(Data #5) 90°, PVC ell – G = 2$\frac{5}{16}$"

(Data #9) 2" × 1½" PVC wye – I = 1$\frac{1}{16}$"

e-e = 20" – (2$\frac{5}{16}$" + 1$\frac{1}{16}$")

e-e = 20" – 3⅜"

e-e = 16⅝"

Solve for pipe K:

c-c = 75" – 20"

c-c = 55"

(Data #9) 2" × 1½" PVC wye – H = 3$\frac{5}{16}$"

(Data #5) 90°, PVC ell – G = 2$\frac{5}{16}$"

e-e = 55" – (3$\frac{5}{16}$" + 2$\frac{5}{16}$")

e-e = 55" – 5⅝"

e-e = 49⅜"

Solve for pipe L:

(Data #20) diagonal = 1.414 × offset

c-c = 1.414 × 29"

c-c = 41.006" or 41" (Data #1)

(Data #5) 45°, PVC ell – G = 1⅛"

(Data #9) 2" × 1½" PVC wye – J = 3$\frac{7}{16}$"

e-e = 41" – (1⅛" + 3$\frac{7}{16}$")

e-e = 41" – 4$\frac{9}{16}$"

e-e = 36$\frac{7}{16}$"

Solve for Pipe M:

(Data #20) rise = offset

rise = 29"

c-c = 75" – (20" + 29")

c-c = 26"

(Data #5) 45° PVC ell – G = 1⅛"

(Data #5) 90° PVC ell – G = 1¾"

e-e = 26" – (1⅛" + 1¾")

e-e = 26" – 2⅞"

e-e = 23⅛"

Exercise

Solve for c-c and e-e lengths for pipes J, K, L, and M in each problem.

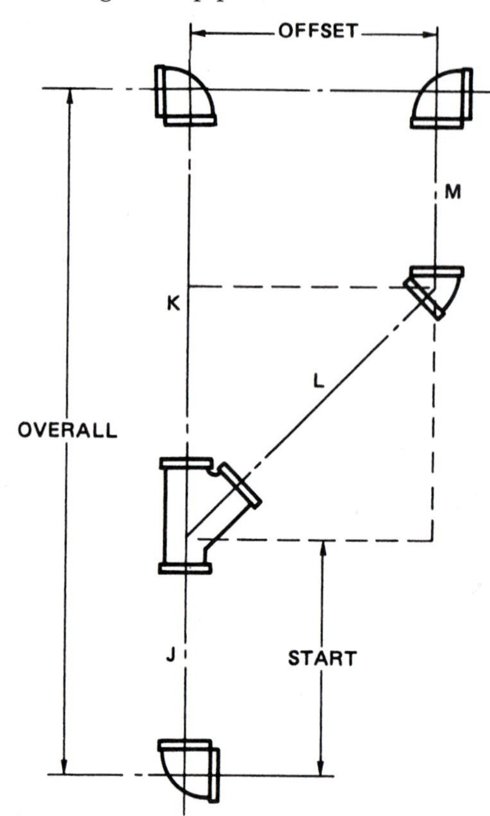

	Overall	Offset	Start	Pipe
1.	45½"	23"	8"	1½" threaded
2.	5' 7"	2' 6"	15"	1½" PVC
3.	4' 5"	2' 0"	12"	2" PVC
4.	4' 8"	9¾"	22"	2" threaded
5.	26"	8½"	11¾"	2" copper
6.	5' 10"	29"	19"	2" (1½" branch) threaded

Unit 23 Wye and Tee-Wye Assemblies

Objectives

- Learn flow patterns of the drainage fittings.
- Study drain and vent solutions with wye assemblies.

The tee wye is a 90° drainage fitting (DWV) that has an inside construction designed to cause as little turbulence as possible. Unlike a standard tee, which has three equal allowances, the tee wye has two long allowances and one short allowance.

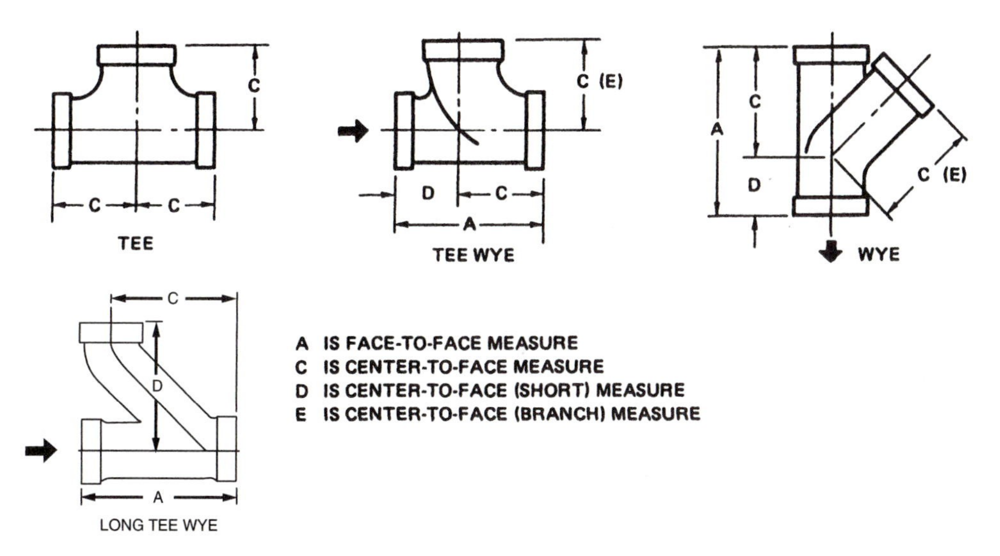

The wye and tee wye illustrations show the proper direction of the main flow with a dark arrow. You can see that the branch flow in the drainage fittings is directed smoothly into the main flow. Correct installation of drainage fittings regarding flow is vital. Another consideration is that a ¼-inch per foot slope is built into the branch of the fitting. Installing a DWV tee or tee wye upside down would result in a reverse flow situation. In the case of the simple wye shown previously, the (E) illustrates that a different allowance would apply if the branch size differed from the run size, i.e., a reducing wye was involved in the calculations. We will use drainage (DWV) fittings exclusively for threaded pipe in this unit. Data #8 is used for regular wyes. Data #9 is used for reduced branch wyes. Data #10 is used for regular tee wyes. Data #11 is used for reduced branch tee wyes.

The wye and tee wye fittings, both regular and reduced branch, are used in various types of pipe assemblies with 45° diagonals.

Sample Problem 1

Solve for c-c and e-e lengths of threaded pipes J, K, L, and M shown in the following pipe diagram.

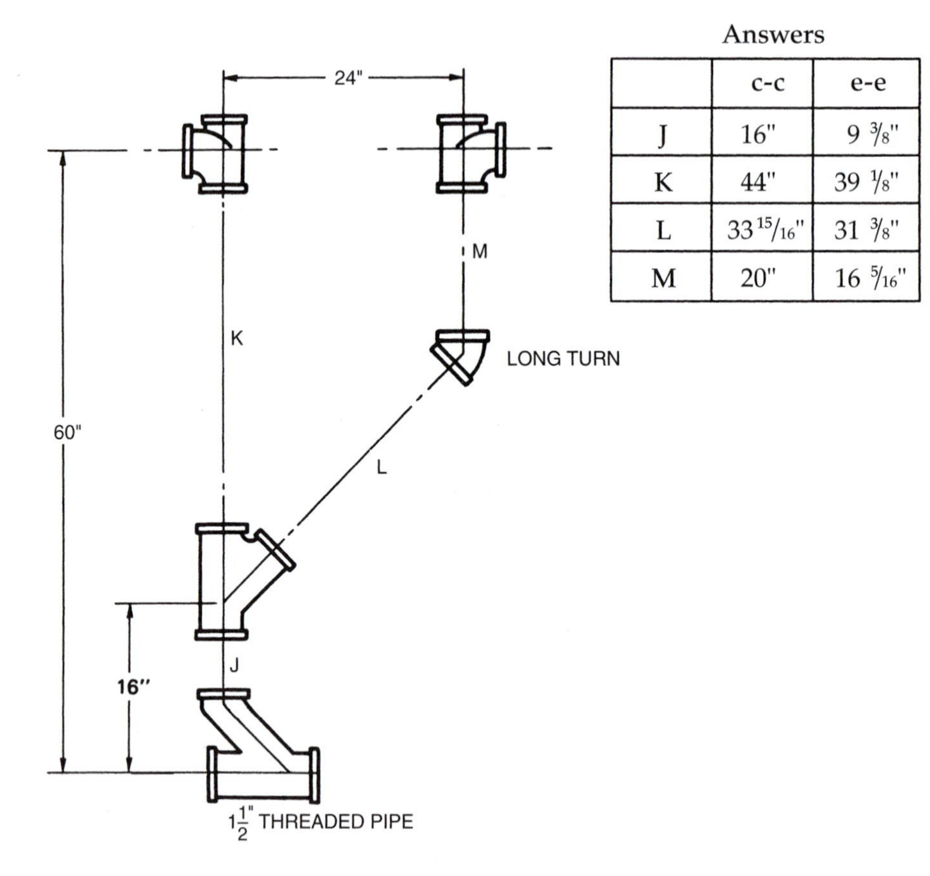

Answers

	c-c	e-e
J	16"	9 3/8"
K	44"	39 1/8"
L	33 15/16"	31 3/8"
M	20"	16 5/16"

Solution

Solve for pipe J: c-c = 16"

(Data #10) D = 4 3/16", B = 1/2"

(Data #8) D = 3 7/16", B = 1/2"

e-e = 16" – [(4 3/16" + 3 7/16") – 2(1/2")] = 9 3/8"

Solve for pipe K:

c-c = 60" – 16" = 44"

(Data #8) C = 3 7/16", B = 1/2"

(Data #10) C = 2 7/16", B = 1/2"

e-e = 44" – [(3 7/16" + 2 7/16") – 2(1/2")] = 39 1/8"

Solve for pipe L: offset = 24"

(Data #20) diagonal = 1.414 × offset

c-c = 1.414 × 24"

c-c = 33.936" = $33\frac{15}{16}$" (Data #1)

(Data #8) D = $3\frac{7}{16}$", B = $\frac{1}{2}$"

(Data #5) A = $1\frac{7}{8}$", B = $\frac{1}{2}$"

e-e = $33\frac{15}{16}$" – [($3\frac{7}{16}$" + $1\frac{7}{8}$") – 2($\frac{1}{2}$")] = $31\frac{3}{8}$"

Solve for pipe M:

(Data #20) rise = offset

rise = 24"

c-c = 60" – (16" + 24") = 20"

(Data #5) A = $1\frac{7}{8}$", B = $\frac{1}{2}$"

(Data #10) C = $2\frac{7}{16}$", B = $\frac{1}{2}$"

e-e = 20" – [($1\frac{7}{8}$" + $2\frac{7}{16}$") – 2($\frac{1}{2}$")] = $16\frac{5}{16}$"

Sample Problem 2

Solve for c-c and e-e of threaded pipes N, O, and P.

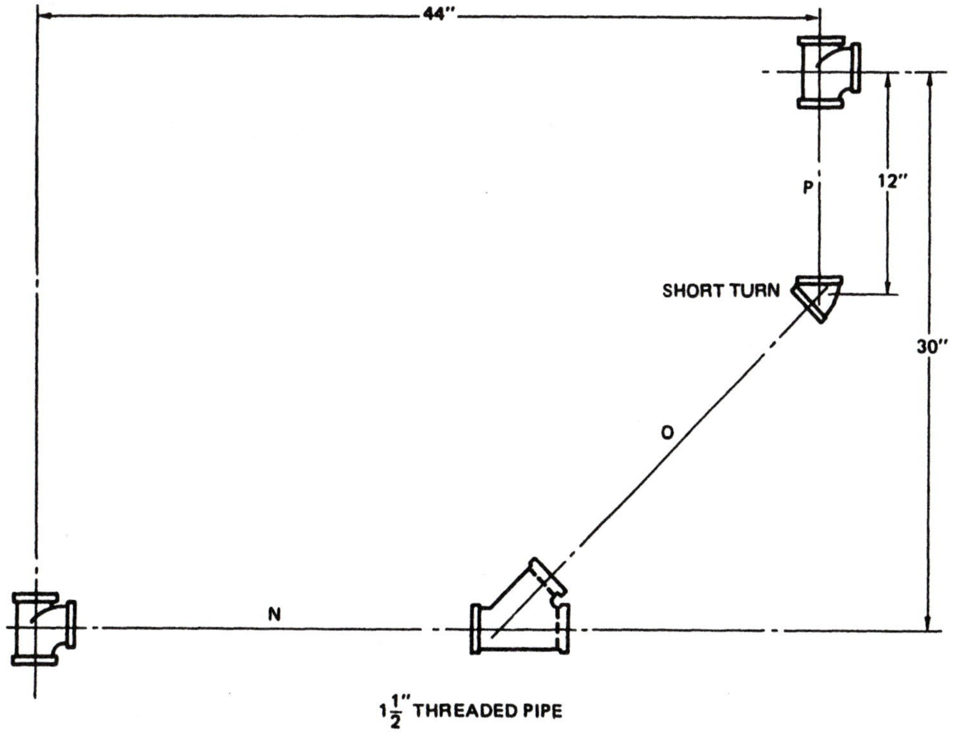

44"
12"
30"
P
O
N
SHORT TURN
$1\frac{1}{2}$" THREADED PIPE

Answers

	c-c	e-e
N	26"	$22\frac{11}{16}$"
O	$25\frac{7}{16}$"	21"
P	12"	$9\frac{1}{16}$"

Solution

Solve for pipe N:

rise of pipe O = 30" – 12" or 18"

(Data #20) offset = rise = 18"

c-c = 44" – 18" = 26"

(Data #10) D = 2½", B = ½"

(Data #8) A – C = 1$\frac{13}{16}$", B = ½"

e-e = 26" – [(2½" + 1$\frac{13}{16}$") – 2(½")] = 22$\frac{11}{16}$"

Solve for pipe O:

(Data #20) diagonal = 1.414 × offset

c-c = 1.414 × 18"

c-c = 25.452" or 25$\frac{7}{16}$"

(Data #8) D = 3$\frac{7}{16}$", B = ½"

(Data #5) A = 1½", B = ½"

e-e = 25$\frac{7}{16}$" – [(3$\frac{7}{16}$" + 1½") – 2(½")] = 21"

Solve for pipe P: c-c = 12"

(Data #5) A = 1½", B = ½"

(Data #10) C = 2$\frac{7}{16}$", B = ½"

e-e = 12" – [(1½" + 2$\frac{7}{16}$") – 2(½")] = 9$\frac{1}{16}$"

Sample Problem 3

Solve for c-c and e-e of threaded pipe Q. (Procedures for other pipes are covered in previous problems.)

Answers

	c-c	e-e
Q	34"	29¾"

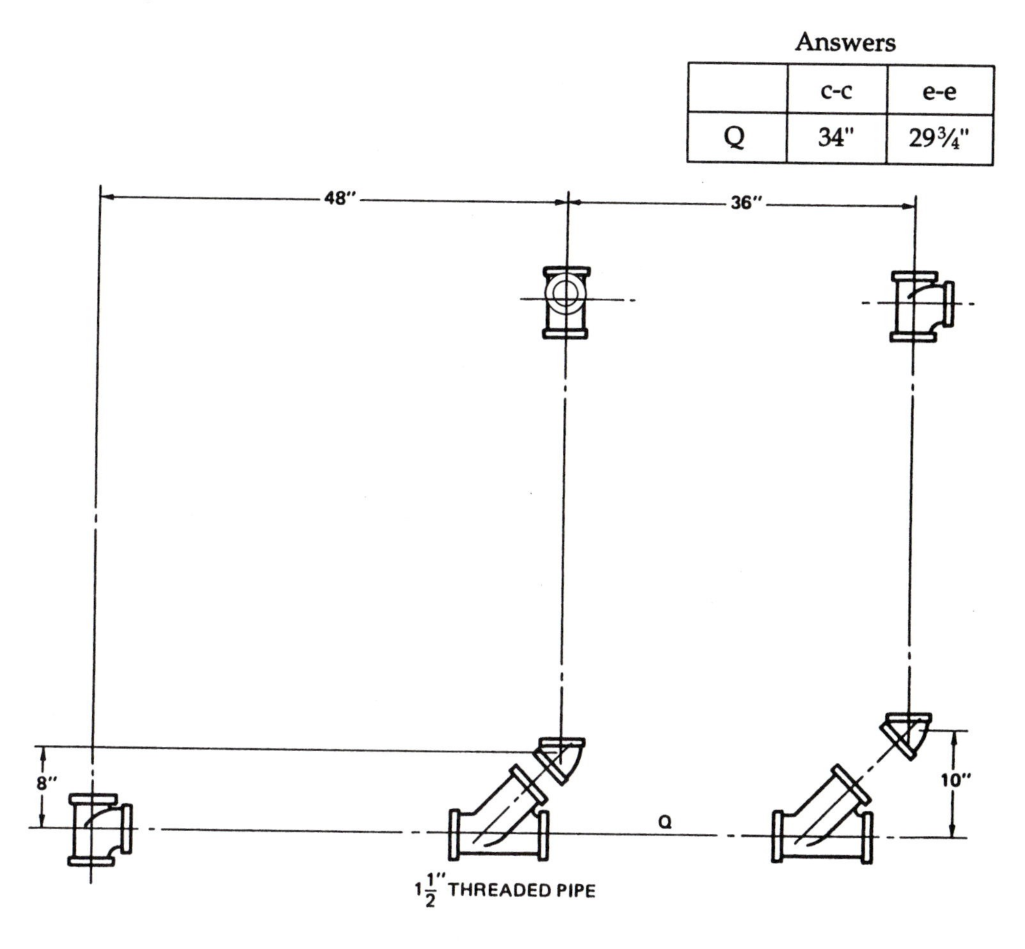

Solution

Solve for pipe Q:

(Data #20) run = rise

outside run = 10"; inside run = 8"

c-c = overall – outside run + inside run

c-c = (36" – 10") + 8" = 34"

(Data #8) C = $3\frac{7}{16}$", B = ½"

A – C = $1\frac{13}{16}$", B = ½"

e-e = 34" – [($3\frac{7}{16}$" + $1\frac{13}{16}$") – 2(½")] = 29¾"

Exercise

Solve for c-c and e-e of each pipe, using dimensions as given for each problem.

Note: Use this illustration for problems 1–5.

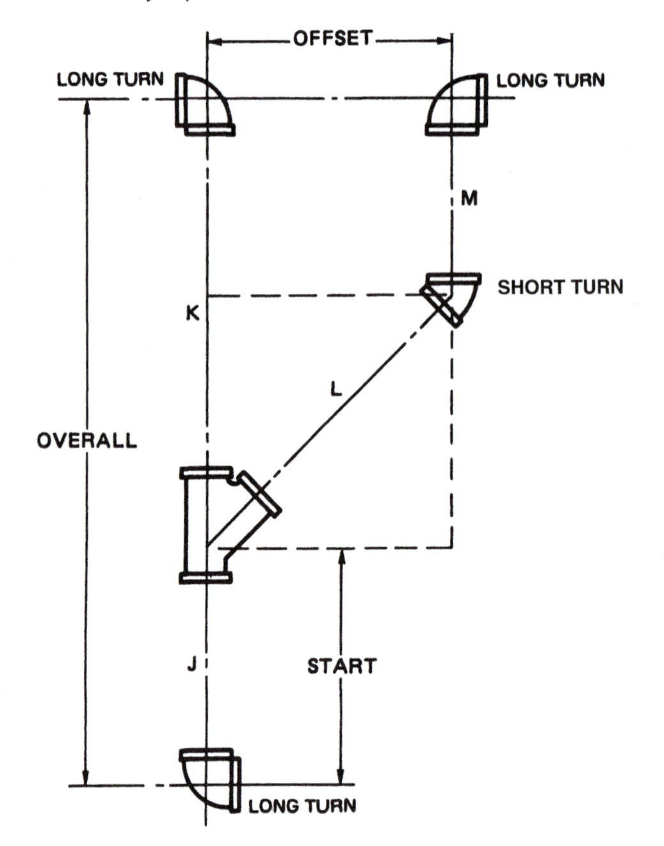

	Overall	Start	Offset	Pipe
1.	62"	14"	27"	1½" threaded
2.	5' 10"	19½"	2' 7"	1½" copper
3.	6' 5½"	21¾"	2' 9½"	2" threaded
4.	7' 4"	2' 3"	3' 1¼"	2" PVC
5.	9' 1½"	2' 10½"	3' 7"	2" (1½" branch) copper

Note: Use this illustration for problems 6–10.

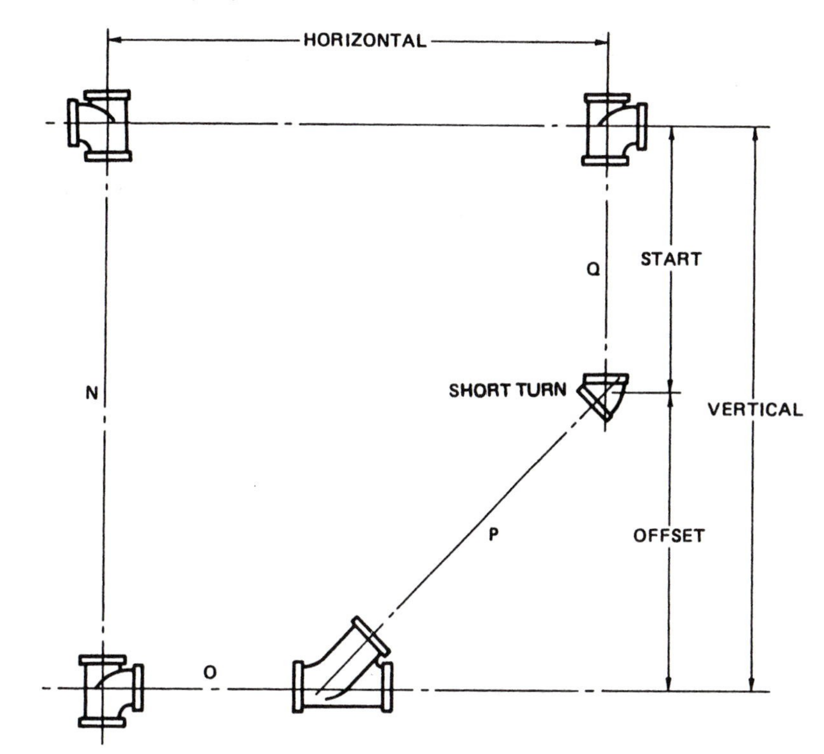

	Horizontal	Vertical	Start	Offset	Pipe
6.	40"	50"	30"		1½" PVC
7.	46"	32"		12½"	1½" threaded
8.	3' 7¾"	5' 3"	2' 11"		2" copper
9.	5' 1½"	4' 10¼"		2' 5"	2" threaded
10.	3' 2"	4' 1½"	1' 9½"		2" (1½" branch) copper

Note: Use this illustration for problems 11–15.

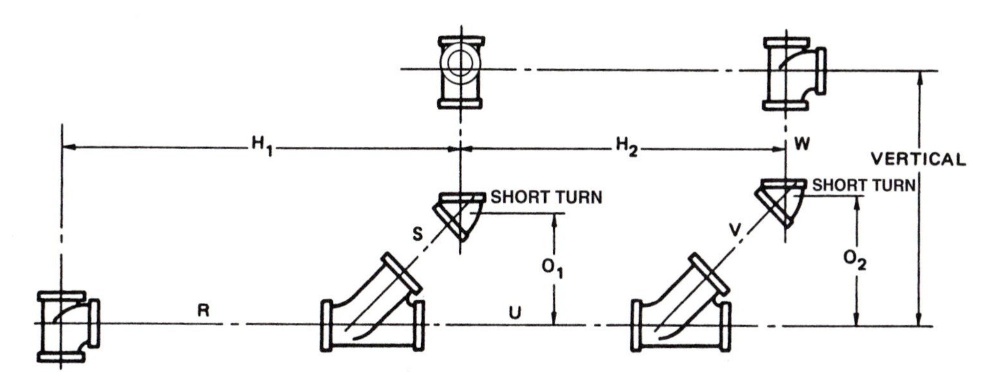

	H_1	H_2	Vertical	O_1	O_2	Pipe
11.	38"	25"	40"	8"	8"	1½" threaded
12.	42"	31"	53"	12"	9"	1½" PVC
13.	3' 7½"	2' 4"	4' 10½"	10"	10"	2" threaded
14.	3' 9"	2' 5¾"	5' 0"	1' 2"	1' 7"	2" copper
15.	4' 7¾"	3' 5"	6' 2½"	1' 5½"	1' 9"	2" PVC

16. This is an assembly for under the floor instead of in the wall. Solve for c-c and e-e lengths of threaded pipes A, B, C, D, and E.

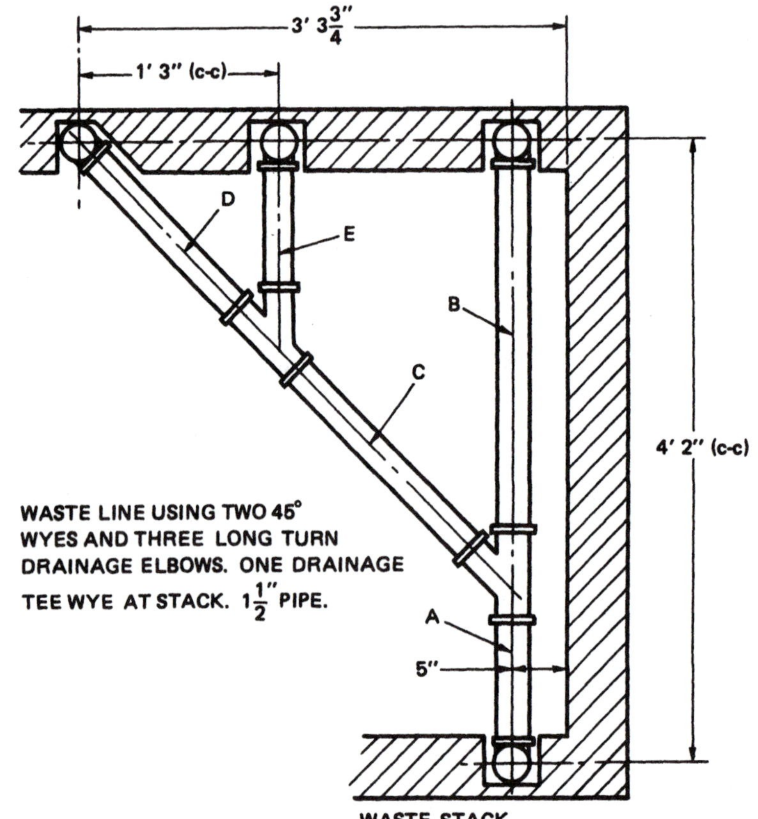

Unit 24

The 0.707 Constant in 45° Pipe Assemblies

Objective

- Find the offset using the length of the diagonal.

The use of a 45° ell and a wye to turn a corner works equally well with any diagonal length. The pipe fitter can use a stock nipple or other threaded pipe length to make up the diagonal. It is then necessary to determine the c-c or diagonal length and the offset or run.

In preceding units, the c-c length has been found by first determining the offset and multiplying by 1.414. Now the diagonal or c-c length is found and multiplied by 0.707 to find the offset. The rise (or run) length for 45° diagonals is equal to the offset.

Note: Constants for 45° fittings are given in Data #20 in the Appendix.

Sample Problem Solve for c-c and e-e lengths of threaded pipes J, K, and L.

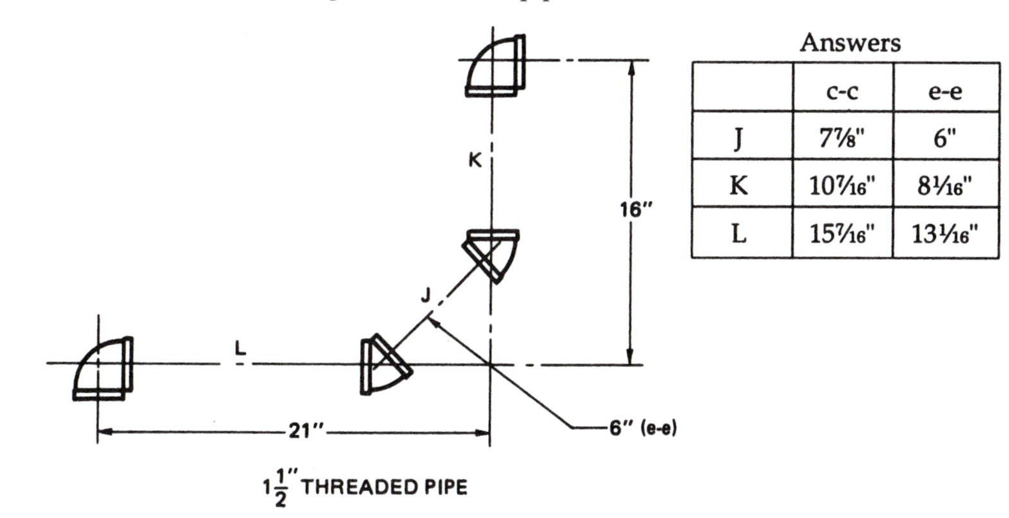

Answers

	c-c	e-e
J	7⅞"	6"
K	10 7/16"	8 1/16"
L	15 7/16"	13 1/16"

Solution

Solve for pipe J: e-e = 6"

(Data #5) A = $1\frac{7}{16}$", B = $\frac{1}{2}$"

c-c = 6" + [($1\frac{7}{16}$" + $1\frac{7}{16}$") – 2($\frac{1}{2}$")] = 6" + $1\frac{7}{8}$" = $7\frac{7}{8}$"

Solve for pipe K:

(Data #20) offset = 0.707 × diagonal

offset = 0.707 × $7\frac{7}{8}$" = $5\frac{9}{16}$"

c-c = 16" – $5\frac{9}{16}$" = $10\frac{7}{16}$"

(Data #5) A = $1\frac{15}{16}$", B = $\frac{1}{2}$"

A = $1\frac{7}{16}$", B = $\frac{1}{2}$"

e-e = $10\frac{7}{16}$" – [($1\frac{7}{16}$" + $1\frac{15}{16}$") – 2($\frac{1}{2}$")] = $10\frac{7}{16}$" – $2\frac{3}{8}$" = $8\frac{1}{16}$"

Solve for pipe L:

(Data #20) run = offset

run = $5\frac{9}{16}$"

c-c = 21" – $5\frac{9}{16}$" = $5\frac{7}{16}$"

The fitting allowance is the same as for pipe K.

e-e = $15\frac{7}{16}$" – [($1\frac{7}{16}$" + $1\frac{15}{16}$") – 2($\frac{1}{2}$")] = $15\frac{7}{16}$" – $2\frac{3}{8}$" = $13\frac{1}{16}$"

Exercise

Solve for c-c and e-e lengths for each lettered pipe in the diagrams and find the offset for each diagram.

1.

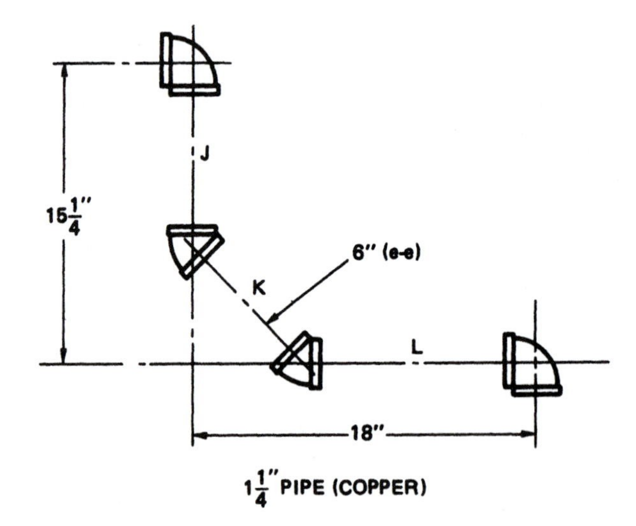

2. Use threaded drainage fittings for exercise 2.

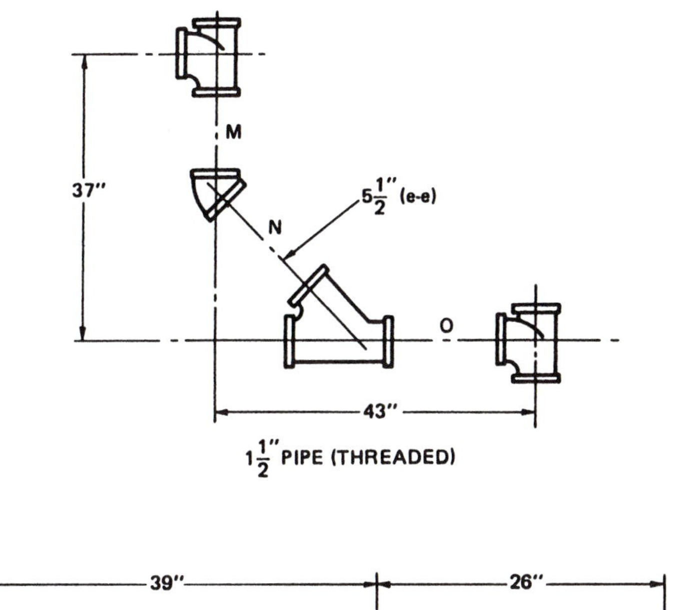

3.

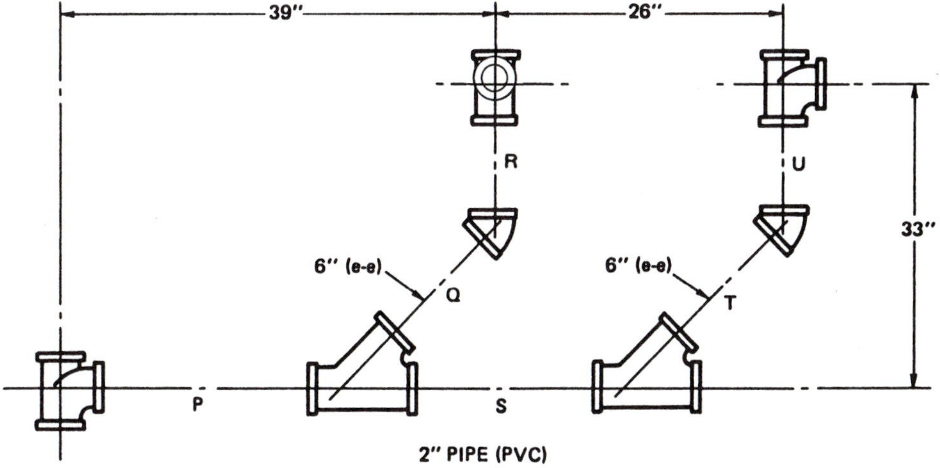

Unit 25

Solving for Other Angles

Objectives

- Review offset terminology.
- Solve for offsets of any angle.

As has been shown in previous units, the lengths of the *offset, diagonal,* and *run* of an offset are rigidly proportional to each other and dependent on the angle of the bends or the fittings used. For a review of some of the terminology, see Figure 25-1.

For any given angle (see Figure 25-1), the run and set will have the same proportion to each other. For instance, if the angle were 30° and the offset were 2, the run will be 3.464. This is equal to 1.732 times 2 (the offset). If you were dealing with many offsets using 30° angle elbows, the length of each run will always be equal to 1.732 times whatever the offset happens to be.

RUN
angle
DIAGONAL
OFFSET or SET
OFFSET or SET
angle
RUN

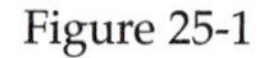

Figure 25-1

Another way of looking at this: If we were to construct a rectangle that had one side of 2 feet and the other side of 3.464 feet, and we bent a piece of tubing to fit the rectangle, as in Figure 25-1, the angles of the bends would be forced to be 30°.

Note that when an offset is in a vertical line of pipe, some will call the RUN the RISE. They mean the same thing.

So perhaps the student is beginning to see that for any given angle, the sides of the triangle within the rectangle, the offset, run, and diagonal are locked together in the relationship of the sides to each other. We could actually construct a table.

To Find Any Side of a 30° Offset

To find this side	when this is known	multiply this	by this
diagonal	offset	offset	2.000
offset	diagonal	diagonal	0.500
run	offset	offset	1.732
offset	run	run	0.577
diagonal	run	run	1.155
run	diagonal	diagonal	0.866

As you might imagine, it's important that you memorize the names for the different parts of an offset. *Take the time, now, to do this.*

Imagine the offset in Figure 25-1 with the pipe ends outside of the rectangle running vertically instead of horizontally. One of the names would stay the same. The other two names or labels would switch places. Try drawing a few offsets with vertical and horizontal pipe ends (these are called pipe tangents) and label the parts or sides without looking at Figure 25-1. Now look.

You may be wondering where these magic numbers in the right column of the table come from. These are trigonometric functions, or *constants* for short. These functions have names like **sine**, **cosine**, **tangent**, **cotangent**, **secant,** and **cosecant**. A table of these constants can be found in Appendix A, Data #27. We will demonstrate how these names would enter into the same chart. Notice that the only thing that changes is the right-most column. Now you, the student, should have a kind of template for making a chart for an offset of any angle.

To Find Any Side of a 30° Offset

To find this side	when this is known	multiply this	by this for 30°
diagonal	offset	offset	(cosecant)
offset	diagonal	diagonal	(sine)
run	offset	offset	(cotangent)
offset	run	run	(tangent)
diagonal	run	run	(secant)
run	diagonal	diagonal	(cosine)

Problem: Make an offset chart for a 60° angle. Use the trigonometric values from the chart provided (Appendix A, Data #27). To make the chart for any angle, go to the line on the trigonometric chart that corresponds to the angle to be used and copy the values from this line to the chart below in the same positions as the right-most column of the chart above.

To Find Any Side of a ___° Offset

To find this side	when this is known	multiply this	by this for ___°
diagonal	offset	offset	
offset	diagonal	diagonal	
run	offset	offset	
offset	run	run	
diagonal	run	run	
run	diagonal	diagonal	

The Answer: When you have completed your chart for the 60° angle it should look like the chart below.

To Find Any Side of a 60° Offset

To find this side	when this is known	multiply this	by this for 60°
diagonal	offset	offset	1.1547
offset	diagonal	diagonal	0.8660
run	offset	offset	0.5774
offset	run	run	1.7320
diagonal	run	run	2.0000
run	diagonal	diagonal	0.5000

Note: There are many pocket calculators with function keys which provide the trigonometric values. Be careful that these give the same values as those shown in the chart. If your calculator does not give the same values as those in the chart, you will find that the function keys on your calculator are assuming that the angle you are entering is in an angle style called *radians* or *grads* and not in *degrees.* If this is the case with your calculator, you will probably have a key on it somewhere to shift the device into *degree mode.*

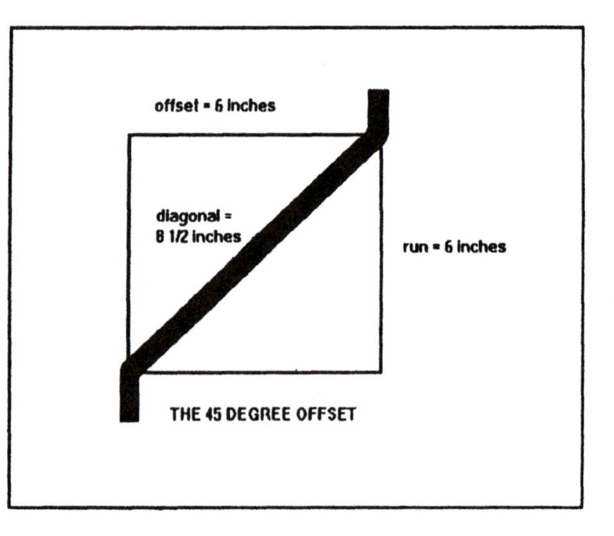

Figure 25-2

The most common offsetting angle, other than 90°, is the 45° offset. This gives an advantage insofar as calculations go because *the offset and the run are always the same.* The diagonal is always equal to 1.414 times either the offset or the run. If the diagonal of the 45° offset happens to be known, both the offset and the run can be obtained by dividing this distance by 1.414. See Figure 25-2.

The visual aid that we use for the offset in this case is actually a square. You may recall that all sides of a square are equal in length. Notice that the tangents of the pipe leaving the square are vertical rather than horizontal, as was the case in Figure 25-1. The side of the square that was the offset in Figure 25-1 is the run in Figure 25-2. One way of remembering this is to say, "**The run runs with the main pipeline.**" In other words, if the pipe entering and leaving the offset rectangle is vertical, then the run of the offset is the same as the vertical side of that rectangle.

Other Uses for Offset Mathematics

Notice that the diagonal bisects the rectangle that is the visualization aid for a pipe offset. This divides the rectangle into two similar and equal right triangles. The diagonal is equal to the *hypotenuse* of either of the triangles. The other two sides of the triangles are called the *adjacent* side and the *opposite* side in geometry. With a little reflection, the student should be able to use pipe offset math to solve for the length of the sides of any right triangle where one of the non-90° angles and the length of one of the sides are known.

Sample Problem

Solve for c-c and e-e lengths of threaded pipes J, K, and L.

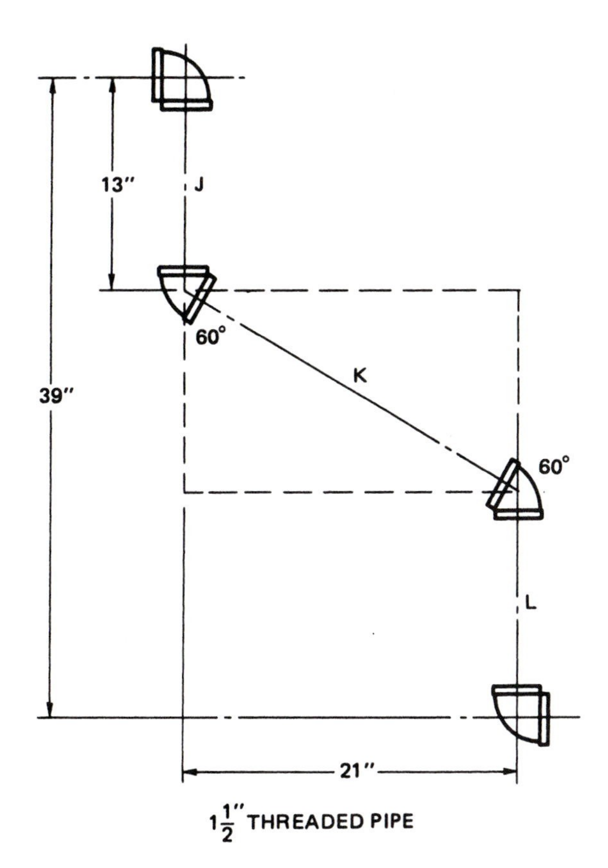

Answers

	c-c	e-e
J	13"	10⁵⁄₁₆"
K	24¼"	21¾"
L	13⅞"	11⁹⁄₁₆"

Solution

Solve for pipe J: c-c = 13"

(Data #5) A = $1\frac{15}{16}$", B = $\frac{1}{2}$"

(Data #6) A = $1\frac{3}{4}$", B = $\frac{1}{2}$"

e-e = 13″ – [($1\frac{15}{16}$" + $1\frac{3}{4}$") – 2($\frac{1}{2}$")] = $10\frac{5}{16}$"

Solve for pipe K: (from 60° table we have constructed previously)

diagonal = 1.55 × offset

c-c = 1.155 × 21" = 24.255" or $24\frac{1}{4}$"

(Data #6) A = $1\frac{3}{4}$", B = $\frac{1}{2}$"

e-e = $24\frac{1}{4}$" – [($1\frac{3}{4}$" + $1\frac{3}{4}$") – 2($\frac{1}{2}$")] = $21\frac{3}{4}$"

Solve for pipe L:

set or rise = 0.577 × offset

set or rise = 0.577 × 21" = 12.117" or $12\frac{1}{8}$"

c-c = 39" – (13" + $12\frac{1}{8}$") = 39" – $25\frac{1}{8}$" = $13\frac{7}{8}$"

The fitting allowance is the same as for pipe J.

e-e = $13\frac{7}{8}$" – [($1\frac{15}{16}$" + $1\frac{3}{4}$") – 2($\frac{1}{2}$")] = $11\frac{3}{16}$"

Exercise

Solve for c-c and e-e dimensions of all lettered pipes in the diagram. Use dimensions as given on the drawing.

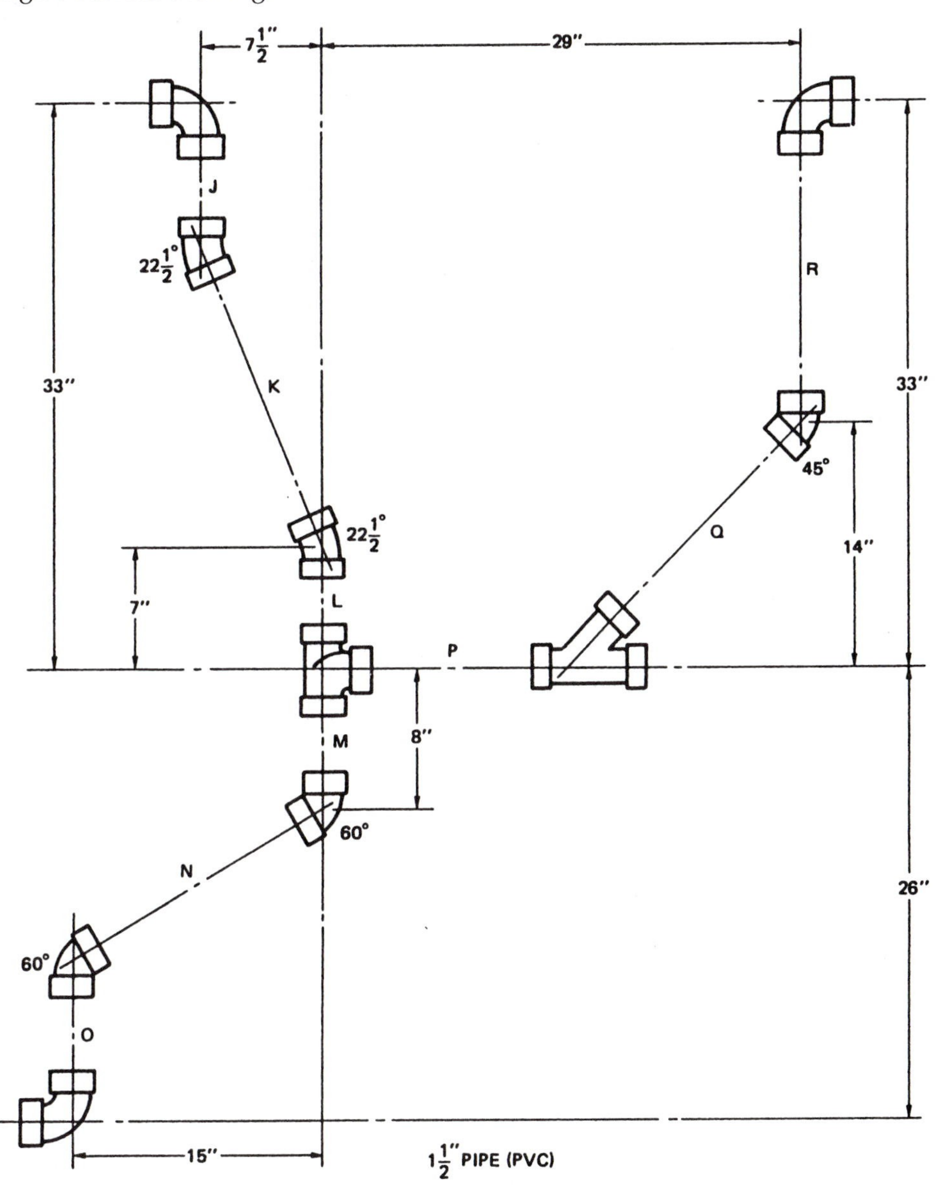

Unit 26 Pipe Bending

Objectives

- Learn when pipe bending is used.
- State the methods of bending.
- Do bending calculations.

Every fitting is a possible source for a future leak. Vibration, pressure, corrosion, and chemical breakdown of the fitting materials are causes of fitting failure. In some circumstances pipefitters and plumbers wish to have as few fittings and welds as possible to avoid future sources of leakage. This can be important in heating pipes which are to be buried in concrete, boiler piping inside boiler casings, hydraulic piping, and others. The best solution to these situations is often to substitute pipe bends in place of fittings wherever possible.

A pipe bend can be visualized as an arc of a circle (see Figure 26-1). When a pipe is bent, the outside of the bend is stretched and the inside is compressed. The stretched outside has safety considerations. When the pipe wall is stretched it becomes thinner. The general practice, with standard weight pipe, is to minimize this thinning by limiting the radius of the bend to *no less than five times the pipe size.* A bending wheel is a steel or cast iron circle that is pressed into the pipe to make the bend. A groove is machined into its perimeter to a depth of one-half the pipe diameter. If the bending wheel is measured across its greatest width and this is divided by 2, the radius of the bend that this particular bending wheel will make is obtained. Bending on too short a radius will flatten the pipe and reduce its carrying capacity.

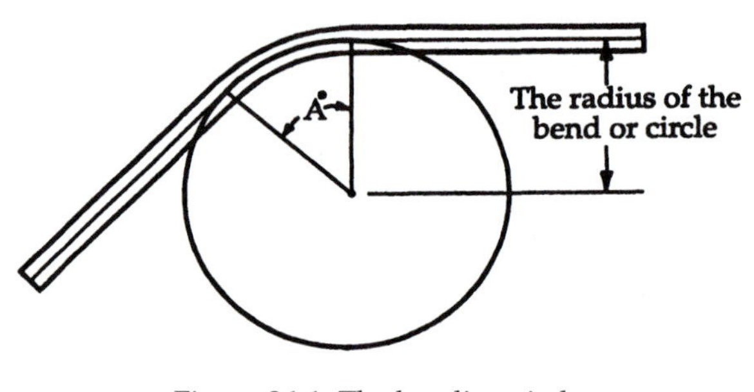

Figure 26-1. The bending circle
A is the bend angle

Making the Bend

There are two kinds of bending methods. In the first, the bending wheel is pressed into the pipe, which is held in position by two "bending blocks" (see Figure 26-2). The beginning contact point between the machine and the pipe is the center of the arc of the finished bend. The other kind of bending machine grasps the pipe to the wheel at the beginning of the bend and then rotates the wheel, forcing the pipe to bend with the wheel as it rotates (see Figure 26-3). This makes all calculations start at the beginning of the arc of the bend. The mathematics is slightly different in each situation. This is the type process found in small hand benders used for hydraulic, gas, and air pressure lines, but it is also used in heavy pipe fabrication shops where pipe sizes 8" IPS and more are bent on a machine weighing many tons.

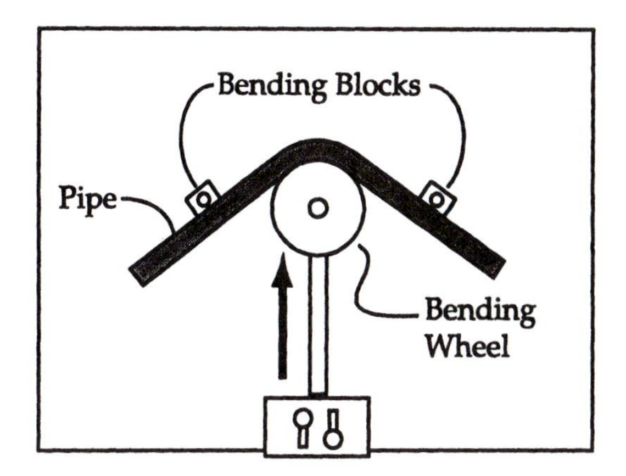

Figure 26-2. A bending table

Visualizing the Math

If we make a 90° bend in the center of a piece of pipe 24 inches long, how long will the pipe tangents be? We can't be exactly sure, except to say that they *will not be* 12 inches long. We can't be certain because the size of the pipe is not known and, therefore, the radius of the bending wheel is not known. But the reason that the length of the pipe tangents added together is greater than 24 inches is that *the pipe takes a short-cut* across the angle of the bend. This short-cut results in what pipefitters call *gain*. Gain is illustrated in the following paragraph.

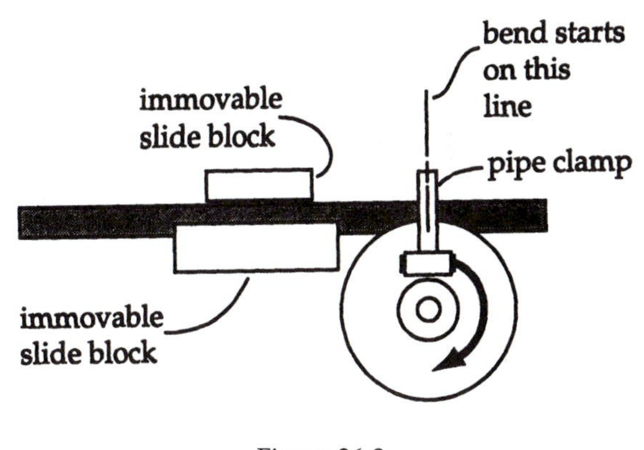

Figure 26-3

To make this calculation, calculate the difference between the straight line length of the pipe from where the bend starts and stops and the length of the arc made in the pipe by the bending wheel. In the case illustrated by Figure 26-4, a piece of 2-inch pipe is bent at a 90-degree angle. The pipe follows the 20-inch diameter bending wheel (5 times 2 inches times 2). The pipe length consumed by the bending arc is ¼ of the perimeter of the 20-inch circle. The perimeter of a 20-inch circle is 20π or 62.832 inches, ¼ of which is 15.708 inches. So if an abrupt 90-degree angle were possible, as shown in Figure 26-4, it would consume 20 inches (10 plus 10) between the intersection points with the bending wheel. The pipe, however, follows the wheel and consumes only 15.708 inches, a difference of 20 – 15.708 or 4.292 inches.

Note: *The radius of the bending wheel can be specified to be anything larger than 5 times the pipe size. If the radius is not specified it will be 5 times the pipe size.*

A Formula for Calculating Pipe Involved in a Bend

P = pipe size. D = number of degrees in the bend.

Pipe length = the length of the pipe actually incorporated in the bend.

$$\textit{Pipe length} = P \times 5 \times 2 \times \pi \times \frac{D}{360}$$

This calculation is helpful in determining how close together bends can be made for a particular angle.

The number 5 represents the standard bending radius multiplier. This would be changed to suit for special cases. The number 2 is the multiplier necessary to change the radius into the diameter of the bending wheel. This is so that the circumference of the bending wheel can be calculated using the formula $C = \pi \times D$, i.e., the circumference of a circle is equal to the value of pi times the diameter of that circle. If a bending radius different from 5 times the pipe diameter is *not* going to be used, the formula can be shortened to

$$\textit{Pipe length} = P \times 31.4159 \times \frac{D}{360}, \text{ or}$$

$$\textit{Pipe length} = P \times D \times 0.0873$$

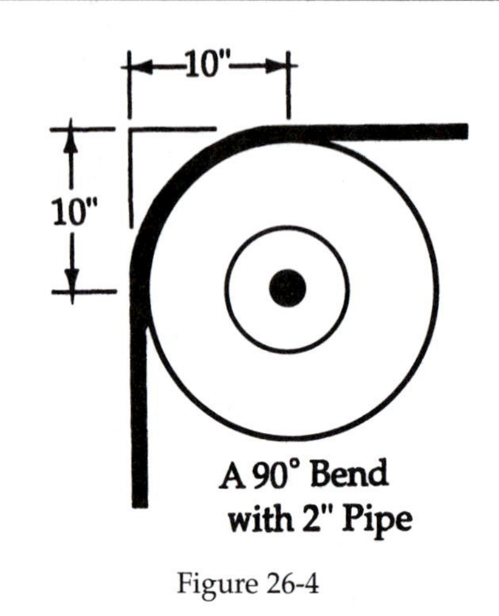

Figure 26-4

A Common Pipe Bending Problem

Problem: Make a 10-inch offset in 2-inch pipe using 20° bends.

Visualize: Figure 26-5 shows the centerline of the 2-inch pipe offset. In practical bending practice with the push-type bending wheel, two strips of metal are joined at one end with a clamp screw. This creates a tool that can be fastened at any degree setting. We use a protractor to set this tool at 20°. Then we press the bending wheel into the pipe until the desired degree of bend is reached by comparing the bend with our tool as the work progresses.

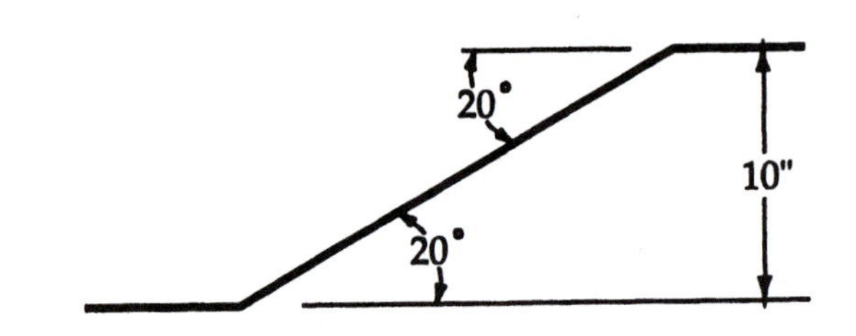

Figure 26-5

Figure 26-6 shows the piece of pipe that is going to be used to make the proposed offset. If the clamp wheel is used, the center of the wheel is fastened to the points marked "B" (for beginning). If the push-wheel type of bending machine is used, the center of the wheel is pressed into the points marked "M" (for middle). The wheel is pressed in at the first mark and then the pipe is turned over and the wheel is pressed in at the other mark from the opposite direction, forming the offset. The distances B-B and M-M are exactly the same. With either type of bending machine it should be obvious that the student must first calculate the diagonal (travel) of the offset.

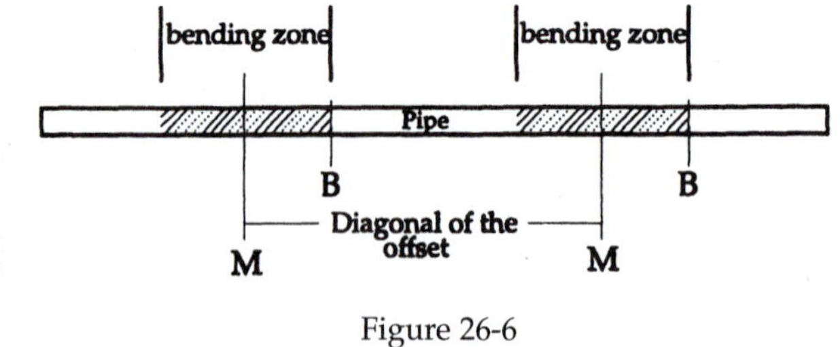

Figure 26-6

The Process:

1. Consult Unit 25 and calculate the diagonal for a 20°, 10-inch offset.
   a. Look at the example offset chart to find out which function to use when the offset is known and the diagonal is needed. In this case, it will be the cosecant of the bend angle.
   b. Look up the constant in the trigonometric function chart in the Appendix. The cosecant of 20 degrees is 2.9238.
   c. The example offset chart also tells us to multiply the function value by the offset. 2.9238 × 10 = 29.238 inches. This answer is the length of the diagonal in our offset.
   d. Convert the decimal fraction to an English measure fraction. 29.238" becomes 29¼ inches.
2. Put lines on the pipe indicating the point at which the bending wheel makes first contact with the pipe.
3. Bend it, using some kind of gauge to check progress. You will have to overbend somewhat because there will be a certain amount of spring-back in the pipe.

Problems

Calculate the length of pipe that will be needed for the following pipe sizes and angles using the standard minimum bending radius.

	Pipe size (inches)	Bend angle (degrees)	Length (inches)
1.	1	15	
2.	1	25	
3.	1½	15	
4.	1½	25	
5.	2	15	

How far apart should the bends be made for the following offsets?

	Offset (inches)	Bend angle (degrees)	Between bends (inches)
6.	8	9	
7.	10	11	
8.	13	14	
9.	15	17	
10.	16½	22	
11.	8	32	
12.	10	9	
13.	13	11	
14.	15	14	
15.	16½	17	
16.	8	22	
17.	10	32	
18.	13	9	
19.	15	11	
20.	16½	14	

Unit 27 Cast Iron Pipe

Objectives

- Learn standard pipe lengths of cast iron pipe.
- Allow for hubbed pipe.
- Allow for No-Hub pipe.

Cast iron (C.I.) pipe is very durable in most cases and is preferred to wrought iron or steel pipe when long service with little maintenance is expected. Connections are made by one of the three methods shown in the half-section views. The lead and oakum joint can be used either with beaded or plain spigot end pipe. The compression gasket uses plain spigot end pipe. The No-Hub connection is available for 2", 3", and 4" cast iron pipe and fittings. No-Hub is service-weight C.I. pipe and fittings. The service-weight C.I. pipe is also made with hubs.

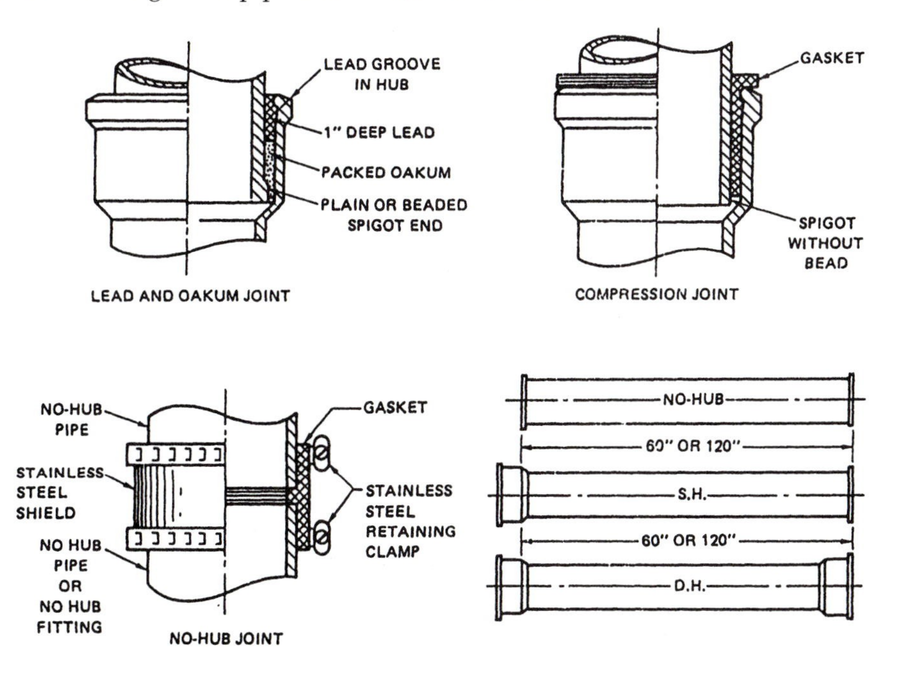

Pipe is cast in 5-foot and 10-foot lengths. For No-Hub pipe this is the actual, or laying, pipe length. Single hub (S.H.) pipe has a laying length of 5 or 10 feet; this is the pipe length without measuring the hub. Any part cut off a single hub length has no hub. Such "bald" pieces are often scrap and are as short as planning can allow.

The double hub piece can be a saving as it will make two shorter pieces each with a hub. The laying length of double hub (D.H.) pipe is shorter than 60 inches or 120 inches by one hub depth. The laying length for 4" D.H. cast iron pipe is 57 inches for 5-foot pipe and 117 inches for 10-foot pipe. The overall lengths of the single hub and the double hub pipe are alike, probably as an aid to the molding and casting of pipe.

Note: *The problems in this text are based on the 5-foot cast iron pipe but are easily usable with 10-foot lengths.*

Sample Problem

Using 5-foot lengths of cast iron pipe, find the following:

a. Find the number of full lengths of single hub pipe needed for a cast iron line that is 16' 2".

16' 2" ÷ 5' = 3 full lengths

b. Find the length of the additional piece needed.

3 × 5' = 15'

16' 2" – 15' 0" = 1' 2"

c. How much is left (scrap) after cutting the additional piece needed from a length of single hub pipe?

5' = 60"

1' 2" = 14"

60" – 14" = 46"

d. Find the laying length of the remaining piece after cutting the additional piece needed from double hub pipe.

(60" – 3") – 14" = 43"

Exercise

Using 5-foot lengths, determine for each of the given total lengths:

a. Number of cast iron single hub pipes needed

b. Length of any piece less a full pipe

c. How much is left (scrap) after cutting the piece from an S.H. pipe

d. Find the laying length of the remaining piece after cutting the piece from a D.H. pipe with 3″ hub depth

Total length	a. Number of full lengths needed S.H.	b. Length of additional piece needed	c. Length of remaining piece if cut from S.H.	d. Laying length of remaining piece if cut from D.H.
1. 10' 10"	2	10"	50"	47"
2. 13' 7"				
3. 22' 5"				
4. 49' 2"				
5. 63' 8"				

Unit 28

Cast Iron Assemblies with Various Bends

Objectives

- Solve for e-e lengths of cast iron pipe.
- Compute lead amounts for various sizes.

Allow 12 ounces of lead for each inch of pipe size, e.g., a 4-inch pipe joint consumes $4 \times 12 = 48$ ounces. Since there are 16 ounces in one pound, one 4-inch cast iron pipe joint consumes 3 pounds of lead ($48/16 = 3$).

The center-to-center lengths for cast iron assemblies are the same as for threaded pipe. It is the allowance for fittings that is somewhat different. The pipe fitter must get used to the fact that a $\frac{1}{6}$ bend is 60°, a $\frac{1}{8}$ bend is 45°, and a $\frac{1}{16}$ bend is $22\frac{1}{2}$°. Because these angles are the same as for threaded fittings, the same corresponding constants are used.

Sample Problem 1

Solve for c-c and e-e lengths of No-Hub pipes J, K, and L.

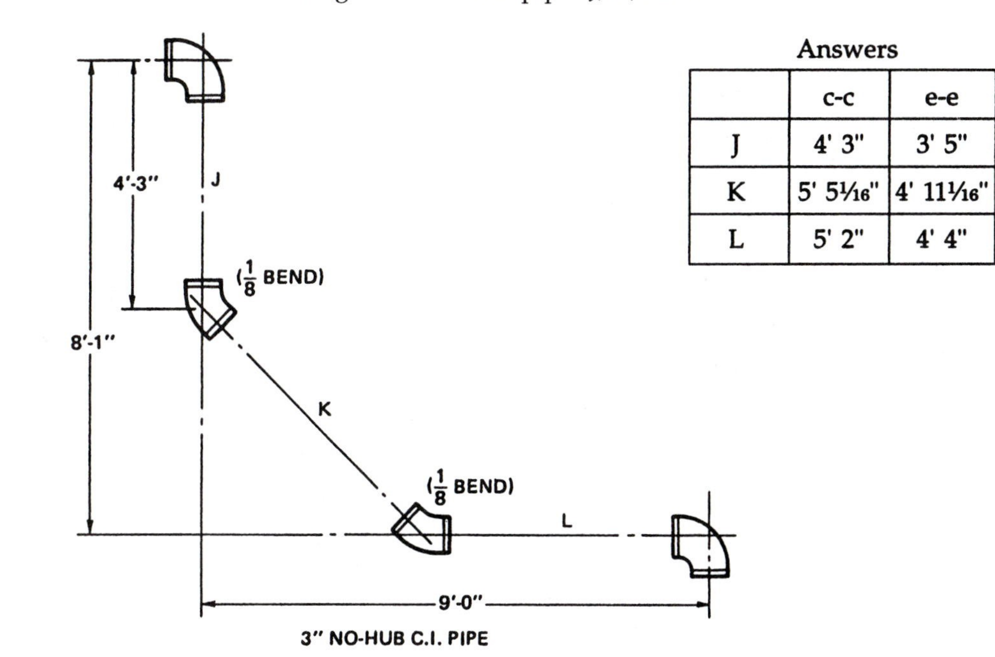

Answers

	c-c	e-e
J	4' 3"	3' 5"
K	5' 5$\frac{1}{16}$"	4' 11$\frac{1}{16}$"
L	5' 2"	4' 4"

Solution

Solve for pipe J: c-c = 4' 3"

(Data #12) D = 7"

D = 3"

e-e = 51" – (7" + 3") = 41" = 3' 5"

Solve for pipe K:

(Data #20) diagonal = 1.414 × rise

c-c = 1.414 × (97" – 51")

c-c = 65.044 = 65$\frac{1}{16}$" = 5' 5$\frac{1}{16}$"

(Data #12) D = 3"

e-e = 65$\frac{1}{16}$" – (3" + 3") = 59$\frac{1}{16}$" = 4' 11$\frac{1}{16}$"

Solve for pipe L:

(Data #20) offset = rise

offset = 97" – 51" = 46"

c-c = 108" – 46" = 62" = 5' 2"

The fitting allowance is the same as for pipe J.

e-e = 62" – (7" + 3") = 52" = 4' 4"

Sample Problem 2

a. Solve for c-c and e-e lengths for service weight cast iron (SWCI) pipes J, K, and L.

b. Find the number of hubs to pour for pipes J, K, and L.

c. Compute pounds of lead required.

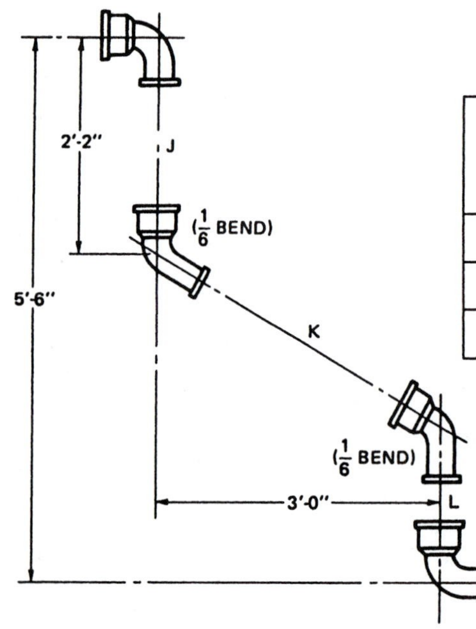

Answers

a.	c-c	e-e	b. Number of Hubs
J	2' 2"	1' 4"	2
K	3' 5$\frac{9}{16}$"	2' 10$\frac{13}{16}$"	2
L	1' 7$\frac{1}{4}$"	9$\frac{1}{4}$"	2

Solution

a. Solve for pipe J: c-c = 26"

(Data #12) D = 8"

(Data #13) C = 2"

e-e = 26" – (8" + 2") = 16" = 1' 4"

Solve for pipe K:

(Data #21) diagonal = 1.155 × offset

c-c = 1.155 × 36"

c-c = 41.58" = 41 9/16" = 3' 5 9/16"

(Data #13) D = 4 3/4", C = 2"

e-e = 41 9/16" – (4 3/4" + 2") = 34 13/16" = 2' 10 13/16"

Solve for pipe L:

(Data #21) rise = 0.577 × offset

rise = 0.577 × 36"

rise = 20.772" = 20 3/4"

c-c = 66" – (26" + 20 3/4") = 19 1/4" = 1' 7 1/4"

(Data #13) D = 4 3/4"

(Data #12) C = 5 1/4"

e-e = 19 1/4" – (4 3/4" + 5 1/4") = 9 1/4"

b. Pipe J: use one 16" length – 2 hubs

Pipe K: use one 34 13/16" length – 2 hubs

Pipe L: use one 9 1/4" length – 2 hubs

2 hubs + 2 hubs + 2 hubs = 6 hubs

c. 12 oz/in × 2 in × 16 = 144 oz

144 oz ÷ 16 oz/lb = 9 lb

Exercise

For each of the problems:

a. Solve for c-c and e-e lengths if XHCI pipe is used.

b. Solve for c-c and e-e lengths if No-Hub pipe is used.

c. How many pounds of lead are needed to pour the hubs for the XHCI pipes?

d. How many connectors are needed for the No-Hub pipes?

Note: Use this illustration for problems 1–4.

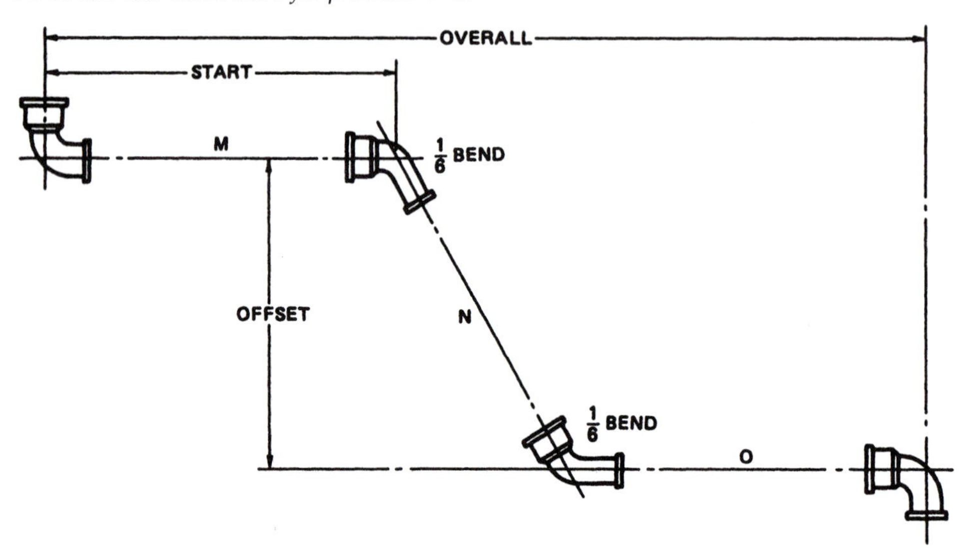

	Overall	Start	Offset	Pipe Size
1.	60"	22"	18"	2"
2.	8' 0"	2' 0"	4' 0"	3"
3.	3' 10"	1' 3"	1' 4"	3"
4.	14' 6"	7' 0"	2' 9"	4"

Note: Use Data #13

Note: Use this illustration for problems 5–8.

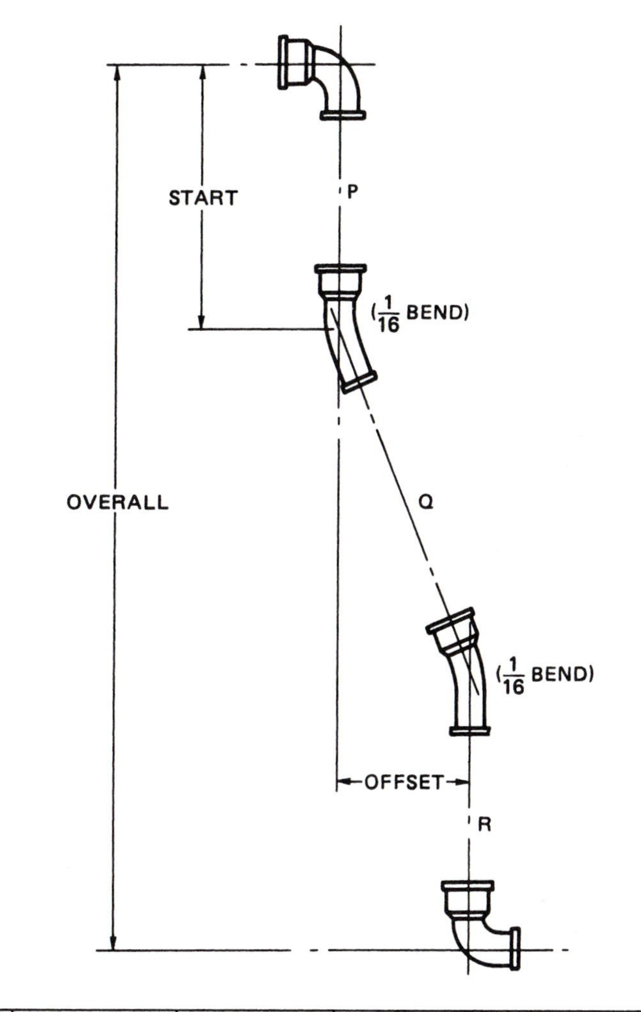

	Overall	Start	Offset	Pipe Size
5.	5' 5"	20"	7"	2"
6.	70"	24"	10½"	3"
7.	7' 8"	2' 6"	1' 0"	4"
8.	12' 9"	5' 2"	1' 9"	4"

Note: Use Data #14

Unit 29 Bend Offsets

Objectives

- Calculate an offset with fittings only.
- Allow for the rare 72° fitting.

Cast iron bends are made with a 72° fitting angle. These are ⅕ bends (360° ÷ 5 = 72°). There is no similar fitting angle in threaded fittings. Also, the 11¼° fitting angle available in threaded fittings is not available in cast iron. Each style of fittings has one angle not available to both kinds of pipe.

It is necessary to use a new set of constants to work with ⅕ bends. Constants are given in Data #24 in the Appendix.

Sample Problem 1

Solve for c-c and e-e lengths of pipes J, K, and L.

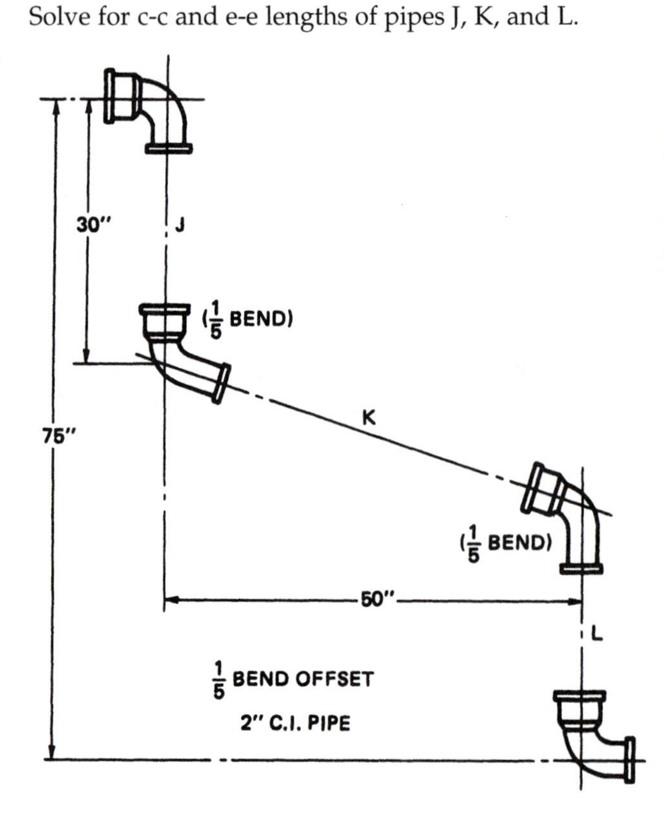

Answers

	c-c	e-e
J	30"	19½"
K	52⅝"	44⅞"
L	28¾"	18¼"

Solution

Solve for pipe J: c-c = 30"

(Data #12) D = 8"

(Data #13) C = 2½"

e-e = 30" – (8" + 2½") = 19½"

Solve for pipe K:

(Data #24) diagonal = 1.052 × offset

c-c = 1.052 × 50"

c-c = 52.6" = 52⅝"

(Data #13) D = 5¼"

C = 2½"

e-e = 52⅝" – (5¼" + 2½") = 44⅞"

Solve for pipe L:

(Data #24) rise = 0.325 × offset

rise = 0.325 × 50" = 16.25" = 16¼"

c-c = 75" – (30" + 16¼") = 28¾"

(Data #13) D = 5¼"

(Data #12) C = 5¼"

e-e = 28¾" – (5¼" + 5¼") = 18¼"

Sample Problem 2

What is the shortest diagonal that can be made with 2 two-inch ⅕ bends? What is the offset? What is the rise?

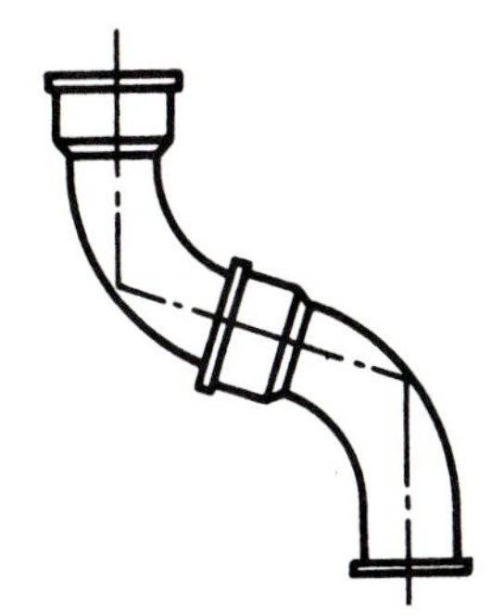

Solution

Diagonal:

(Data #13) D = 5¼"

C = 2½"

diagonal = 5¼" + 2½" = 7¾"

Offset:

(Data #24) offset = 0.951 × diagonal

offset = 0.951 × 7¾" = 7.37025" = 7⅜"

Rise:

(Data #24) rise = 0.309 × diagonal

rise = 0.309 × 7¾" = 2.39475" = 2⅜"

Exercise

1. Solve for c-c and e-e for each lettered pipe.

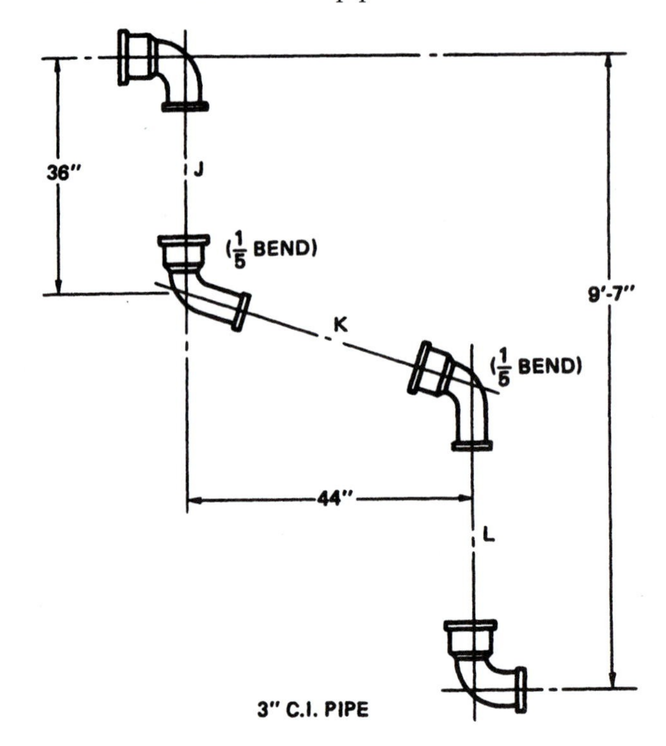

2. Solve for c-c and e-e for each lettered pipe.

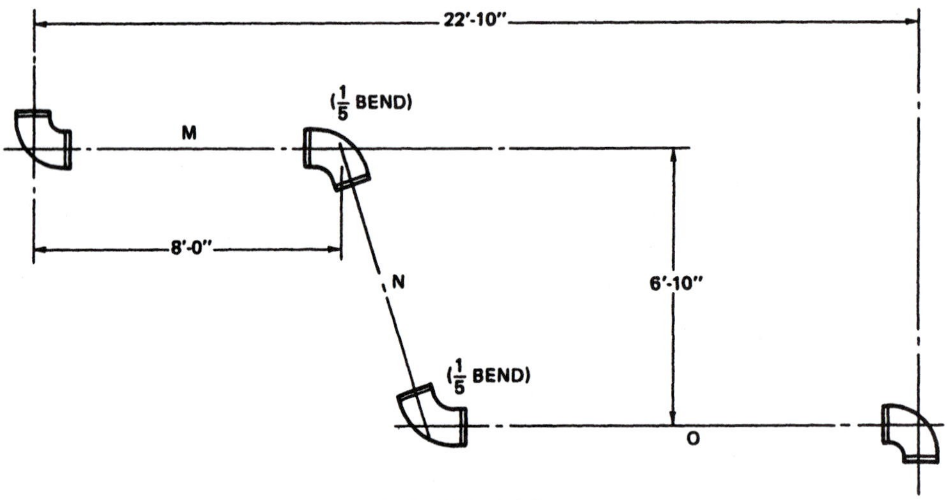

3. Solve for c-c and e-e for each lettered pipe.

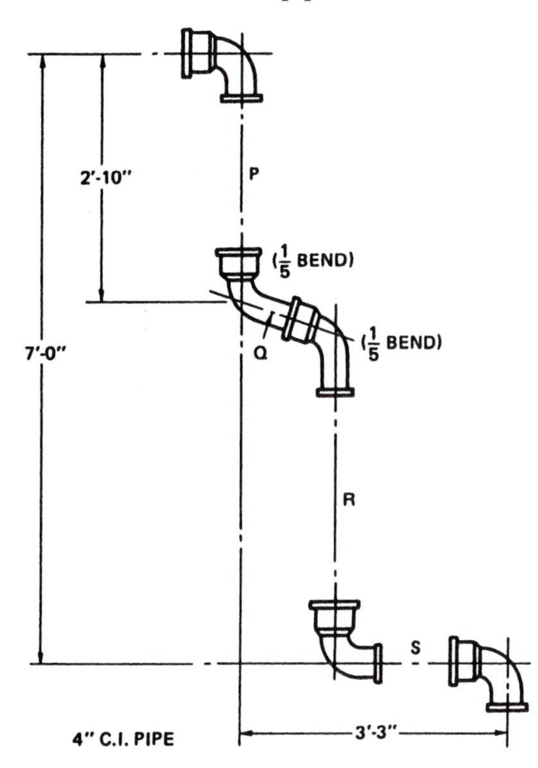

Unit 30

Assemblies with Cast Iron Wyes and Tee-Wyes

Objective

- Combine various fitting angles into practical plumbing assemblies.

Cast iron pipe assemblies use bends, wyes, and tee wyes as well as some other fittings. The wyes and tee wyes have three measurements to the center point. These lengths do not have the consistent pattern of threaded fittings. However, most manufacturers of cast fittings use the dimensions shown in Data #12 through Data #17 in the Appendix.

The new problem is to make the fittings allowances correctly, as the calculations for c-c are the same as in previous units showing similar diagrams with threaded pipe.

Exercise

For diagrams 1–3:

a. Solve for c-c and e-e lengths of each lettered pipe.

b. Compute the pounds of lead required to assemble.

1.

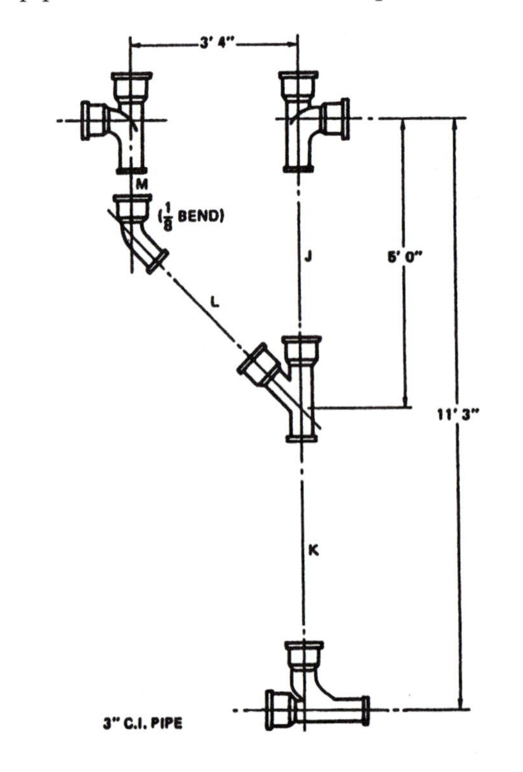

2.

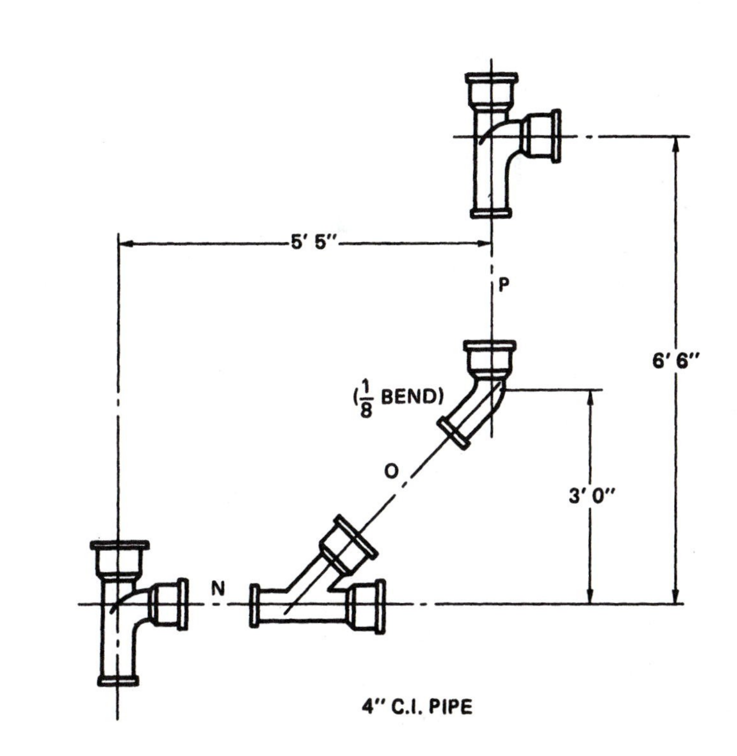

3.

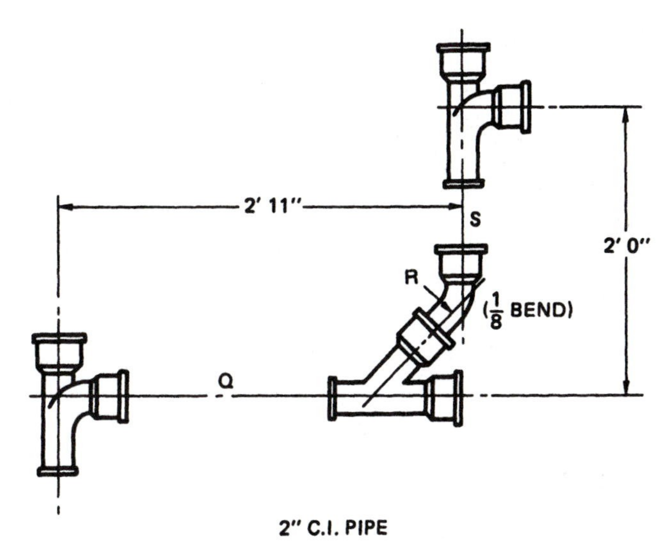

4. a. Solve for c-c and e-e lengths of each lettered pipe.
   b. Find the number of connectors.

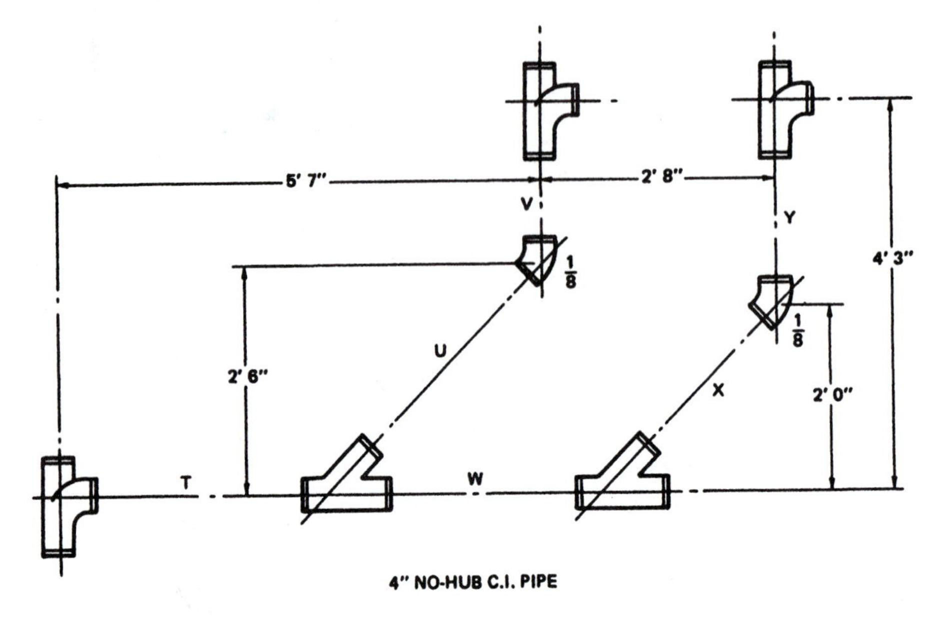

Unit 31

Single Loop Back-Venting

Objectives

- Learn waste and vent loop layout.
- Combine different fitting materials into one assembly.

Stacks are pipes that generally run vertical. There are *soil stacks,* which are stacks that receive the discharge of at least one water closet; *drain stacks,* which receive the discharge of fixtures that are not water closets; and *vent stacks,* which exist to provide air to the plumbing sanitary system. Stacks can be plastic, copper, cast iron, or steel pipe. The example and exercise on the following pages show cast iron main stacks with galvanized steel branches.

Traps must be installed with provisions for a sufficient supply of air to prevent trap siphonage. When traps are to be installed long distances from the venting stack, additional piping must be installed to simply supply air (Figure 31-1). This often takes the form of back-venting or loop venting.

In loop venting, the vent stack is continued upward in the building until it is higher than the highest fixture. Then it is connected back into the main house stack.

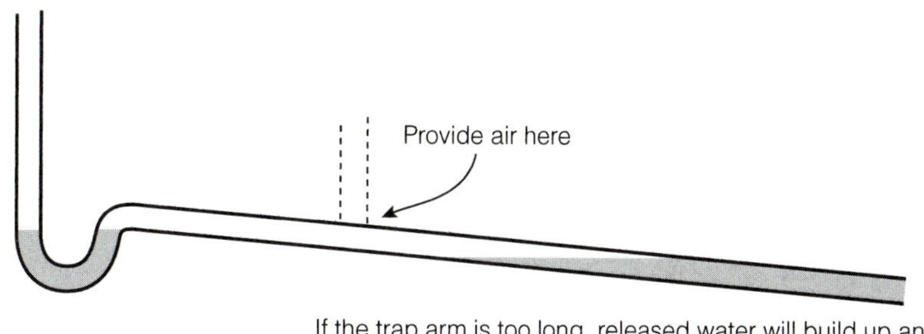

Figure 31-1 Preventing the loss of the trap's seal

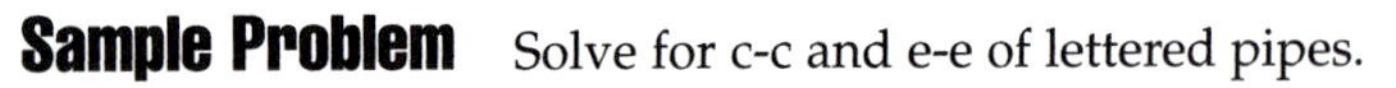

Sample Problem

Solve for c-c and e-e of lettered pipes.

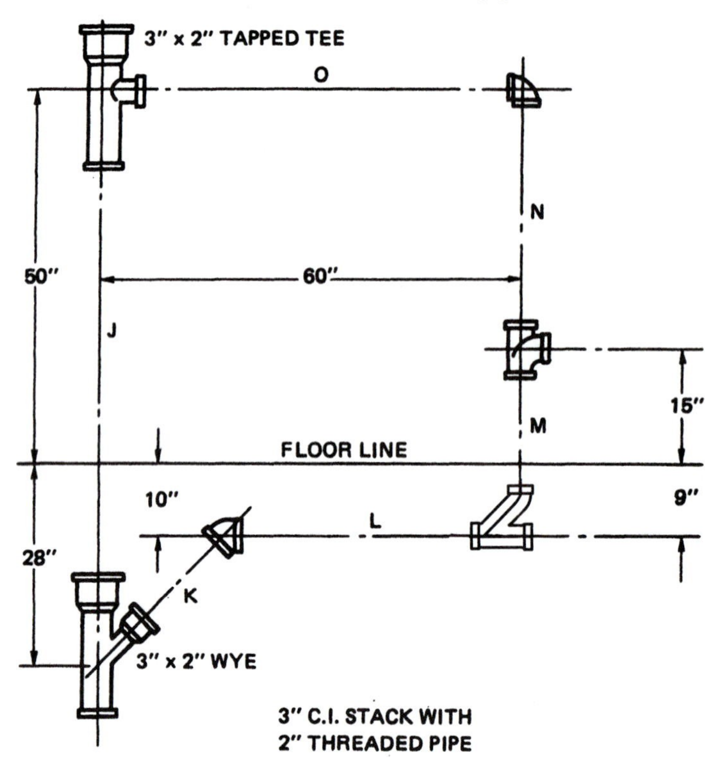

Answers

	c-c	e-e
J	78"	$66\frac{1}{4}$"
K	$25\frac{7}{16}$"	$19\frac{5}{16}$"
L	42"	$34\frac{3}{4}$"
M	24"	$15\frac{3}{16}$"
N	35"	$28\frac{13}{16}$"
O	60"	$53\frac{7}{8}$"

Solution

Solve for pipe J:

c-c = 28" + 50" = 78"

(Data #17) D = 7"

(Data #15) D = $4\frac{3}{4}$"

e-e = 78" – ($4\frac{3}{4}$" + 7") = $66\frac{1}{4}$"

Solve for pipe K: rise = 28" – 10" = 18"

(Data #20) diagonal = 1.414 × rise

c-c = 1.414 × 18"

c-c = 25.452" = $25\frac{7}{16}$"

(Data #15) E = 5"

(Data #5) A = $1\frac{5}{8}$", B = $\frac{1}{2}$"

e-e = $25\frac{7}{16}$" – [(5" + $1\frac{5}{8}$") – $\frac{1}{2}$"] = $19\frac{5}{16}$"

Solve for pipe L:

(Data #20) offset = rise = 18"

c-c = 60" – 18" = 42"

(Data #5) A = $1\frac{5}{8}$", B = $\frac{1}{2}$"

(Data #10) C = $4\frac{5}{8}$", B = $\frac{1}{2}$"

e-e = 42" – [($1\frac{5}{8}$" + $4\frac{5}{8}$") – 2($\frac{1}{2}$")] = $34\frac{3}{4}$"

Solve for pipe M: c-c = 9" + 15" = 24"

(Data #10) D = $5\frac{1}{8}$", B = $\frac{1}{2}$"

(Data #10) C = $3\frac{3}{16}$", B = $\frac{1}{2}$"

e-e = 24" [($5\frac{1}{8}$" + $3\frac{3}{16}$") – 2($\frac{1}{2}$")] = $15\frac{3}{16}$"

Solve for pipe N: c-c = 50" – 15" = 35"

(Data #10) A – C = $2\frac{9}{16}$", B = $\frac{1}{2}$"

(Data #5) A = $2\frac{5}{8}$", B = $\frac{1}{2}$"

e-e = 35" – [($2\frac{9}{16}$" + $2\frac{5}{8}$") – 2($\frac{1}{2}$")] = $28\frac{13}{16}$"

Solve for pipe 0: c-c = 60"

(Data #5) A = $2\frac{5}{8}$", B = $\frac{1}{2}$"

(Data #17) E = $2\frac{1}{2}$", B = $\frac{1}{2}$"

e-e = 60" – [($2\frac{5}{8}$" + $2\frac{1}{2}$") – 2($\frac{1}{2}$")] = $53\frac{7}{8}$"

Exercise

Solve for c-c and e-e for each lettered pipe in the following diagrams.

1.

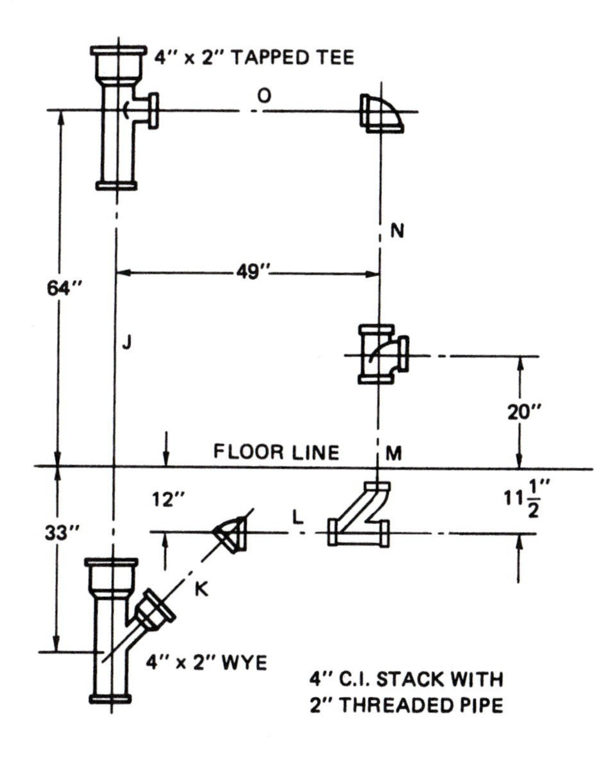

2.

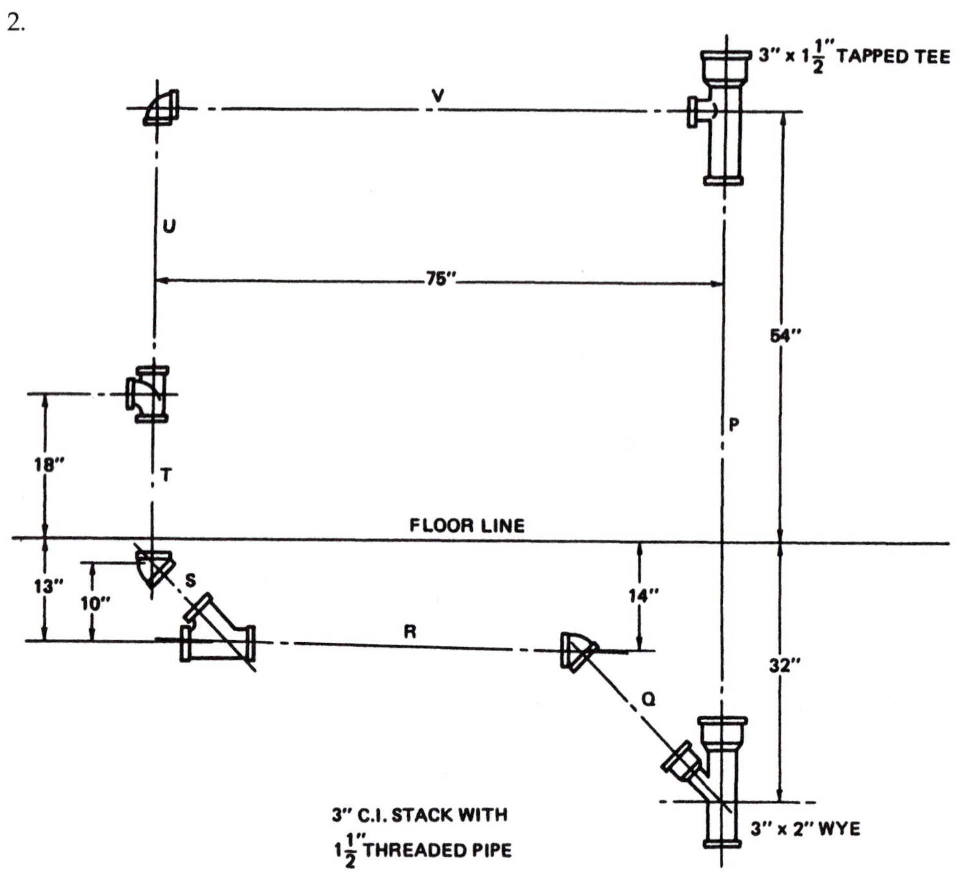
3" x 1½" TAPPED TEE
V
U
75"
54"
P
18"
T
FLOOR LINE
13"
10"
S
14"
R
32"
Q
3" C.I. STACK WITH
1½" THREADED PIPE
3" x 2" WYE

Unit 32 Grade, Percent Grade, Drop, and Run

Objectives

- Define drop, grade, and run.
- Calculate grade.
- Calculate percent grade.
- Calculate drop.

Horizontal waste piping and sewers are installed at a slight slope to carry solids better than would be possible in a true horizontal pipe. The plumber measures grade, sometimes called *pitch,* in inches per foot by use of a level held on the pipe. The level may have a small adjustable foot to give the proper drop for the length of the level.

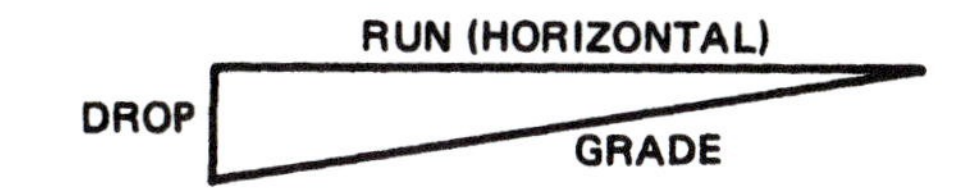

The *slope* of a pipe, sewer, or land may be dimensioned as percent grade. A 1-percent grade is at the rate of 1 foot of drop for each 100 feet of horizontal run. This is the same rate as 0.01 foot of drop for 1 foot of run. Because 0.01 foot is very nearly ⅛ inch, a 1-percent grade is equal to a ⅛ inch per foot grade.

1% grade = 1 foot drop for 100 feet of run =
0.01 foot of drop for 1 foot of run =
⅛" of drop for 1 foot of run

Drainage fittings such as tee wyes and 90° ells are tapped to give a grade of ¼ inch per foot. This means that the horizontal drain line is out of level by ¼ inch for each foot of run. Some plumbing codes specify a minimum grade of ⅛ inch per foot. The *drop* is the total amount out of level. Thus at ¼ inch per foot, the drop would be one inch if continued for four feet.

Formulas

Inches per Foot

Drop is amount out of level in inches.

- Drop = grade × run

Run is horizontal length in feet.

Grade is rate of slope in inches per foot.

- $\text{Grade} = \frac{\text{drop}}{\text{run}}$

Note: Drop is in inches
Run is in feet

Percent Grades

Drop is the amount out of level in the same units of measure as the run.

- $\text{Drop} = \frac{\text{percent grade} \times \text{run}}{100}$

Run is horizontal length.

Percent grade is a ratio number.

- $\text{Percent grade} = \frac{\text{drop}}{\text{run}} \times 100$

Note: Drop and run must be like units; both may be feet, inches, or meters.

Sample Problem 1

a. How much drop is there in 18 feet of run at ⅛ inch per foot of grade?

b. How much drop is there in 18 feet of run at 1-percent grade?

Solution

a. Drop = grade × run

Drop = ⅛ × 18 = 2¼"

b. $\text{Drop} = \frac{\text{percent grade} \times \text{run}}{100}$

$\text{Drop} = \frac{1 \times 18'}{100} = 0.18 \text{ foot}$

Sample Problem 2

a. What grade to the nearest 32nd of an inch is a drop of 5 inches in 22 feet?

b. What percent grade is a drop of 8 inches in 60 feet?

Solution

a. $\text{Grade} = \frac{\text{drop}}{\text{run}}$

Grade = 5/22 = 0.227 = 7/32 inch per foot

b. Percent grade = $\frac{\text{drop}}{\text{run}} \times 100$

Drop = 8 inches = 0.67 foot

Percent grade = $\frac{0.67'}{100} = \times 100 = \frac{67}{60} = 1.1\%$

Exercise

Compute each of the problems for drop, grade, and percent grade as indicated by the blank spaces in the chart.

	Run	Drop	Grade	Percent Grade
1.	12'		¼ in/ft	
2.	27'			2%
3.	5'		⅛ in/ft	
4.	35'		5⁄32 in/ft	
5.	26'	5"		
6.	46'	6½"		
7.	72'	9¾"		
8.	84'		⅛ in/ft	

Unit 33 Elevation and Grade

Objectives

- Define an elevation and a bench mark.
- Convert to engineer's measure.
- Solve problems in elevations.

An *elevation* is a certain distance above or below a fixed point. Land or geographic elevations are distances above or below sea level. The architect may specify building elevations based on sea level but often a number, such as 100.00 feet, is used as a reference or *bench mark.* The bench mark is some nonmoving spot as on the curb. It is called 100.00 feet elevation because the number is easy to add to, or subtract from. A first floor at elevation 103.50 feet would be 3½ feet higher than the bench mark. A basement floor at elevation 94.00 feet would be 6 feet lower than the bench mark.

The surveyor or civil engineer uses elevation measurements in feet and decimal parts of a foot. One hundredth (0.01) foot is the smallest elevation difference. The plumber uses feet, inches, and nearest eighth of an inch in elevation measure. An eighth of an inch is very nearly the same as one hundredth of a foot. It may be necessary for the plumber to convert engineering and building trade elevations because both are used.

A pipe that is at a grade changes elevation. The plumber needs to determine the elevations of points on house sewers and house drains.

Sample Problems and Solutions

1. Convert 8' 4⅝" to engineer's measure.

   8' = 8.00'

   4⅝" = 4.625"

   4.625" ÷ 12 in/ft = 0.38'

   8' 4⅝" = 8.00' + 0.38' = 8.38'

   Also,

   8' = 8.00'

   4" = 4" ÷ 12 in/ft = 0.33'

   ⅝" = 0.01' × 5 = 0.05'

   8' 4⅝" = 8.00' + 0.33' + 0.05' = 8.38'

2. Convert 94.28' to builder's measure.

   94' = 94'

   0.28' = 0.28' × 12 in/ft = 3.36"

   3.36" = 3⅜"

   94.28' = 94' 3⅜"

   Also,

   94' = 94'

   0.28' = 28/8" = 3½"

   94.28' = 94' 3½"

(Note: *Using the approximation ⅛" = 0.01' results in a slightly different answer.)*

3. The bench mark is 100.00 feet elevation.

   The basement floor is 94.70 feet elevation.

   The first floor is 102.95 feet elevation.

   a. How far below the bench mark is the basement floor?

   b. How far above the bench mark is the first floor?

   c. How far is it from basement floor to first floor?

   a. 100.00' – 94.70' = 5.30 feet or 5' 3⅝" below bench mark

      (0.30' = 0.25' + 0.05'; 0.25' = 3"; 0.05' = ⅝")

   b. 102.95' – 100.00' = 2.95 feet or 2' 11⅜" above bench mark

      (0.95' is 1'-0.05'; thus, ⅝" less than 12")

   c. 102.95' – 94.70' = 8.25 feet or 8' 3" between basement and first floors

4. A house drain has a run of 30 feet at a grade of ⅛ inch per foot. The low end has elevation of 93.50 feet. What elevation is the high end?

   30 × ⅛" = 3¾" difference in elevation

   93.50' = 93' 6"; 93' 6" + 3¾" = 93' 9¾"

   Also, ⅛" is about 0.01 foot. Therefore,

   0.01' × 30 = 0.30 foot, difference in elevation

   93.50' + 0.30' = 93.80 feet

(Note: *The two answers are not exactly equal because ⅛ inch is a little larger than 0.01 foot. Either answer is acceptable.)*

Exercise

A. Convert each measurement to engineer's measure.

1. 18' 6"
2. 15' 5½"
3. 32' 9"
4. 33' 10"
5. 35' 11¼"

B. Convert each measurement to builder's measure.

1. 84.27'
2. 93.08'
3. 106.71'
4. 123.44'
5. 156.92'

C. Some building elevations follow: bench mark 100.00', basement floor 94.20', first floor 103.75', second floor 113.00'.

1. How far below the bench mark is the basement floor?
2. How far above the bench mark is the first floor?
3. What is the distance from the basement floor to the first floor?
4. What is the distance from the first floor to the second floor?

D. The low end of a house drain has an elevation of 91.55 feet. The grade is ⅛ inch per foot.

1. What is elevation of a fitting 10 feet from the low point?
2. What is the elevation of a fitting 25 feet from the low point?

Unit 34 Elevation in a Plan View Pipe Diagram

Objectives

- Define the invert.
- Do problems with pipe in horizontal position.
- Calculate elevation.

Most of the pipe diagrams shown in this book have been drawn as elevation views. Some of them would be installed only in walls and are diagramed in the proper position. However, pipe is run under ground or under floors and is then best shown in a plan view. House sewers and house drains are drawn as plan view pipe diagrams.

Because the pipe has grade, there is a change in elevation. The elevation does not show in a plan view so a note is made of the elevation of points along the pipe.

Most pipes are measured to a centerline but sewers and house drains show elevations measured to the invert of the sewer. The *invert* is the inside and bottom of the pipe. The advantages of measuring to the invert of the pipe, which is a solid surface, can be seen.

Sample Problem Determine the c-c lengths for pipes J, K, L, and M and the elevations of points A, B, C, and D. These points are on the invert of the pipe directly under the intersection of the centerlines.

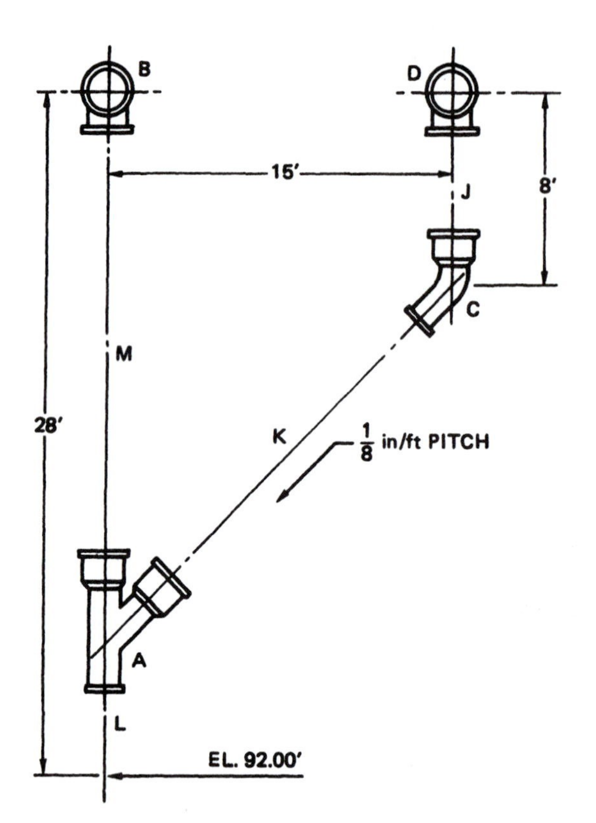

Answers

	c-c
J	8'
K	21' 2½"
L	5'
M	23'

Elevations	
A	92.05'
B	92.29'
C	92.27'
D	92.35'

Solution

Solve for pipe J: c-c = 8'

Solve for pipe K:

(Data #20) diagonal = 1.414 × offset

c-c = 1.414 × 15'

c-c = 21.21' = 21'-2½"

Solve for pipe L:

(Data #20) rise = offset

rise = 15'

c-c = 28' – (8' + 15') = 5'

Solve for pipe M: c-c = 28' – 5' = 23'

elevation A: c-c of L = 5'

5 × ⅛" = ⅝" or 0.05'

92.00' + 0.05' = 92.05'

elevation B:

(c-c of L) + (c-c of M) = 5' + 23' = 28'

28 × ⅛" = 3½" = 0.29'

92.00' + 0.29' = 92.29'

Elevation C:

(c-c of L) + (c-c of K) = 5' + 21' = 26'

26 × ⅛" = 3¼" = 0.27'

92.00' + 0.27' = 92.27'

Elevation D:

(c-c of L) + (c-c of K) + (c-c of J) = 5' + 21.21' + 8' = 34.21'

34.21 × ⅛" = 4¼" = 0.35'

92.00' + 0.35' = 92.35'

Exercise

1. Determine the c-c lengths of pipes J, K, and L and the elevation for points A, B, and C in the illustrated plan of a house drain.

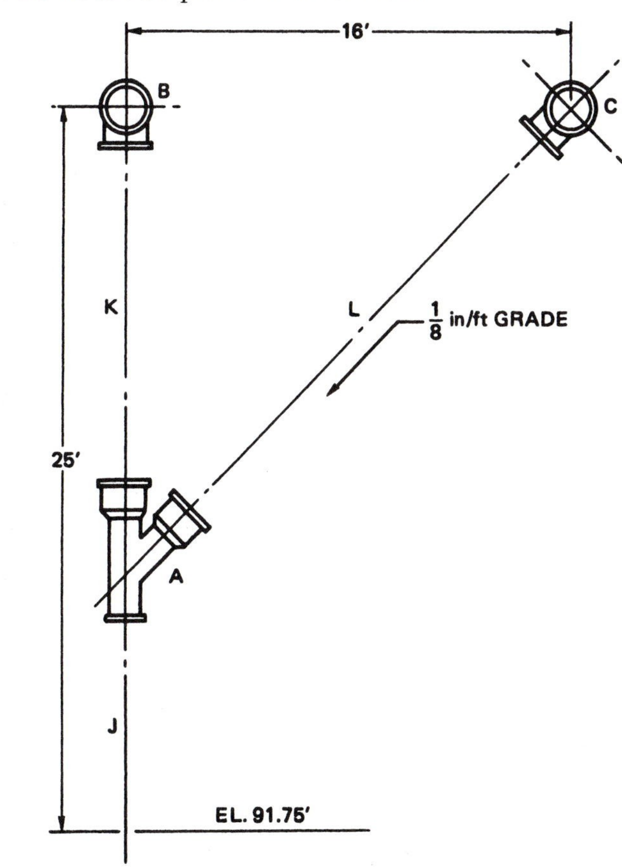

2. Determine the c-c lengths for pipes J, K, L, M, and N. Compute the elevations at points A, B, C, and D.

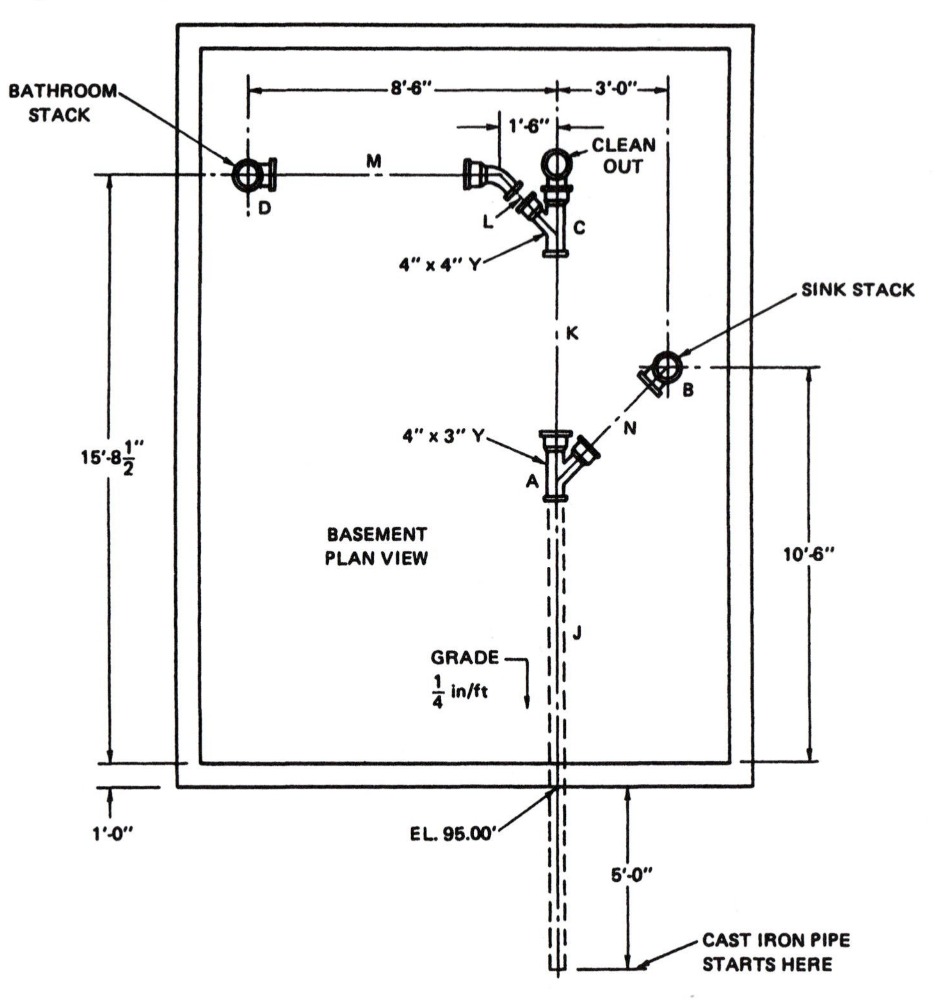

Unit 35 Jumper Offsets

Objectives

- Define the jumper offset.
- Visualize the secondary square or rectangle.
- Use constants to effect a solution.

A jumper offset is one that goes around a stack, tank, or other cylindrical object. The diagonals are usually at 45° angles but the on-the-job measurements change the solution from the usual.

There are two different patterns of jumper offsets. One is made with two 45° ells and one 90° ell. The other uses four 45° ells.

The four different solutions used in connection with various 45° diagonals are as follows:

- Diagonal = 1.414 × offset
- Offset = 0.707 × diagonal
- Diagonal = 2 × side of secondary square
- Use of 0.414 constant for 22½° angle in some 45° fitting diagrams.

In calculation of the two patterns of jumper offsets, all four solutions are used.

Sample Problem 1

Determine the c-c lengths of J and K.

For the offset use two 45° ells and one 90° ell. On this problem the centerline of the cylindrical obstacle coincides with the centerline of the pipe.

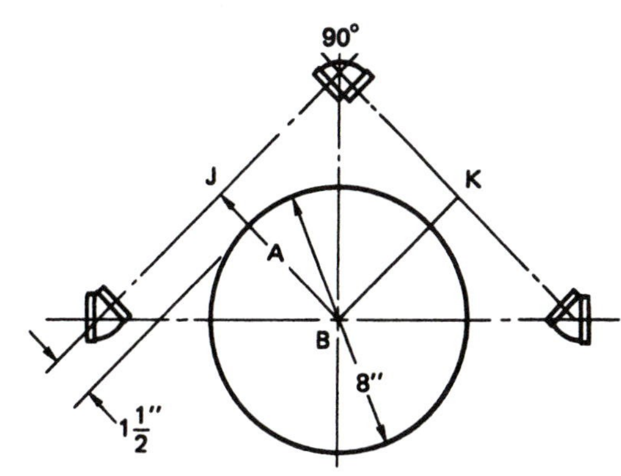

Answers

	c-c
J	11"
K	11"

Solution

Solve for J:

side of secondary square = A

$A = 4" + 1\frac{1}{2}" = 5\frac{1}{2}"$

$\text{c-c} = 2 \times 5\frac{1}{2}" = 11"$

Solve for K:

side of secondary square = $5\frac{1}{2}"$

$\text{c-c} = 2 \times 5\frac{1}{2}" = 11"$

the offset (B-B) = $0.707 \times 11" = 7\frac{3}{4}"$

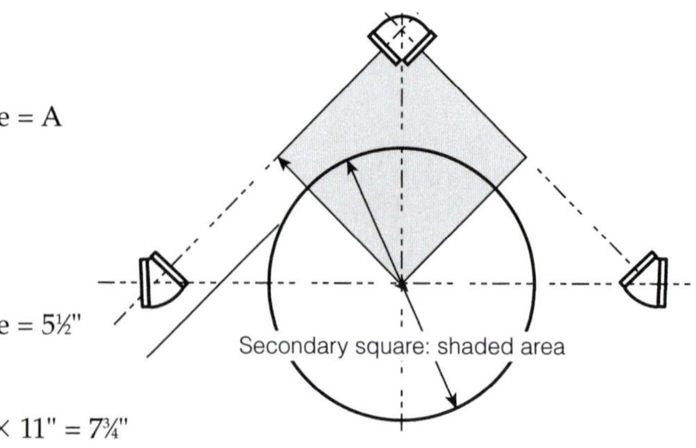

Sample Problem 2

Determine c-c lengths of pipes J, K, L, and M. In this problem, the centerline of the obstacle does not coincide with the centerline of the pipe.

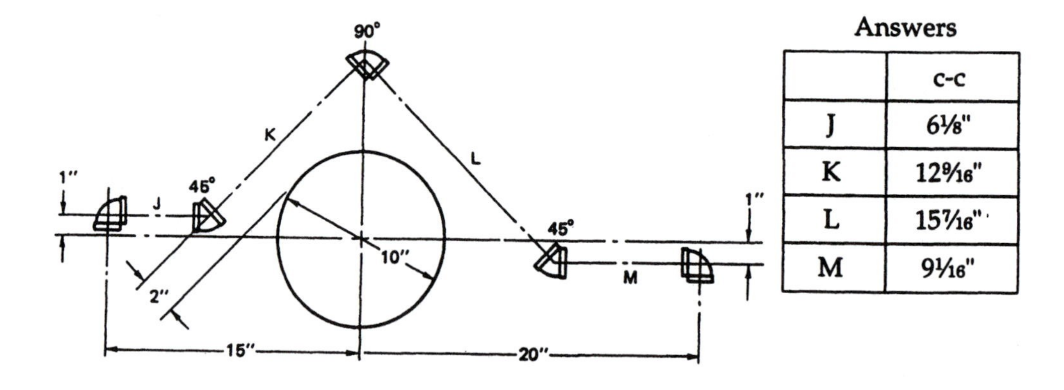

Answers

	c-c
J	$6\frac{1}{8}"$
K	$12\frac{9}{16}"$
L	$15\frac{7}{16}"$
M	$9\frac{1}{16}"$

Solution

Find the side of the secondary square:

$5" + 2" = 7"$

Find the diagonal to the centerline:

$2 \times 7" = 14"$

Find the diagonal distance above and below the centerline for each 45° elbow.

diagonal = $1.414 \times$ rise

diagonal = $1.414 \times 1" = 1\frac{7}{16}"$

Solve for pipe K:

$\text{c-c} = 14" - 1\frac{7}{16}" = 12\frac{9}{16}"$

Solve for pipe L:

$\text{c-c} = 14" + 1\frac{7}{16}" = 15\frac{7}{16}"$

Solve for pipe J:

offset for K = $0.707 \times 12\frac{9}{16}" = 8.88" = 8\frac{7}{8}"$

$\text{c-c} = 15" - 8\frac{7}{8}" = 6\frac{1}{8}"$

Solve for pipe M:

offset for L = $0.707 \times 15\frac{7}{16}" = 10\frac{15}{16}"$

$\text{c-c} = 20" - 10\frac{15}{16}" = 9\frac{1}{16}"$

Sample Problem 3

Determine c-c for pipes J, K, L, M, and N.

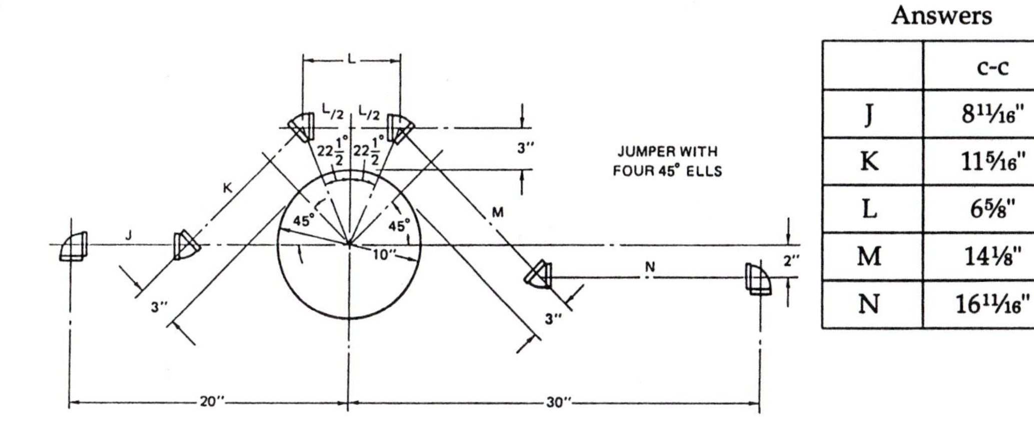

Answers

	c-c
J	8¹¹⁄₁₆"
K	11⁵⁄₁₆"
L	6⅝"
M	14⅛"
N	16¹¹⁄₁₆"

Solution

Find the side of the secondary square.

5" + 3" = 8"

Find the diagonal to the centerline.

2 × 8" = 16"

Find the diagonal distance subtracted for each 45° elbow.

L/2 = 0.414 × 8" = 3.312"

diagonal = 1.414 × 3.312" = 4.683" = 4¹¹⁄₁₆"

Solve for pipe K:

c-c = 16" – 4¹¹⁄₁₆" = 11⁵⁄₁₆"

Solve for pipe L:

c-c = 2 × L/2

c-c = 2 × 3.312" = 6.624" = 6⅝"

Solve for pipe M:

Find diagonal distance below centerline.

diagonal = 1.414 × 2" = 2.828" = 2¹³⁄₁₆"

c-c = 11⁵⁄₁₆" + 2¹³⁄₁₆" = 14⅛"

Solve for pipe J:

diagonal = 1.414 × 8" = 11.312" = 11⁵⁄₁₆"

20" – 11⁵⁄₁₆" = 8¹¹⁄₁₆"

Solve for pipe N:

offset = rise = 2"

30" – (11⁵⁄₁₆" + 2") = 16¹¹⁄₁₆"

Exercise

Note: *Consider both threaded and copper pipe for all problems.*

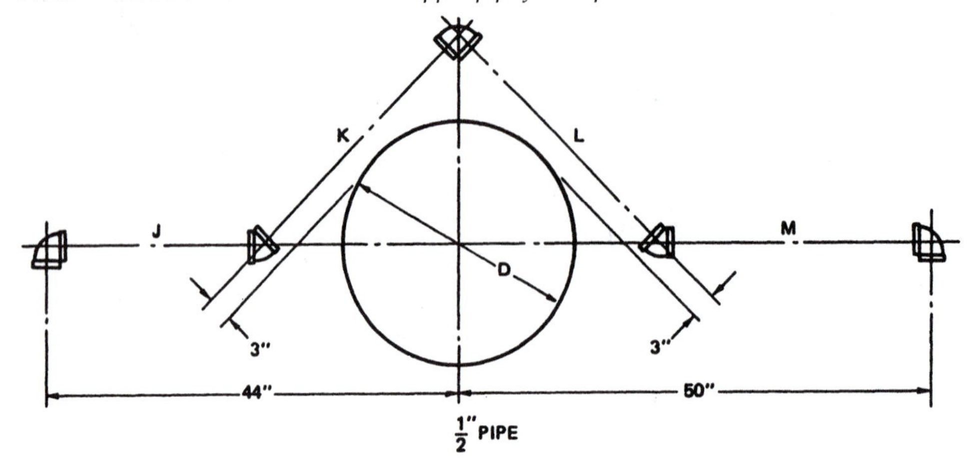

1. Determine c-c and e-e of pipes J, K, L, and M with D = 12".
2. Determine c-c and e-e of pipes J, K, L, and M with D = 20".

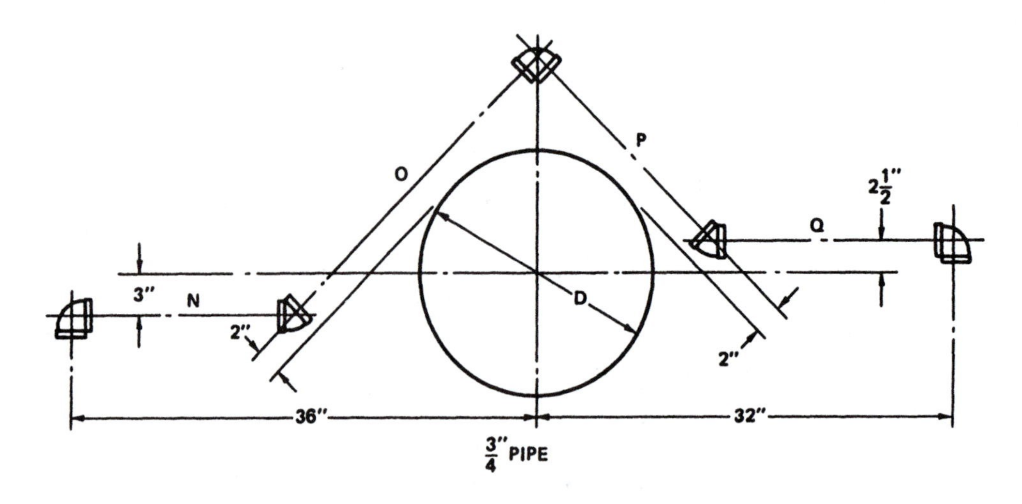

3. Determine c-c and e-e of pipes N, O, P, and Q with D = 9".
4. Determine c-c and e-e of pipes N, O, P, and Q with D = 24".

5. Determine c-c and e-e of pipes R, S, T, U, and V with D = 5".
6. Determine c-c and e-e of lengths of pipes R, S, T, U, and V with D = 15".

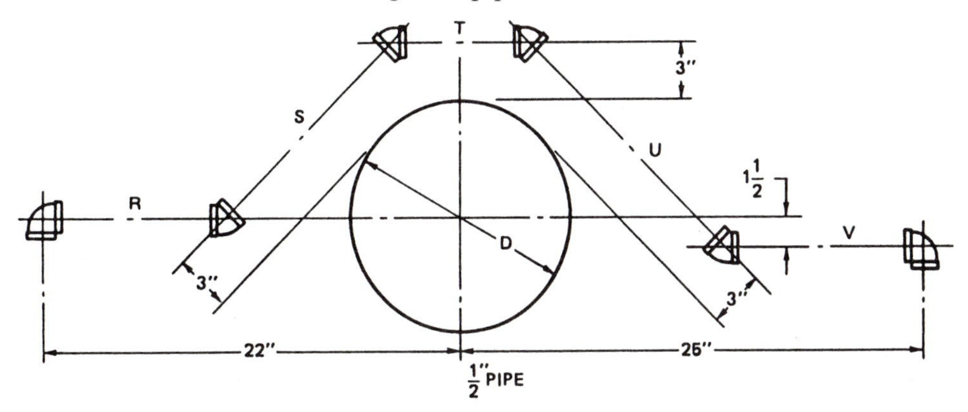

Unit 36 45° Offsets in Parallel

Objectives

- Explain reasons for parallel offsets.
- Determine center-to-center and end-to-end lengths for parallel offsets.
- Use various fittings in parallel offsets.

Parallel offsets are used because of a neat appearance that saves space and allows room for installation of pipe covering. Two or more pipe assemblies can be installed in parallel or equal spread design.

The parallel offsets always have parallel angles one-half of the offset angle. In all but a special case, the diagonals are alike.

One assembly is computed as a simple offset. The second and succeeding assemblies have certain differences in length compared to the first. The difference in length is computed from the spread, or distance between pipe assemblies. Use Data #26 in the Appendix for the constants for parallel offsets.

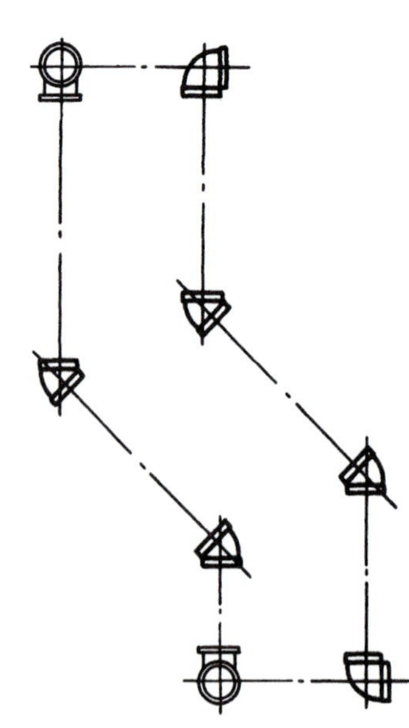

Sample Problem Determine c-c of pipes J, K, L, M, N, and O. (Calculations for e-e are omitted for clarity of new ideas.)

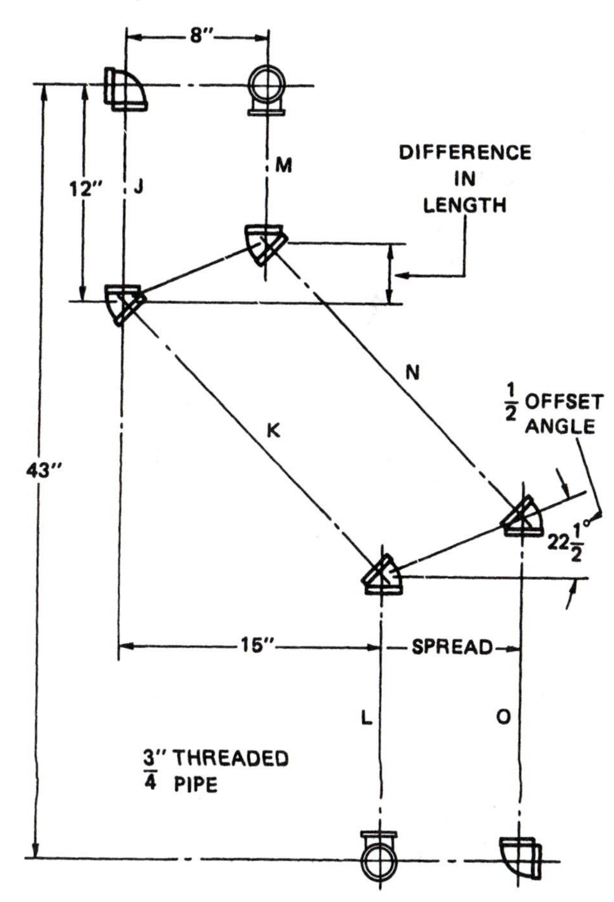

Answers

	c-c
J	12"
K	21 3/16"
L	16"
M	8 11/16"
N	21 3/16"
O	19 5/16"

Solution

Solve for pipe K:

(Data #25) Diagonal = offset × 1.414

c-c = 15" × 1.414 = 21.210" = 21 3/16"

(Data #25) Rise = offset × 1 = 15" × 1 = 15"

Solve for pipe J: c-c = 12"

Solve for pipe L:

c-c = 43" – (12" + 15") = 16"

Solve for pipe M:

(Data #25) Difference in length = 0.414 × spread

Difference in length = 0.414 × 8" = 3.312" = 3 5/16"

c-c = 12" – 3 5/16" = 8 11/16"

Solve for pipe N: c-c = c-c of K = 21 3/16"

Solve for pipe O:

c-c = 16" + 3 5/16" = 19 5/16"

Exercise

1. Determine c-c and e-e lengths of each pipe shown in the diagram. Solve for both threaded and copper pipe.

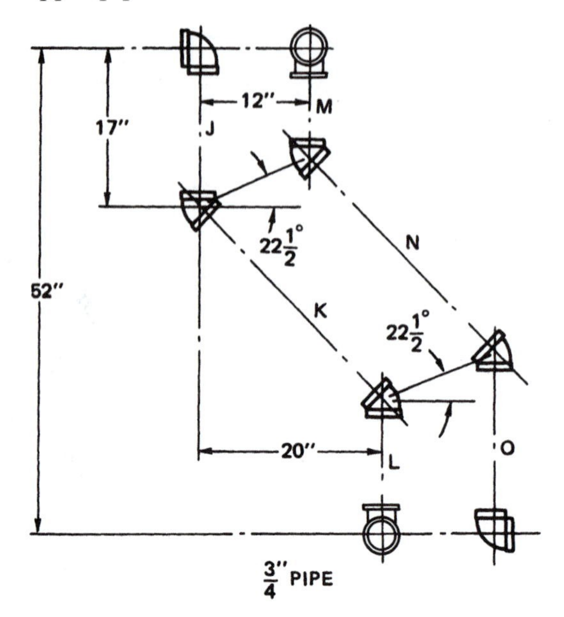

2. Determine the c-c and e-e of lengths of each pipe shown in the diagram. Solve for hub and No-Hub pipe and for threaded and copper pipe.

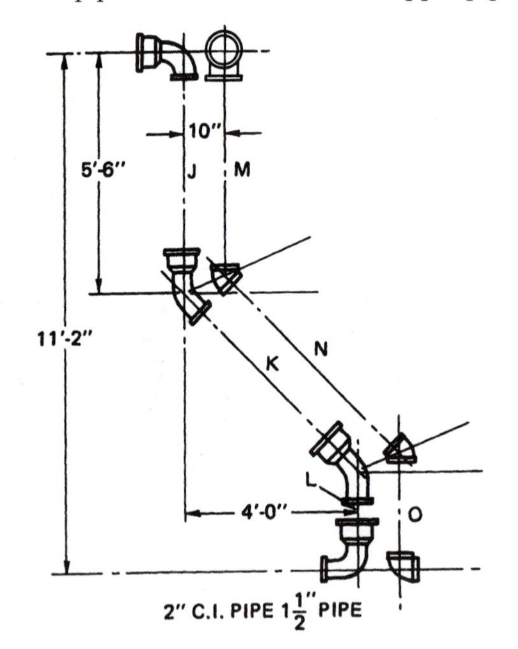

3. Determine c-c and e-e lengths of J, K, L, M, N, O, P, Q, and R shown in the diagram. Consider both threaded and copper pipe.

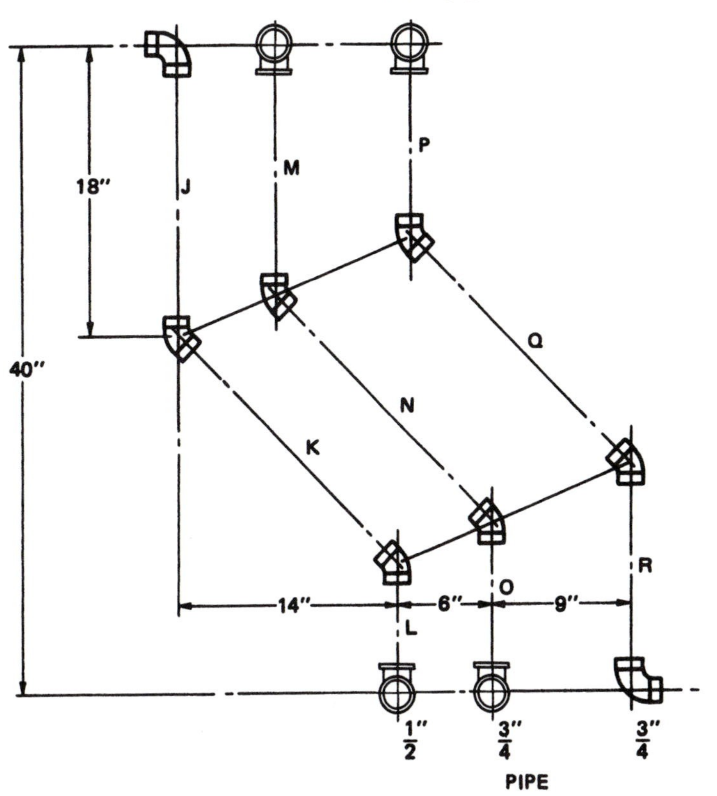

Unit 37 Special Case of 45° Offsets in Parallel

Objectives

- Make 90° turns with parallel offsets.
- Mix sizes and fitting materials.

The special case of 45° parallel offsets uses two 45° or ⅛ bend fittings to make a 90° turn. The pipe assemblies are somewhat like portions of a circle having the same center but different radii. The assembly farther from the center has longer pipes than the nearer assembly.

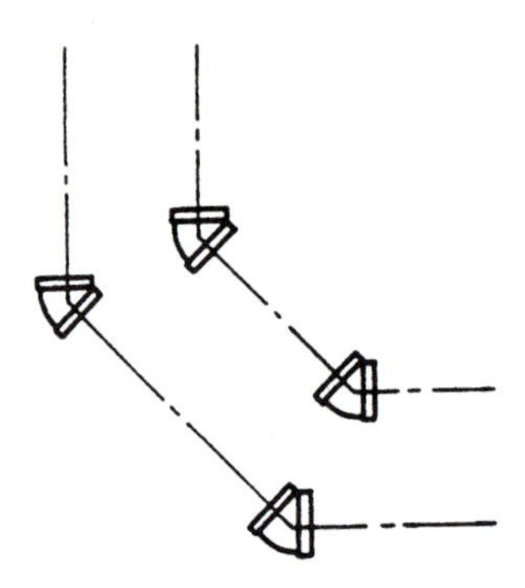

The same parallel angles, 22½°, as for the usual 45° parallel offsets are used. Thus, the 0.414 constant times the spread gives the difference in length. The diagonals, however, are not alike but differ by two times the difference in length.

Sample Problem Determine the c-c of pipes J, K, and L and M, N, and O.

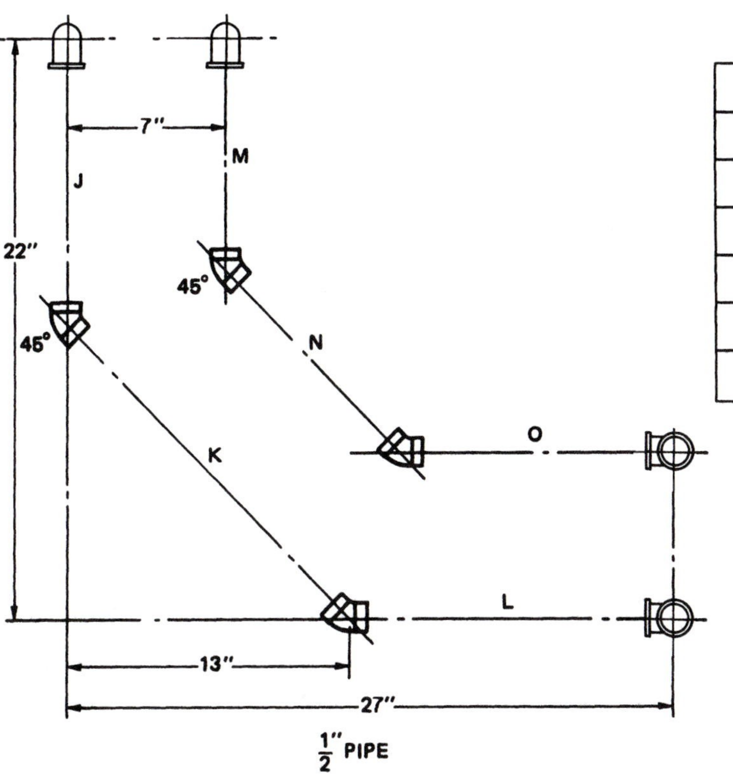

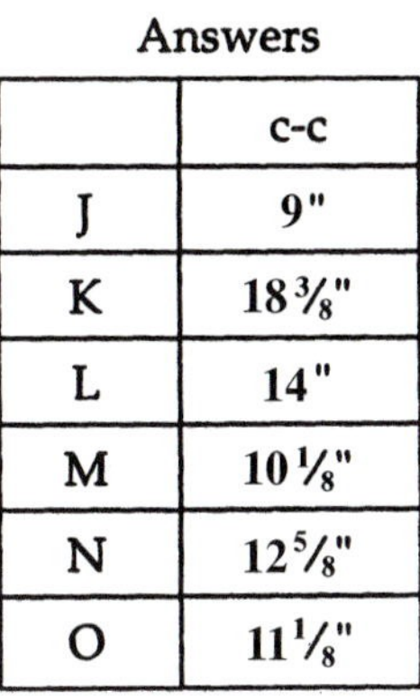

Answers

	c-c
J	9"
K	18⅜"
L	14"
M	10⅛"
N	12⅝"
O	11⅛"

Solution

(Data #25)

Diagonal = 13" × 1.414 = 18⅜"

Rise or offset = 13"

Difference in length = 7" × 0.414 = 2⅞"

J: c-c = 22" – 13" = 9"

K: c-c = 18⅜", diagonal

L: c-c = 27" – 13" = 14"

M: c-c = (c-c of J) – Difference = 13" – 2⅞" = 10⅛"

N: c-c = (c-c of K) – (2 × Difference) = 18⅜" – (2 × 2⅞") = 12⅝"

O: c-c = (c-c of L) – Difference = 14" – 2⅞" = 11⅛"

Exercise

1. Determine the c-c and e-e for lengths for each pipe shown in the diagram. Consider both threaded and copper pipe.

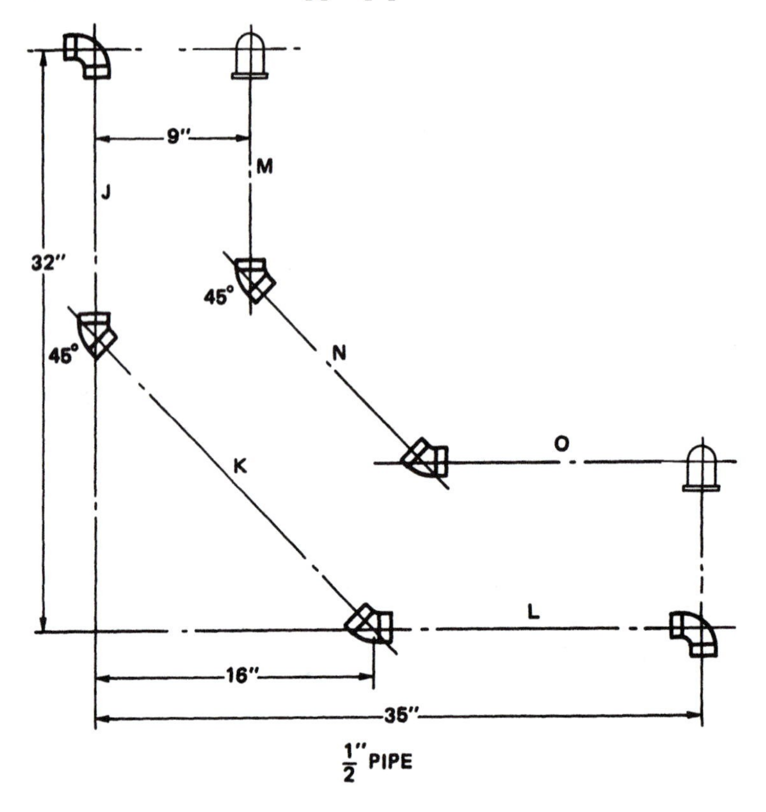

2. Determine c-c and e-e lengths for each pipe shown in the diagram. Consider both threaded and copper pipe.

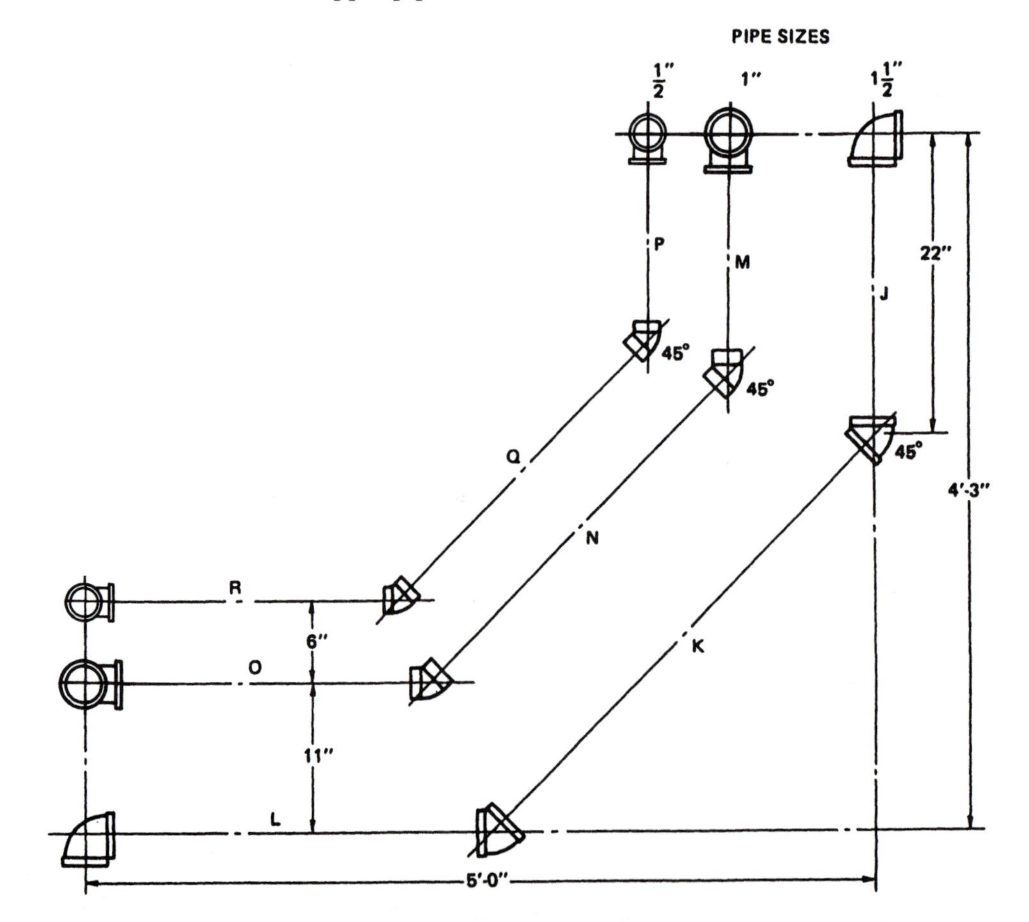

Unit 38 Rolling Offsets

Objectives

- Visualize a rolling offset.
- Define true offset.
- Calculate the travel for any rolling offset.

Up to now we have considered only offsets that were either totally in the horizontal or totally in the vertical plane.

The Horizontal Offset

This can be visualized by imagining the floor being brought up under the diagonal (travel) of the offset and the diagonal then touching the floor at all points along its length.

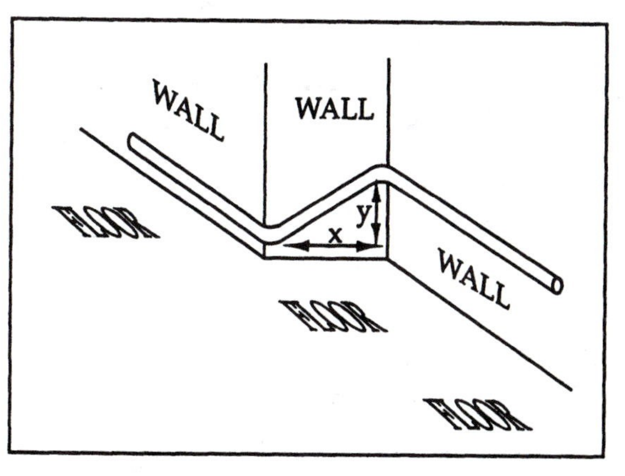

Figure 38–1

The Vertical Offset

This can be visualized by imagining the wall being built tightly against the offset and the diagonal (travel) of the offset then touching the wall at all points along its length.

The Rolling Offset

Occasionally an offset must offset both vertically and horizontally. This could be accomplished by using an offset in the horizontal plane to move the pipeline "over" and another in the vertical plane to move it "up" or "down." This would require more fittings or bends than are actually necessary, perhaps twice as many (see Figure 38–1). The "y" arrow shows the vertical component of the offset and the "x" arrow shows the horizontal component of the offset.

A Rectangular Prism or Box

Think of the rolling offset as a pipe that enters one corner of a box and exits from the farthest possible corner from the entry point (see Figure 38–2).

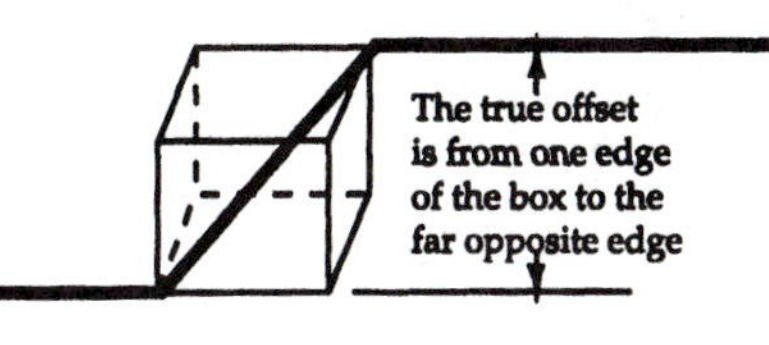

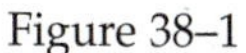

Figure 38–2

The entire solution of the rolling offset depends upon combining the rise and offset into a single dimension called the *true offset*. The true offset and the setback become the two sides of the square, or rectangle, around the diagonal. The true offset is multiplied by the constants under the fitting angle in Data #26 in the Appendix to determine the diagonal and setback. None of the constants are different from those used in simple and parallel offsets.

The square root solution for the true offset gives a little better accuracy than the usual layout solution. However, the layout solution should be practiced and is preferred because it gives a usable answer without the use of square root.

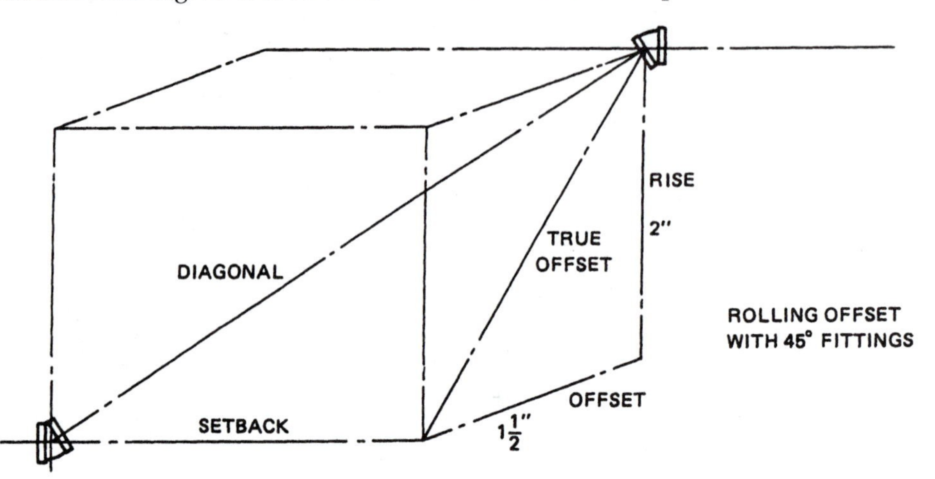

Procedure

1. On the job, measure offset and rise.
2. Determine true offset by formula or by layout method.

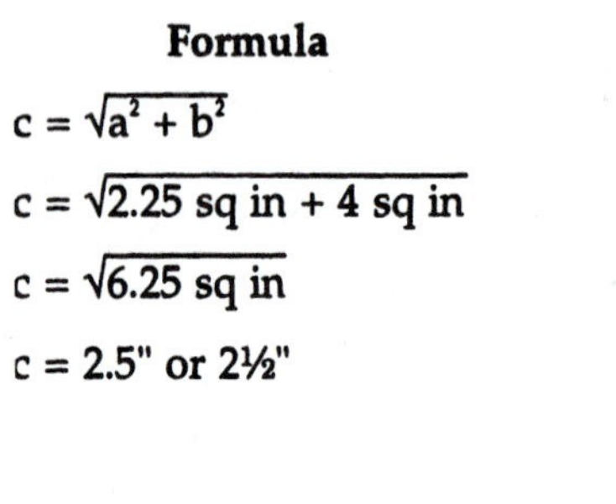

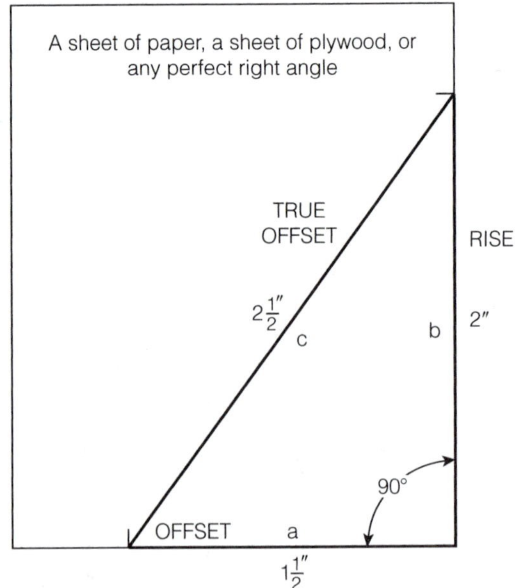

3. Multiply the true offset by the constants under the fitting angle in, Appendix Data #26 to determine the diagonal and setback.

Sample Problem

Determine c-c lengths for pipes J, K, and L.

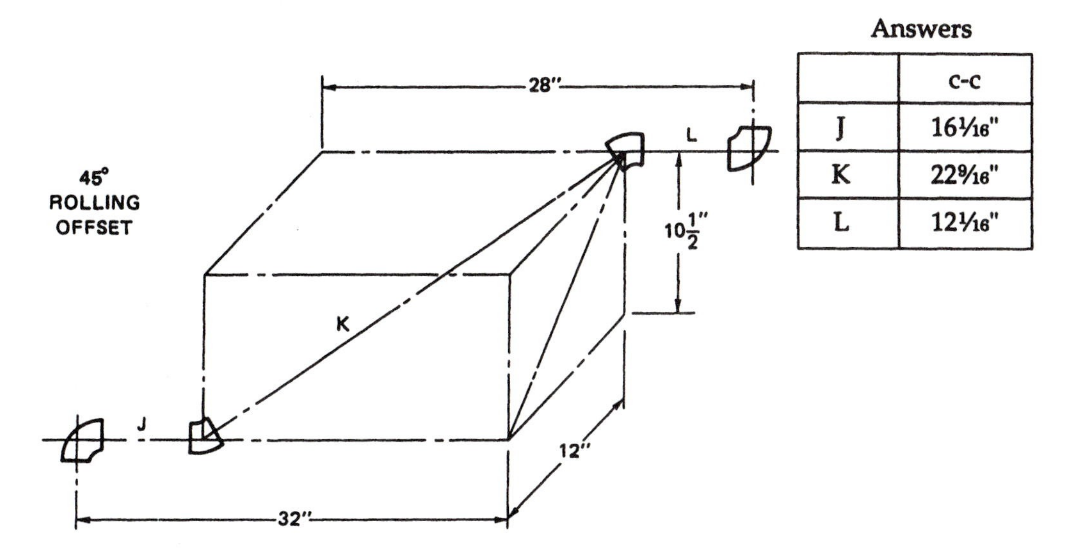

Answers

	c-c
J	16 1/16"
K	22 9/16"
L	12 1/16"

Solution

True offset = $\sqrt{\text{offset}^2 + \text{rise}^2} = \sqrt{(12")^2 + (10.5")^2}$ = 15.95 or 15 15/16"

(Data #26) Diagonal = 1.414 × 15.95" = 22 9/16"

Setback = 1 × 15.95" = 15.95" = 15 15/16"

J: c-c = 32" – 15 15/16" = 16 1/16"

L: c-c = 28" – 15 1/16" = 12 1/16"

K: c-c = diagonal = 22 9/16"

Exercise

Note: *Constants for all fitting angles are listed in Data #26 in the Appendix.*

Note: *Use this diagram for problems 1 and 2.*

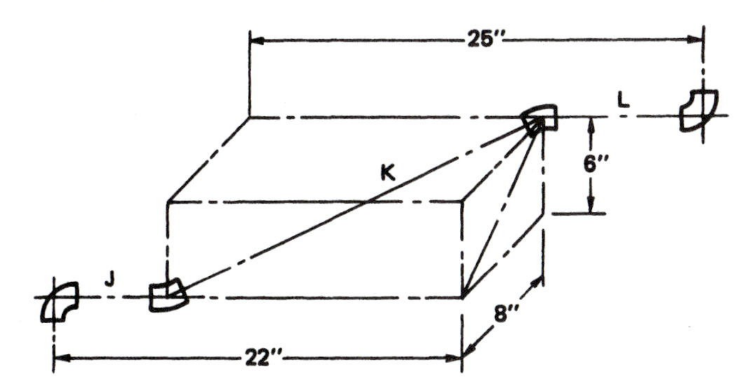

1. Solve for c-c and e-e lengths of each pipe shown in the diagram of a 45° rolling offset using ¾" threaded pipe.
2. Solve for c-c and e-e lengths of each pipe using same diagram and pipe size as in problem 1, but with 11" offset and 9" rise.

3. Solve for c-c and e-e lengths of each pipe shown in the following diagram of a 60° rolling offset using 1¼" threaded pipe.

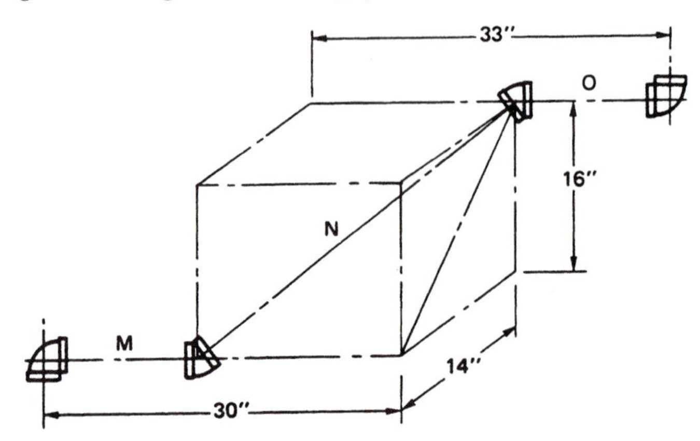

4. Solve for c-c and e-e lengths of each pipe shown in the following diagram of a 22½° rolling offset using 2½" threaded pipe.

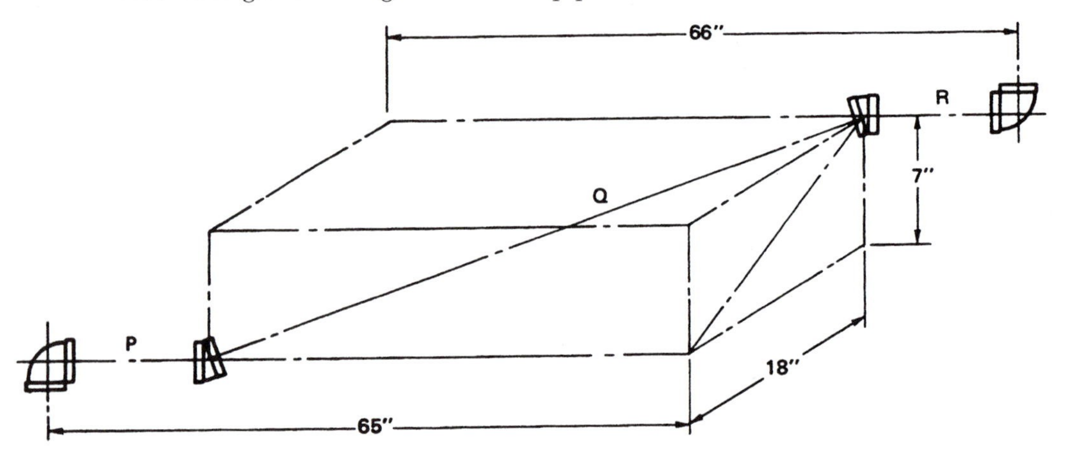

Unit 39

Combination Offsets

Objectives

- Compare combination offsets to rolling offsets.
- Study the 45°–60° combination offset.
- Do case studies with the combination offset.

The combination offset is like a rolling offset in that both are three-dimensional. Therefore, each represents a diagonal in a box.

The combination offset and the rolling offset differ in two ways. The combination connects pipes that are at right angles to each other while the rolling offset connects parallel pipes. Also, the combination offset gets its name from the use of fittings of different angles, usually a 45° and a 60° fitting, while the rolling offset uses any two fittings of the same angle.

There is only one angle at which combination offsets are practical in most plumbing. So it is unnecessary to solve for a true offset, as each length in the box of the diagonal can be found by multiplying the offset by the proper constant.

The alternate directions shown in the following figures mean that a 60° fitting can be rotated to the new direction without any change in the diagonal or the box around the diagonal.

The three figures shown do not represent all the possible combination offsets. Others have been devised for special purposes where the pipes connected by the diagonal are not at right angles to each other.

This figure shows a 45° and a 60° combination, the most used of the combination offsets.

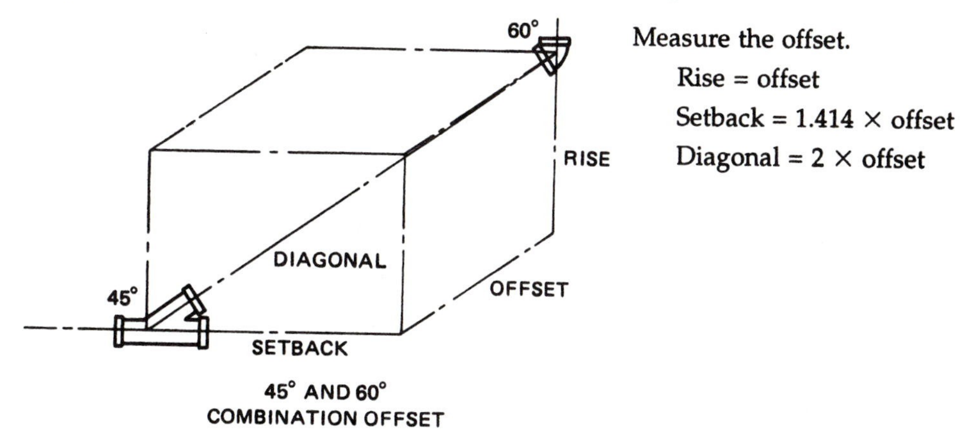

45° AND 60°
COMBINATION OFFSET

A different position or the *alternate direction* is shown in this figure.

Measure the offset.

Rise = 1.414 $\times$ offset

Setback = offset

Diagonal = 2 $\times$ offset

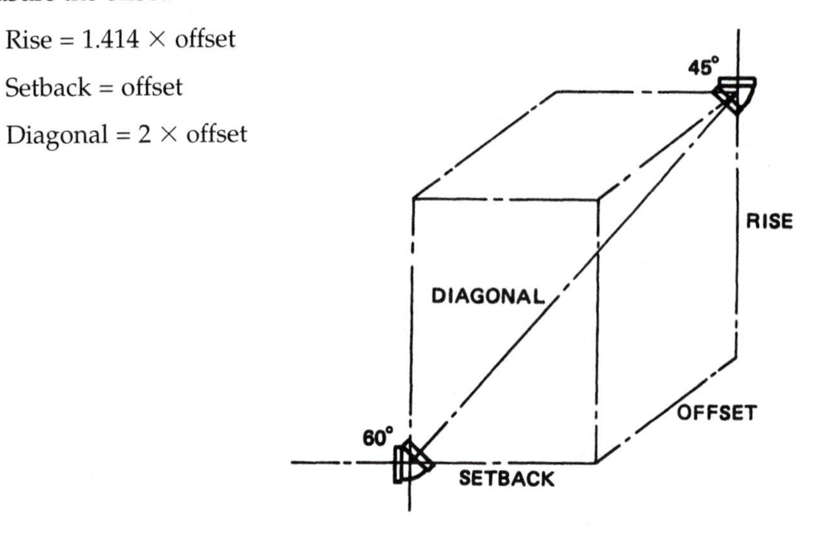

A double combination connecting lines on two faces of a column or on two walls at a corner is shown in this figure.

Measure O_1 and O_2

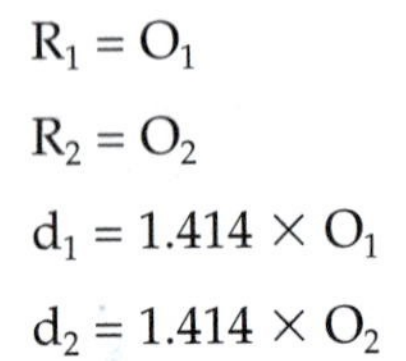

$R_1 = O_1$

$R_2 = O_2$

$d_1 = 1.414 \times O_1$

$d_2 = 1.414 \times O_2$

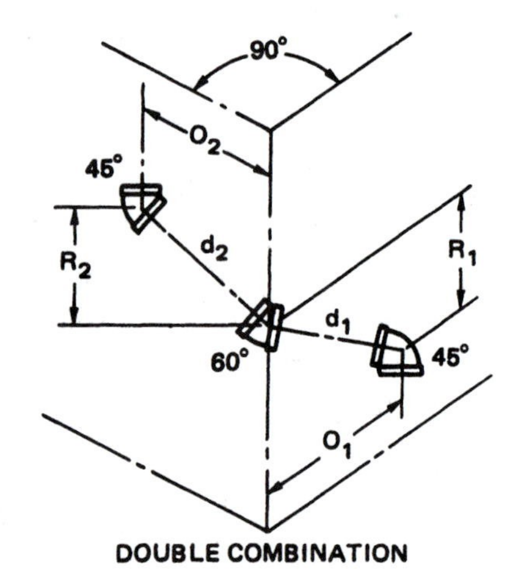

DOUBLE COMBINATION

Sample Problems

1. Determine the c-c length of each pipe in the diagram.

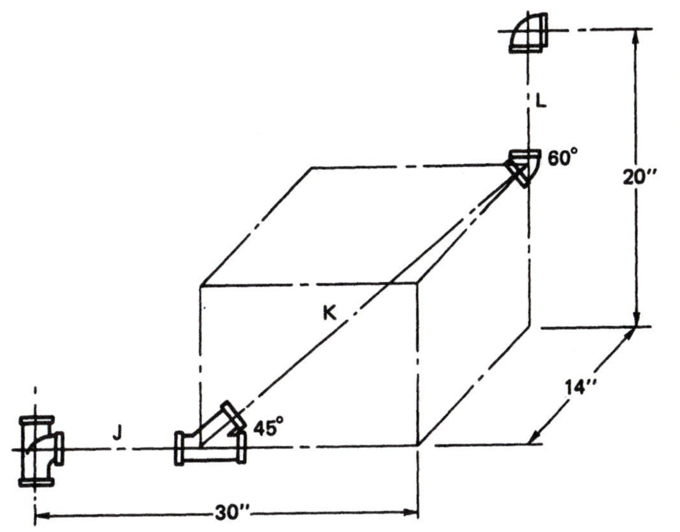

Answers

	c-c
J	10$\frac{3}{16}$"
K	28"
L	6"

Solution

Rise = offset = 14"

L: c-c = 20" – 14" = 6"

K: c-c = 2 × 14" = 28"

Setback = 1.414 × 14" = 19$\frac{13}{16}$"

J: c-c = 30" – 19$\frac{13}{16}$" = 10$\frac{3}{16}$"

2. Determine the c-c length of each pipe in the diagram shown.

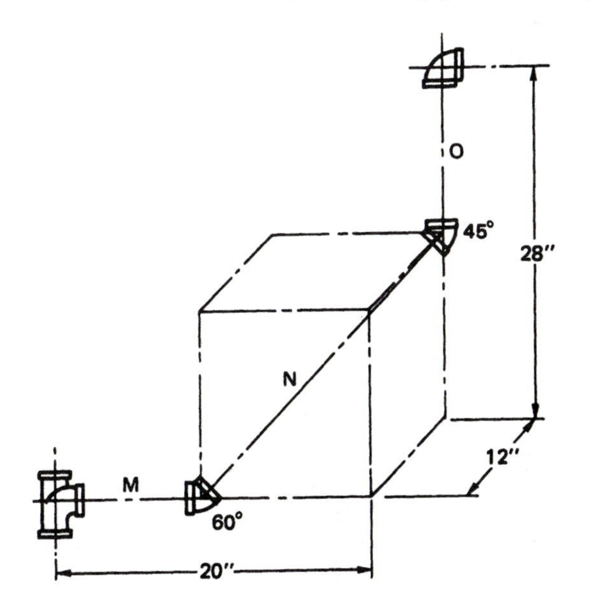

Answers

	c-c
M	8"
N	24"
O	11"

Solution

Rise = 1.414 × 12" = 16$\frac{15}{16}$"

O: c-c = 28" – 16$\frac{15}{16}$" = 11$\frac{1}{16}$"

N: c-c = 2 × 12" = 24"

Setback = offset = 12"

M: c-c = 20" – 12" = 8"

3. Determine the c-c length of each pipe in the diagram shown.

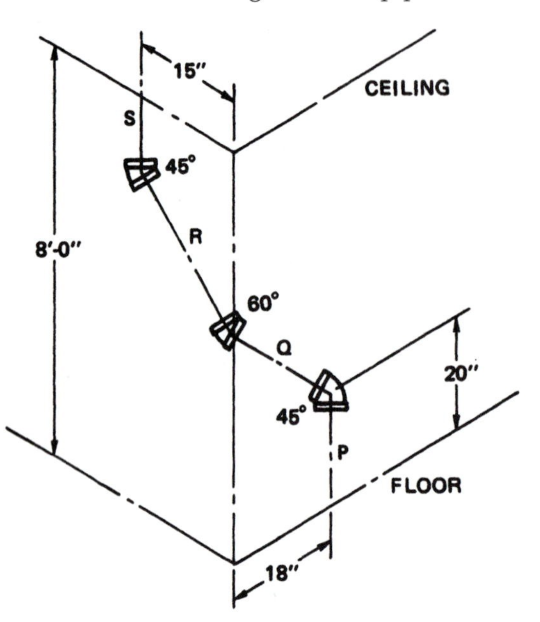

Answers

	c-c
P	20"
Q	25 7/16"
R	21 3/16"
S	43"

Solution

P: floor-to-center = 20"

Q: c-c = 1.414 × 18" = 25 7/16"

R: c-c = 1.414 × 15" = 21 3/16"

S: center-to-ceiling = 8' 0" – (15" + 18" + 20") = 96" – 53" = 43"

Exercise

Note: Use this diagram for problems 1 and 2.

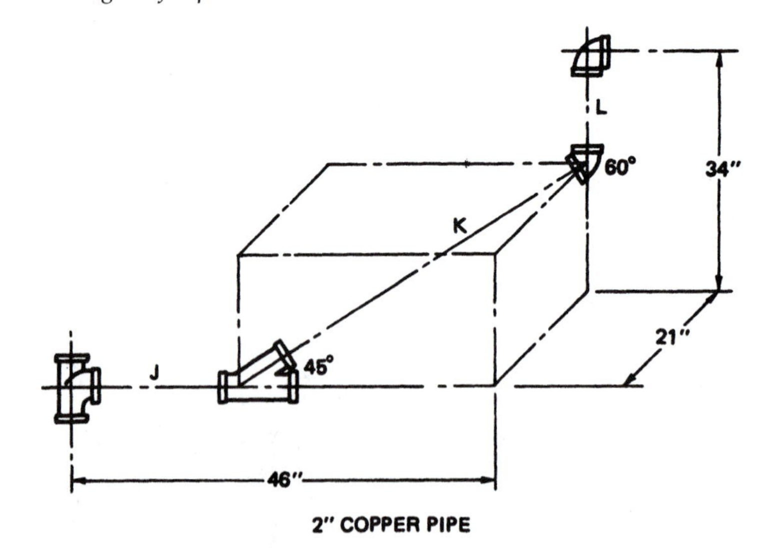

1. Solve for c-c and e-e lengths of each pipe shown.
2. Solve for c-c and e-e lengths using 14½" offset.

Note: Use this diagram for problems 3 and 4.

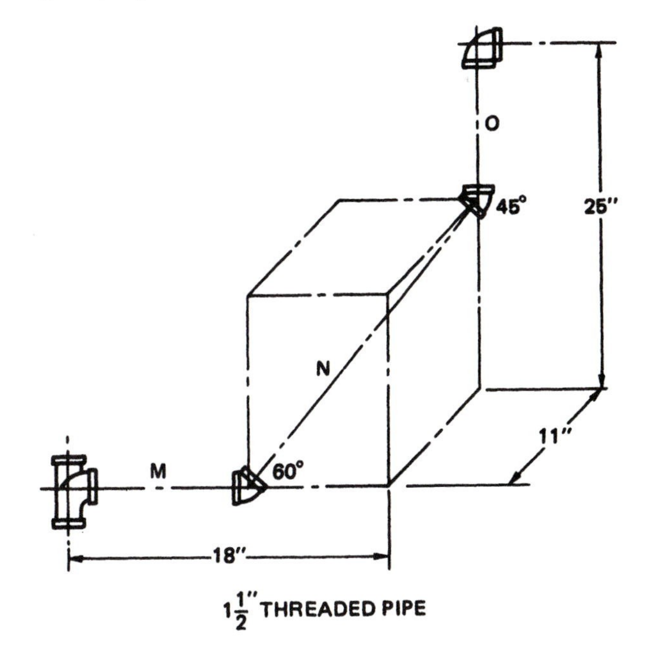

3. Solve for c-c and e-e lengths of each pipe shown.
4. Solve for c-c and e-e lengths using 9½" offset.
5. Solve for c-c and e-e lengths of each pipe shown in this diagram.

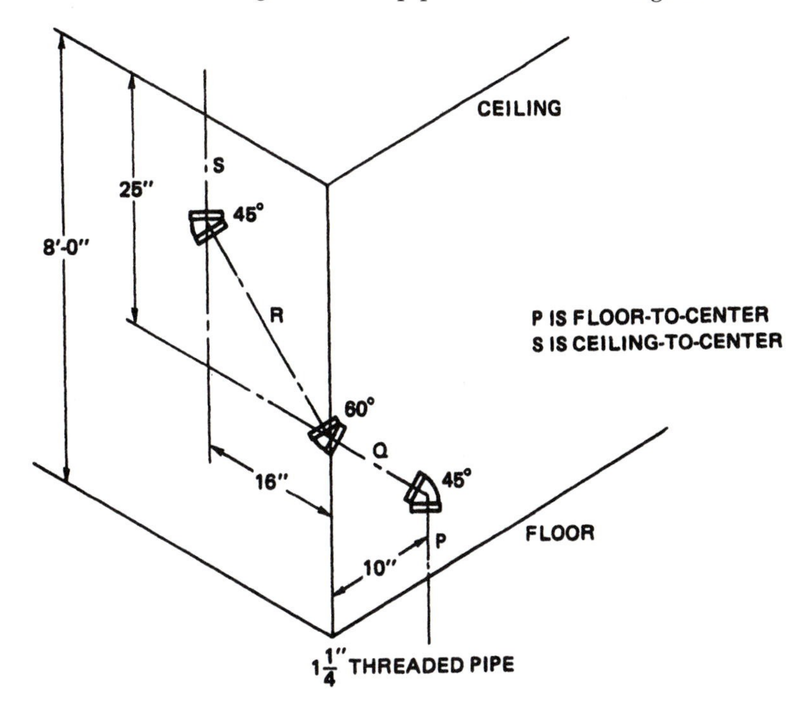

Unit 40

Pipe Length by Layout

Objectives

- Compare different layout methods.
- State the advantages and disadvantages of the layout methods.
- Use graph paper to do some simple layouts.

Many of our calculations can be done quickly and easily by the layout method. When this is done either on the floor, at full scale, or on a piece of paper, to scale, or on a computer screen with a computer aided drafting program (CAD) to scale, approximate answers can be quickly found.

Advantages of the Layout System of Pipe Calculation

1. When done on the job, in the location where the pipe is to be installed, problems are more easily avoided. Out-of-plumb walls and corners that are not good right angles are easily spotted and allowed for.
2. Mistakes, when they occur, are usually small and easily corrected.
3. If conditions are right (clean floor areas, easy-to-get-at measurements) the work usually goes rapidly.
4. Complex piping systems can be easily visualized when laid out in chalk, full scale, on the floor of the installation area.

Disadvantages of the Layout System

1. The craftsman must be on the job to do a full-scale floor layout. Questions invariably arise after the worker is back at the shop and ready to begin pipe cutting.
2. Often the installation area is littered with machinery or storage or may be on more than one level. This will make full-scale layout on the job much less reliable.
3. Jobs that are very large or cover a large area also do not lend themselves well to this method.

Computer or on Paper—"To Scale" Layout

Computer drafting programs allow the scaled layout of complex piping systems. The plumbing draftsman can use the computer to measure between any two points within the system. In this way measurements that become necessary when the job site is far away are easily determined. Much the same thing can be done with pencil and paper. It is easy, however, to make incorrect assumptions that walls are plumb and corners are square. If they are not, mistakes can be made.

Drawing Layouts to Scale

The layout is simply a drawing with correctly measured angles and scale line lengths. The diagrams in this book have used measured angles to obtain the proper shapes; a layout adds scale length to what has already been practiced. To be effective, the layout should be made to relatively large scale, ½" = 1' 0" or larger. This figure is drawn to ¼" = 1' 0". Pipe J measured ⅝" and represents 30", c-c. Pipe K is 7/16" and represents 21", c-c. (At ¼" = 1' 0" scale, 1/16" = 3".) Pipe O, however, is difficult to measure.

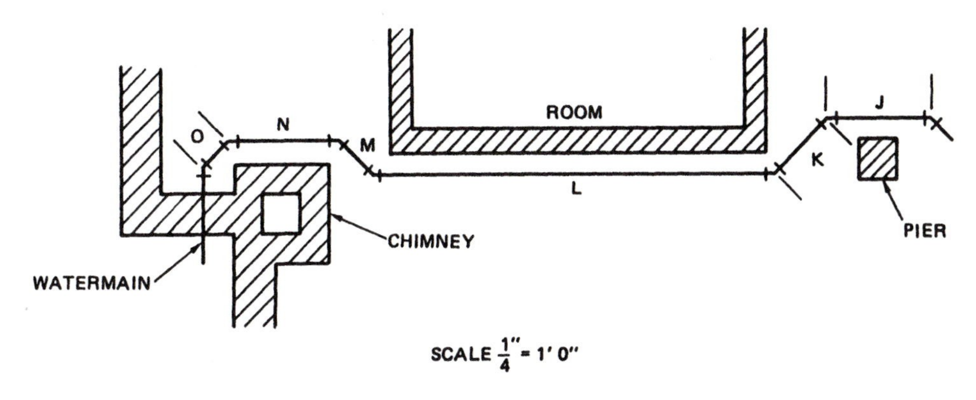

When the drawing is made to ½" = 1' 0", Pipe O is seen to be 7/16" and equals 10½", c-c. (At ½" = 1' 0" scale, 1/16" = 1½").

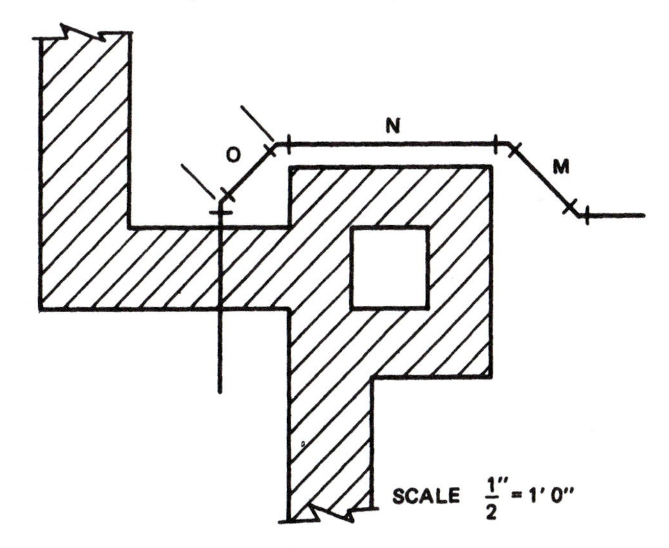

Good drafting techniques give more accurate results than careless drawing. Cross-section paper is sometimes used as an aid in measuring and in locating 45° lines.

In the problems of this unit, a 45° wye is used in a diagonal. The branch from a diagonal saves pipe in some situations.

Exercise

Note: Use cross-section paper for the problems in this unit.

The plan view pipe diagrams used in problems 1, 2, and 3 have a stack at A and risers at B and C. Dimensions are for the inside of the room. Walls are 6" thick.

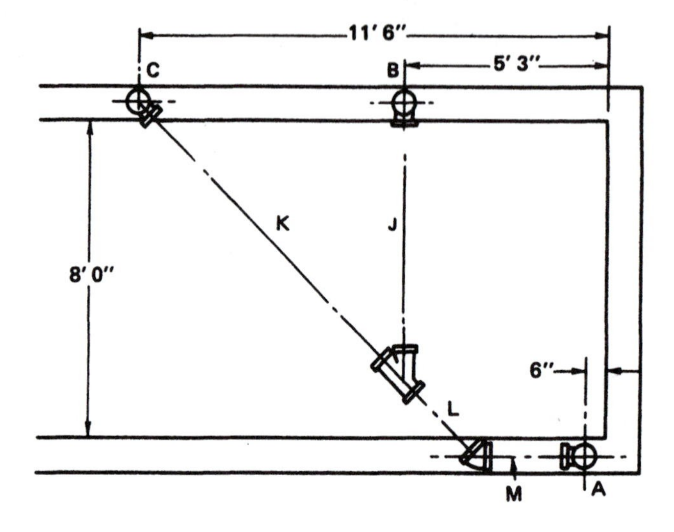

1. Make a layout using a scale of ½" = 1' 0" of the piping shown. Measure each c-c length. Compute each c-c length and compare with the measured lengths.
2. Make a layout of piping according to the diagram shown. Measure each c-c length. Compare the pipe and fittings needed with those for problem 1.

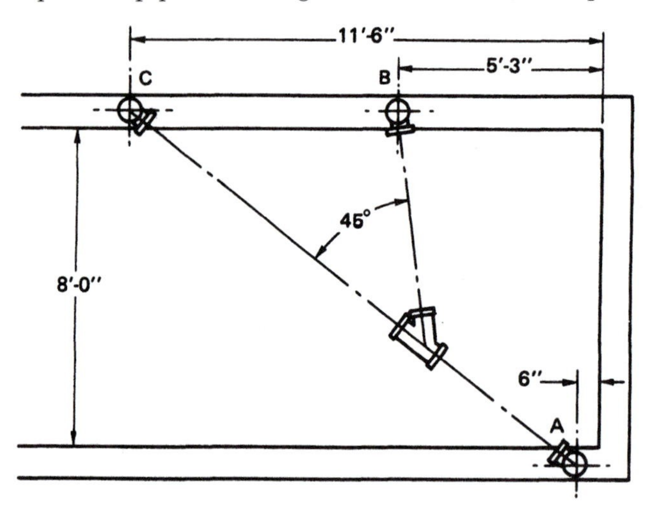

3. Make a layout of the piping shown. Measure the c-c lengths and compare to the pipe that is needed for problem 2.

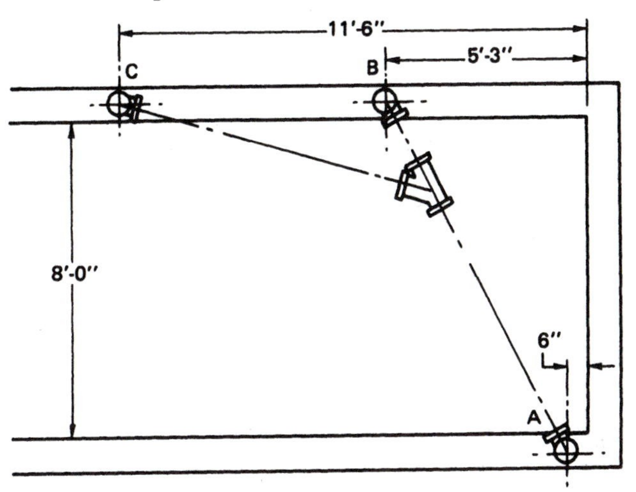

4. Make a layout of the diagram shown, using a scale of 1" = 1' 0". Measure each c-c length. Compute each c-c length and compare with the measured lengths.

Note: *This waste line is to be made up by using two 45° wyes and three 90° drainage elbows. One drainage tee wye installed at stack.*

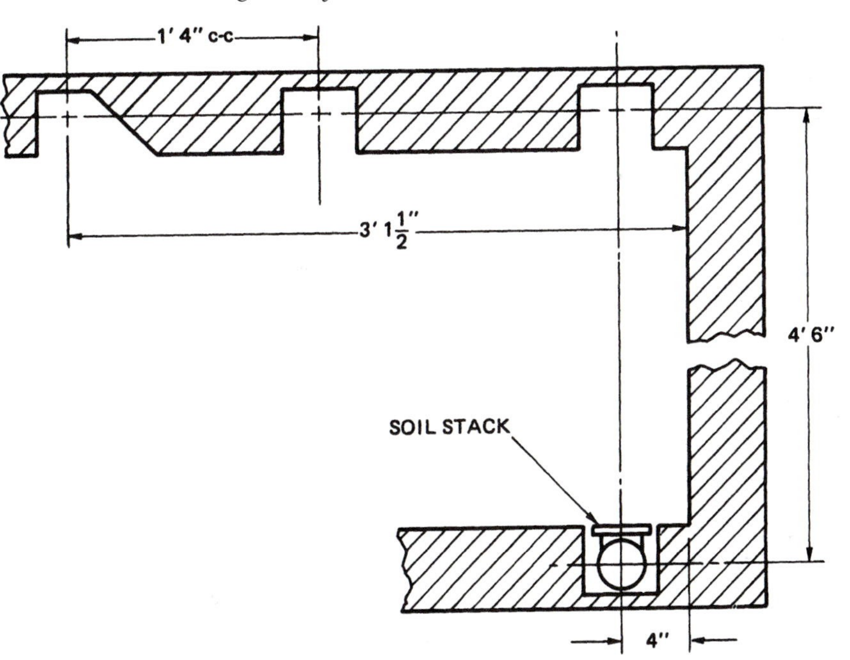

5. Determine the measurements center-to-center by layout using scale ½" = 1' 0" and using the fittings listed. Determine also by calculation. Pipe is to run to the locations shown on the sketch in the best direct manner and is to have a cleanout at the end of the main run.

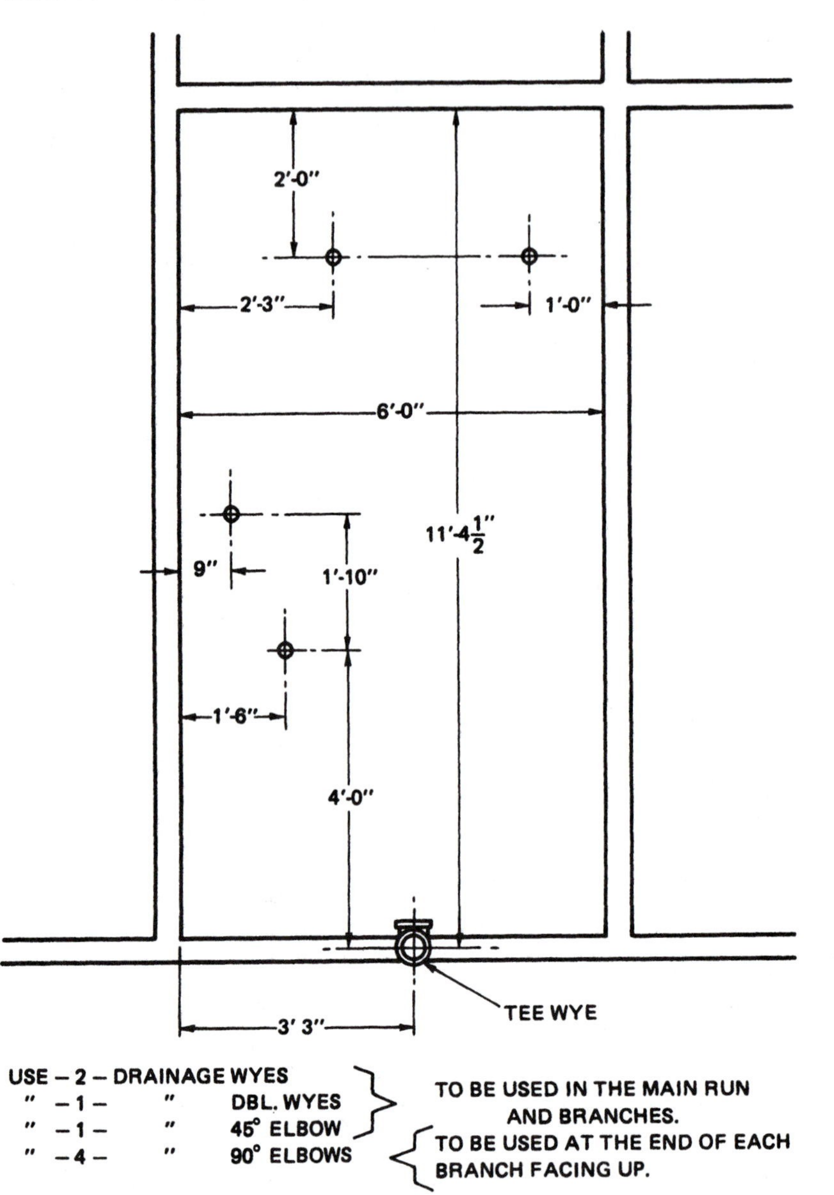

Unit 41

Cast Iron Flanged Fittings

Objectives

- Find end-to-end measurements using the cast iron flanged fitting.

Flanged fittings are useful where a pipe assembly has to be taken apart for cleaning, repair, change of equipment, and other reasons. They may be used on sprinkler fire systems, sewage disposal plants, water pumping stations, and other installations.

One fitting may be bolted directly to another. More often, however, a pipe length with flanges will be used between fittings. The pipe with a flange on each end is called a *spool piece*. The flanges must have bolt holes lined up with the matching flanges on the fittings. With bolt holes aligned, they are called *companion flanges*. A fitting allowance has to be made for each flange threaded on the spool piece pipe.

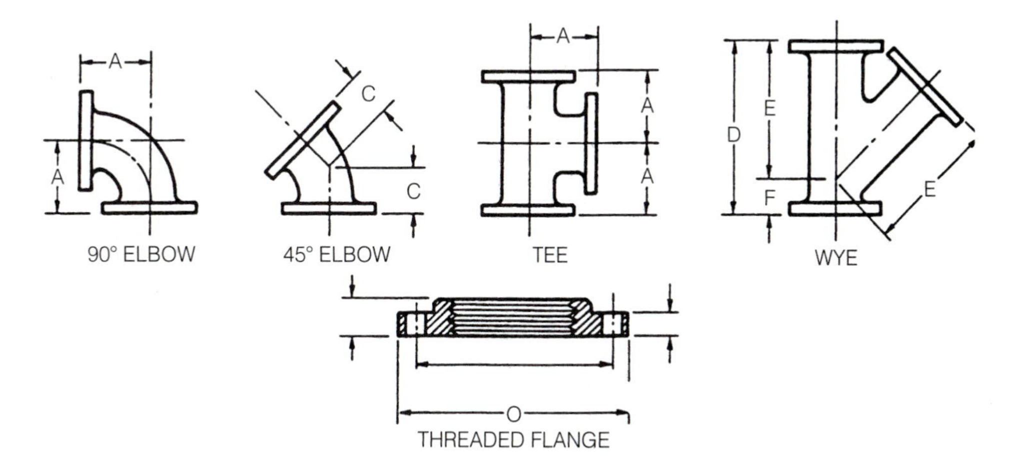

A gasket is used between flanges. The gasket material varies for different fluids and pipe use. Different gasket materials and construction result in a number of gasket thicknesses. Thick gaskets may require a pipe length calculation.

Note: Use Data #19 for finding fitting allowance measures for flanged fittings.

Sample Problem

J is a spool piece to fit between a 4-inch flanged tee and a 90° ell as shown in the sketch.

a. Find the measurement between the tee and the ell.

b. Find the f-f length of the spool piece if the gaskets are $\frac{1}{8}$" thick.

c. Find the e-e length of the spool piece.

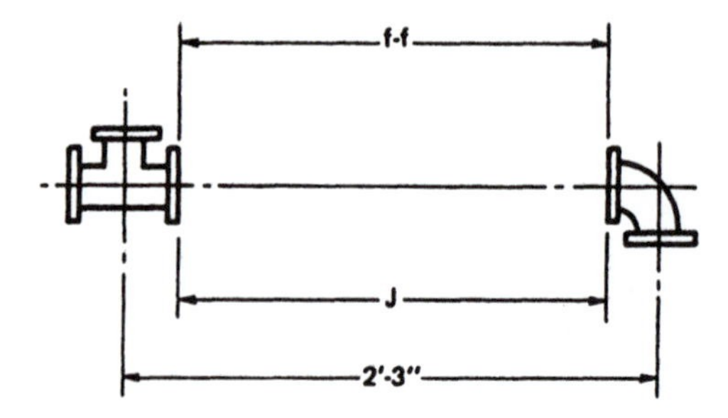

Solution

(Data #19) A = $6\frac{1}{2}$"

A = $6\frac{1}{2}$"

B = $\frac{1}{2}$"

Y = $1\frac{5}{16}$"

a. 27" – ($6\frac{1}{2}$" + $6\frac{1}{2}$") = 14"

b. f-f = 14" – (2 × $\frac{1}{8}$") = $13\frac{3}{4}$"

c. e-e = $13\frac{3}{4}$" – [($1\frac{5}{16}$" + $1\frac{5}{16}$") – 2(1")] = $13\frac{1}{8}$"

Exercise

Compute the measurement between fittings, the f-f length of the spool piece using $\frac{1}{8}$" gaskets, and the e-e length for each spool piece illustrated (K, L, and M).

1. 2. 3.

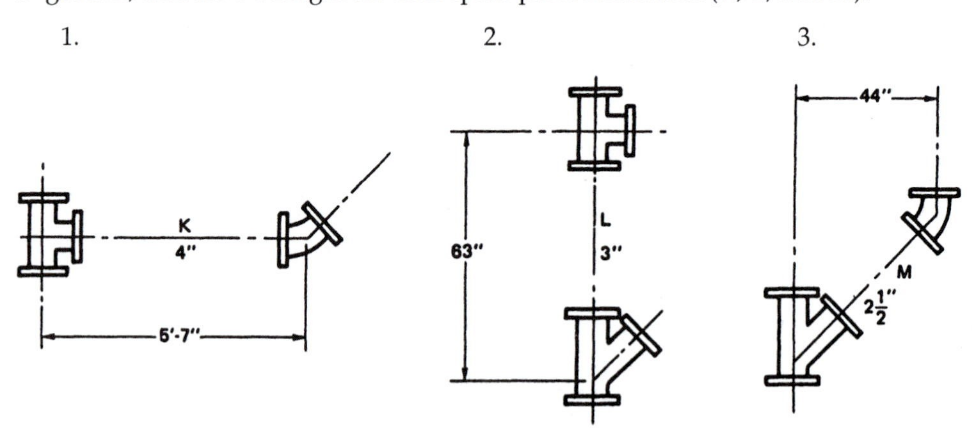

Unit 42

Miter Cuts—Welded Pipe

Objectives

- Learn the anatomy of the miter cut.
- Calculate the setback of any miter cut.

Occasionally it is necessary to change the direction of a line of pipe an unusual number of degrees. Or perhaps a very gradual turn is needed and there are no fittings available that will accomplish the task satisfactorily. The *miter joint* is a solution so long as welding equipment is available. With some very large pipe sizes the miter cut may be the only fitting option. All kinds of fittings may be fabricated using the template stretchout and miter style welded joint, including caps, reducers, tees, and wyes.

If the pipe size is small and large pipe cutting equipment is available like metal cutting, portable bandsaws, or the radial carbide cut-off machine, then the pipe cut may be accomplished in one straight slice. When these tools are not available or the pipe size is too large for this to work, then the miter cut stretchout must be used. Stretchouts will be defined and explained later in this unit.

A Simple Miter Joint

Making one cut at a specific angle of, say 10°, across the pipe line, can result in a pipe bend of 20°, or twice the original cut angle (see Figure 42–1). It can be seen that the miter angle (M.A.) is equal to one-half of the resulting fitting angle (F.A.). If a machine that can cut pipe at an angle in one slice is not available, the pipe cut must be laid out on the outside diameter of the pipe using a *wraparound,* or *stretchout,* and the cut made with a burning torch or similar device. A wraparound, as the name implies, is a band of some non-stretchable material. The sides of the wraparound are parallel; if the sides are lined up

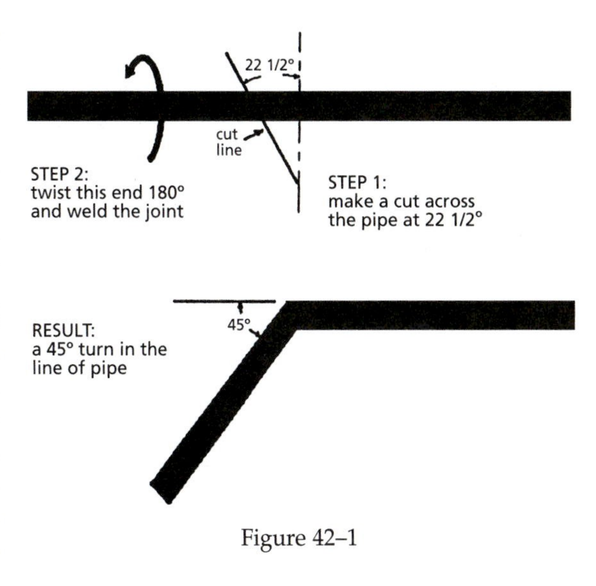

Figure 42–1

after the tool is wrapped around the pipe, the line along the edge of the wrap-around will be "square" (see Figure 42–2). If a line is drawn around the pipe at the edge of the wraparound and a cut made, the pipe joint will result in a straight pipe line.

The Miter Cut

The setback of the cut is equal to the outside diameter of the pipe times the tangent of the miter angle (see Figure 42–3). It should be kept in mind that the miter angle is one-half of the finished fitting angle.

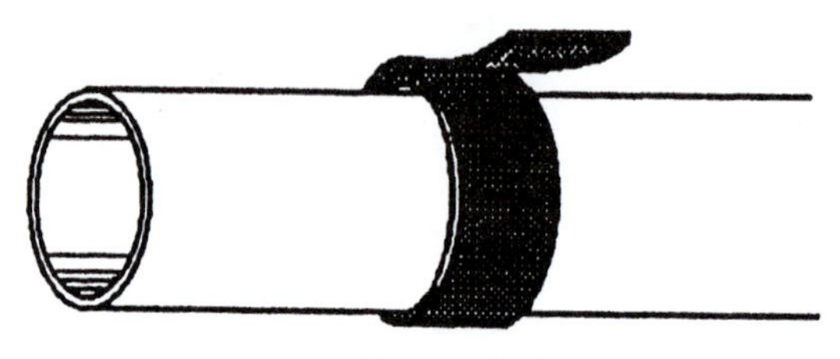

Figure 42–2

Sample Problem

We want to make a 45° turn in a 2" pipe line. What information do we need?

- Outside Diameter—The outside diameter of 2-inch pipe is 2.375 inches (2⅜").
- Miter Angle—Since the desired turn angle is 45°, the miter angle will be 22½° (45 divided by 2).
- Angle Constant—This is the tangent of the miter angle. From the table of trigonometric functions we find that the tangent of 22½° is 0.4142.

The Formula: Setback = OD × TA; the setback is equal to the outside diameter of the pipe times the tangent of the miter angle.

Setback = 2.375 × 0.4142 = 0.984, or 1 inch

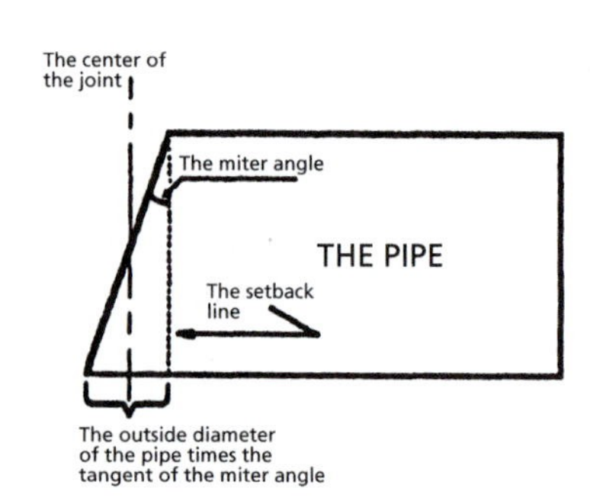

Figure 42–3

Marking the Pipe

Step 1: Use the wraparound to scribe a line around the pipe at the centerline of the cut.

Step 2: Divide the setback in half and use this distance to draw a line around the pipe on both sides of the center line (see Figure 42–4).

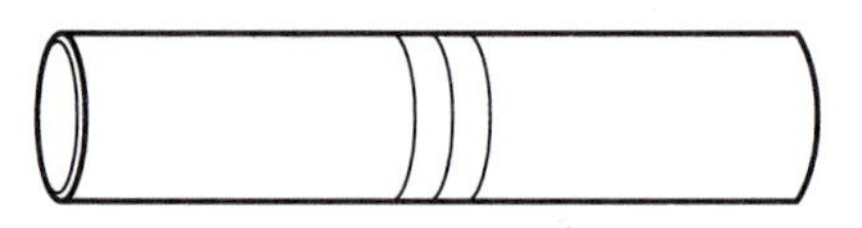

Figure 42–4

Step 3: Cut a piece of paper to go around the pipe precisely. Now fold this paper in half twice. Wrap this around the pipe. The fold lines and the end of the paper when it is wrapped around the pipe will fall at points 90 degrees apart (see Figure 42–5).

Step 4: Use the wraparound at an angle as shown in Figure 42–6 to draw a line from first the top to both sides, turn the pipe upside down, and then draw the other side of the cut line.

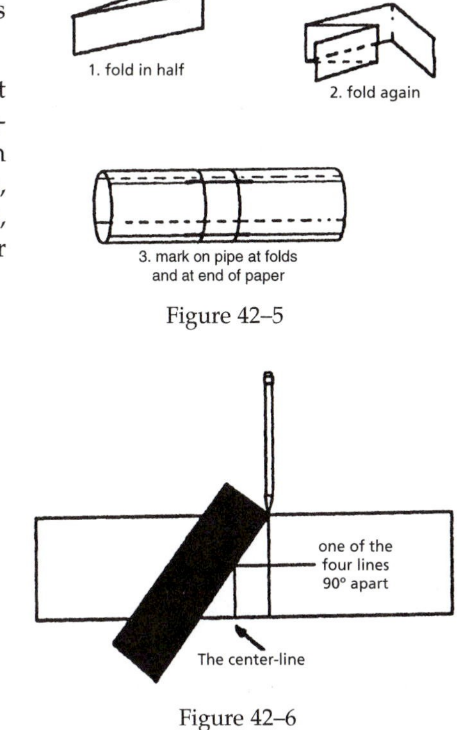

Figure 42–5

Figure 42–6

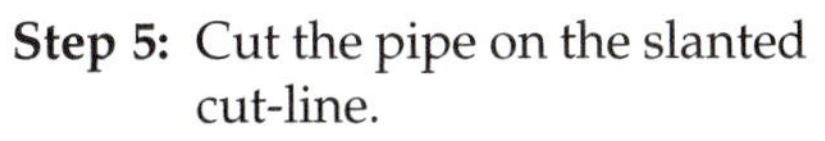

Step 5: Cut the pipe on the slanted cut-line.

Exercise

Find the setback for the problems in the following table. Remember that the outside diameter of the pipe is to be used (use Data #2). The tangent of one-half of the desired bend angle is used. Write the answers in inches and fractional parts of an inch.

Problem	Pipe size I.D.	Bend angle	Setback in inches
1.	2	20	
2.	3	30	
3.	4	40	
4.	5	60	
5.	2	90	
6.	3	20	
7.	4	30	
8.	5	40	
9.	2	60	
10.	3	90	

Section Three
SHEET METAL WORK

Unit 43 Shower Pan or Safe

Objectives

- Learn sheet lead specifications.
- Describe the shower pan stretchout.
- Design shower pans.

Sheet lead work is part of the plumbing trade. The lead sheets are fabricated into various shapes to fit the job. Some lead fabrications are self-supporting but many are liners.

Sheet lead is described by the weight per square foot. Five-pound lead sheet weighs five pounds per square foot. It is also $\frac{5}{64}$ of an inch thick, since a square foot of lead weighs one pound for each 64th of an inch of thickness.

Fiberglass, mat, cloth, and resin are often used instead of lead. The layout work is similar to that for sheet lead.

Sample Problem

A lead shower pan is to have a floor size of 30" × 36". The pan is to be 6" deep.

a. Make a dimensioned sketch of the completed shower pan.

b. Make a stretchout sketch of the layout on flat lead sheet. Use envelope corners.

c. Compute the area of the stretchout in square feet.

d. Compute the weight of lead used if made of 3-pound lead sheet.

Solution

a.

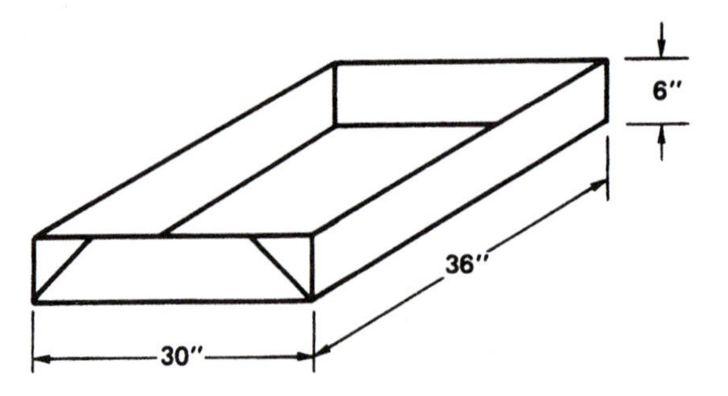

b.

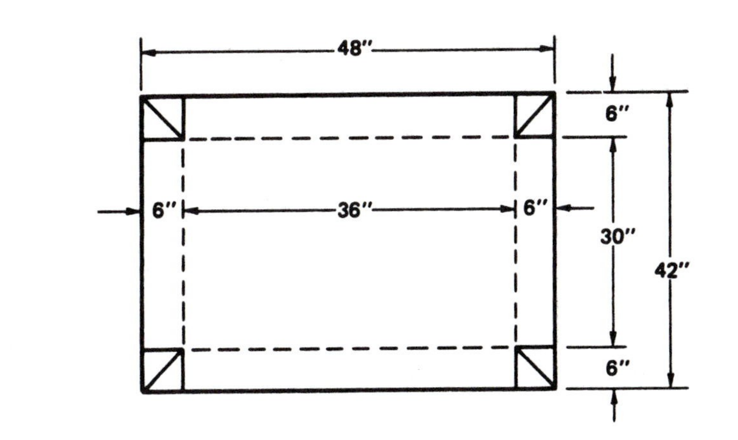

c. $A = LW$

$$A = \frac{48 \text{ in}}{12 \text{ in/ft}} \times \frac{42 \text{ in}}{12 \text{ in/ft}}$$

$A = 4' \times 3.5'$

$A = 14.0$ sq ft

d. 14 sq ft × 3 lb/sq ft = 42 lb of lead

Exercise

For each shower pan size listed:

a. Draw a dimensional sketch of completed plan.

b. Draw a dimensional stretchout sketch.

c. Compute the area of the stretchout in square feet.

d. Compute the total weight of the lead sheet using the thickness given.

1.	Floor Size – 40" × 32"	Depth – 4"	Lead Sheet – 5 lb/sq ft
2.	Floor Size – 35" × 34"	Depth – 5"	Lead Sheet – 3/64" thick
3.	Floor Size – 36" × 36"	Depth – 5"	Lead Sheet – 7 lb/sq ft
4.	Floor Size – 32" × 43"	Depth – 6"	Lead Sheet – 3 lb/sq ft

Unit 44

Rectangular Tank Liner

Objectives

- Define soldered corner tank liners.
- Design tank liners.
- Calculate tank liner materials.

The rectangular tank liner might be for a chemical laboratory sink or an industrial tank. Its construction differs from that of the shower pan in that it is not practical to use the envelope corner. Since the corner material is cut away, the corner seams are soldered or lead-burned. For best corrosion resistance, lead-burning is required. For safety reasons workers must have special skills to do this kind of work. Fiberglass, mat, cloth, and resin are often used instead of lead. The layout work is similar to that for sheet lead.

Sample Problem

A tank 30 inches wide by 50 inches long and 24 inches deep is to have a liner of 7-pound lead sheet.

a. Draw a dimensioned sketch of the completed liner.

b. Draw a dimensioned stretchout sketch of the liner.

c. Compute the square feet of a lead sheet used.

d. Compute the weight of lead used.

a.

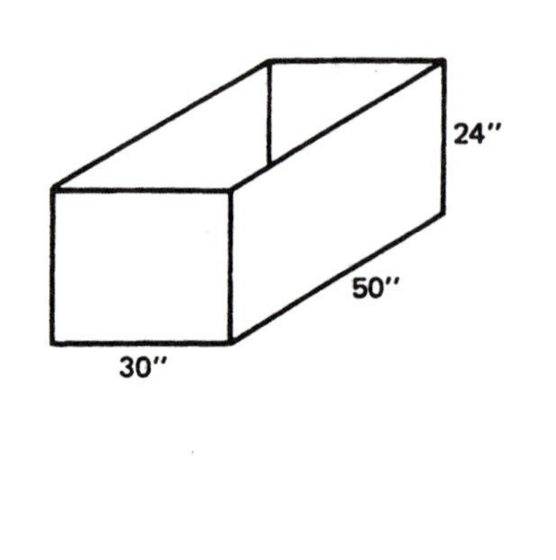

b.

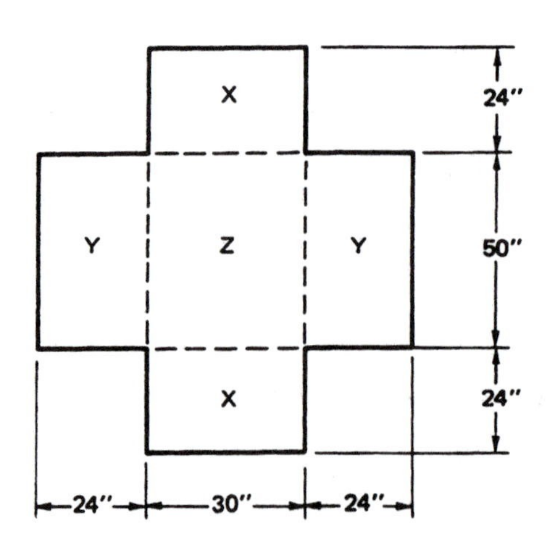

c. Area X

$A = LW$

$A = \frac{30\text{ in}}{12\text{ in/ft}} \times \frac{24\text{ in}}{12\text{ in/ft}}$

$A = 2.5' \times 2'$

$A = 5\text{ sq ft}$

Area Y

$A = LW$

$A = \frac{50\text{ in}}{12\text{ in/ft}} \times \frac{24\text{ in}}{12\text{ in/ft}}$

$A = \frac{25'}{6} \times 2'$

$A = 8.33\text{ sq ft}$

Area Z

$A = LW$

$A = \frac{50\text{ in}}{12\text{ in/ft}} \times \frac{30\text{ in}}{12\text{ in/ft}}$

$A = \frac{25'}{6} \times \frac{5'}{2}$

$A = 10.42\text{ sq ft}$

Total area = 2X + 2Y + Z = 10.00 sq ft + 16.66 sq ft + 10.42 sq ft = 37.08 sq ft

d. Weight = 37.08 × 7 = 259.56 lb

Exercise

For each tank and sheet lead given:

a. Make a dimensioned sketch of the completed liner for the tank.

b. Make a dimensioned stretchout sketch of the liner.

c. Compute the square feet of sheet lead used.

d. Compute the weight of the lead used.

1. Tank Size – 40" × 20" × 20" deep — Lead Sheet – 5/64" thick
2. Tank Size – 22" × 16" × 11" deep — Lead Sheet – 1/16" thick
3. Tank Size – 3' 6" × 2' 9" × 3' 0" deep — Lead Sheet – 3 lb/sq ft
4. Tank Size – 18" × 14" × 10" deep — Lead Sheet – 5 lb/sq ft

Unit 45 Cylindrical Stretchouts

Objectives

- Design cylindrical shapes with sheet metal.
- Define the roof flange.
- Make a roof flange.

A lead tube or cylinder is made by wrapping a rectangular piece of sheet lead into a cylindrical shape. The ends are butted together and soldered or lead-burned. Roof flanges and round pans or liners require the tube construction.

The roof flange consists of a tube and a plate. It is used to make a watertight connection between a vent stack and the roof. In good construction the tube is of sufficient length to be turned into the end of the vent pipe to make a raintight connection between the pipe and the roof flange. The plate is large enough to allow a watertight joint between it and the roofing. Roofing cement is used to seal this connection on a flat roof.

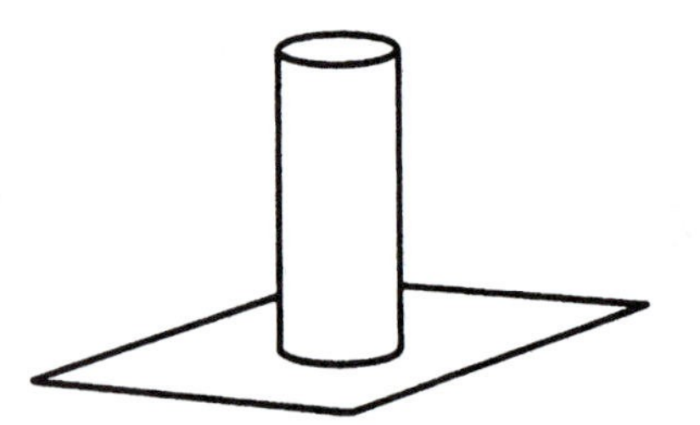

Sample Problem

A roof flange for a flat roof has a tube that is 5 inches in diameter and 22 inches high. The plate is 18" × 18". Both are made of 5-pound sheet lead.

a. Make a dimensioned sketch of the completed roof flange.

b. Make a dimensioned stretchout of the tube.

c. Make a dimensioned stretchout of the plate.

d. Compute the square feet of lead sheet used.

e. Compute the weight of the lead sheet used.

$C = \pi d$

$C = 3.14 \times 5" = 15\frac{11}{16}"$

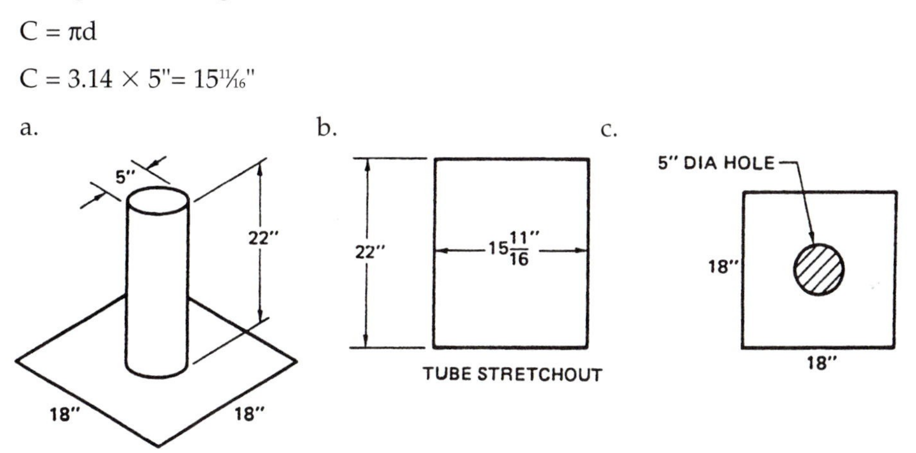

d. **Tube:** $A = LW$

$$A = \frac{22 \text{ in}}{12 \text{ in/ft}} \times \frac{15.7 \text{ in}}{12 \text{ in/ft}} = 2.40 \text{ sq ft}$$

Plate: $A = S^2 - \pi r^2$

$$A = \left(\frac{18}{12 \text{ in/ft}} \times \frac{18}{12 \text{ in/ft}}\right) - \left(\frac{3.14}{1} \times \frac{2.5 \text{ in}}{12 \text{ in/ft}} \times \frac{2.5}{12 \text{ in/ft}}\right)$$

$A = 2.11$ sq ft

Total Area: 2.40 sq ft + 2.11 sq ft = 4.51 sq ft

e. Weight = 4.51 sq ft × 5 lb/sq ft = 22.55 lb

Exercise

For each problem

a. Make a dimensioned sketch of the completed job.

b. Make a dimensioned stretchout of the tube.

c. Make a dimensioned stretchout of the plate or bottom.

d. Compute the square feet of lead sheet used.

e. Compute the weight of the lead sheet used.

1. Job – Roof flange with 6" diameter tube
   20" high and 18" × 18" plate — Lead Sheet – 3 lb/sq ft
2. Job – Roof flange 8½" diameter tube
   24" high with 24" × 24" plate — Lead Sheet – 5 lb/sq ft
3. Job – Circular shower pan
   40" diameter, 6" high — Lead Sheet – 5 lb/sq ft

Unit 46 Roof Pitch

Objectives

- Define roof pitch.
- Apply pitch to a roof flange.
- Design an elliptical roof opening.

Roof pitch is a method of dimensioning the angle or steepness of a roof. Since most roofs are of gable style, pitch is determined from the rise and span of a double slope or *isosceles* triangle. The *rise* is the altitude of the triangle; the *span* is the base; the two sides are equal.

$$\text{Pitch} = \frac{\text{rise}}{\text{span}}$$

$$\text{Rise} = \text{pitch} \times \text{span}$$

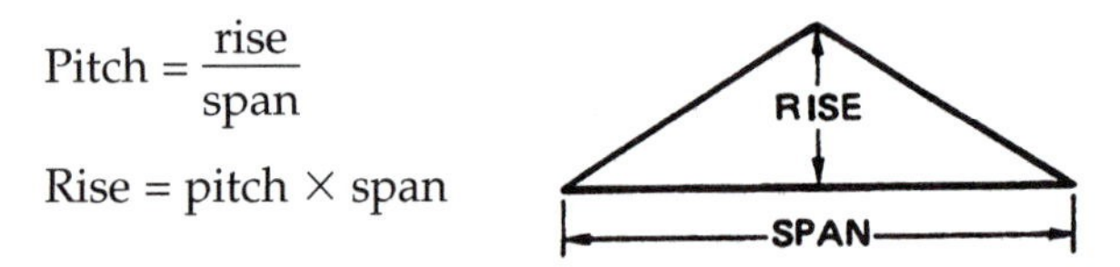

Therefore, a house roof with 5' rise and 15' span would be $\frac{5'}{15'}$ or ⅓ pitch.

The full size of the roof would be difficult to work with for layout of a roof flange or other construction work. For layout, a standard span of 24 inches is used. By using only one slope, a standard run of 12 inches, a right triangle is obtained. The rise is again the altitude; the run, the base. A right triangle with a 12" base and an 8" altitude is exactly ⅓ pitch.

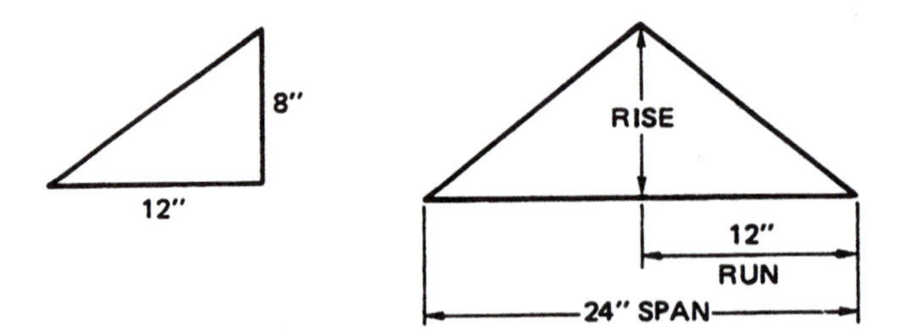

Sample Problem A 6" O.D. lead tube is to be made into a ⅓ pitch roof flange.

a. How can the angle at the bottom of the tube be dimensioned?

The angle of the cut is determined by the altitude, or rise, of the right angle which is formed. The triangle has a base, or run, of 6" (diameter of the tube). The span would be twice the run, or 12".

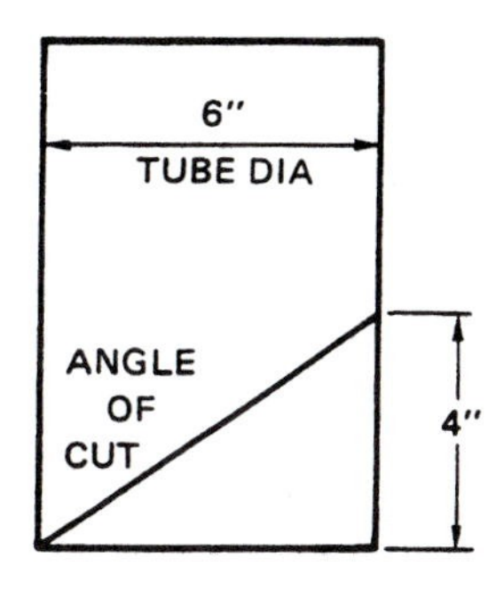

Use the formula: Rise = Pitch × Span

Rise = ⅓ × 12"

Rise = 4" (altitude)

or

Use the formula $\text{Altitude} = \frac{\text{Rise}}{\text{Run}} \times \text{Diameter}$

$$\text{Altitude} = \frac{8''}{12''} \times 6''$$

$$\text{Altitude} = 4''$$

The second formula is based on the proportional relationship that exists between the standard 8" rise to 12" run situation and the new situation in which the run (called diameter) is 6". The ratio of rise to run in the two sets of measurements is equal.

The relationship is expressed as $\frac{\text{Rise}}{\text{Run}} = \frac{\text{Rise (altitude)}}{\text{Run (diameter)}}$.

Therefore, $\text{Altitude} = \frac{\text{Rise}}{\text{Run}} \times \text{Diameter}$.

b. What size hole must be cut in the plate? Construct a right triangle with 6" base and 4" rise. Measure the slope; it will be approximately 7¼". The hole in the plate is an ellipse with major diameter of 7¼" and the minor diameter of 6".

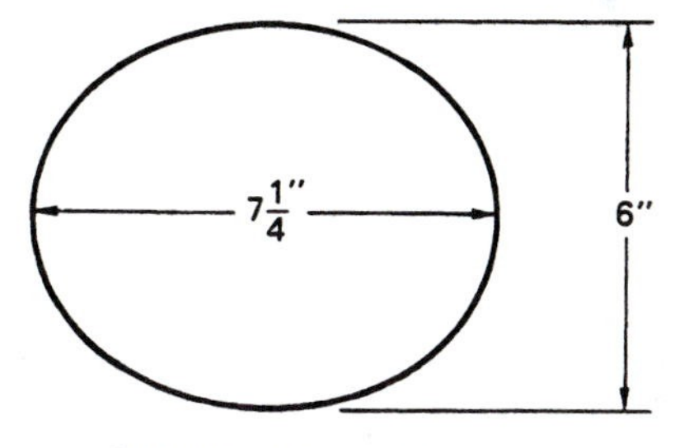

OPENING IN PLATE

Exercise

For each problem

a. Compute the rise for 24" of span.

b. Determine the altitude to use to obtain the proper angle on the tube for a roof flange.

c. Make a dimensioned sketch of the hole through the plate.

1. Pitch ⅓ Tube – 3" diameter
2. Pitch ¼ Tube – 4" diameter
3. Pitch ½ Tube – 5" diameter
4. Pitch ⅛ Tube – 6" diameter

Section Four
VOLUMES, PRESSURES, CAPACITIES

Unit 47
Water Measure

Objectives

- Review volume and weight measure.
- Calculate water weights and volumes.
- Solve problems in metric measure.

Volumes are often computed in cubic inches for small quantities. The U.S. Bureau of Standards has set 231 cubic inches as a U.S. gallon. Cubic inches ÷ 231 = gallons.

Larger volumes are computed in cubic feet. There are nearly 7.5 gallons in one cubic foot. Cubic feet × 7.5 = gallons.

Water weighs nearly 62.5 pounds for a cubic foot or 8.33 pounds for a gallon. Cubic feet × 62.5 = pounds of water; gallons × 8.33 = pounds of water.

Metric volumes are often computed in cubic centimeters. Because 1,000 cubic centimeters is a liter, cubic centimeters ÷ 1,000 = liters.

Larger volumes are calculated in cubic meters. A cubic meter is 1,000 liters.

A liter of water weighs 1 kilogram. One kilogram is about 2.2 pounds. A cubic meter of water weighs 1,000 kilograms, which is a metric ton.

Sample Problems

Problems	Solutions
1. A tank has a volume of 1,000 cubic inches.	
a. What is the volume in gallons?	1,000 ÷ 231 = 4.32 gallons
b. What weight of water can it hold?	4.32 gal × 8.33 = 35.9856 or 35.99 pounds
2. A tank has a capacity of 80 cubic feet.	
a. How many pounds of water can it hold?	62.5 × 80 = 5,000 pounds
b. How many gallons of water can it hold?	7.5 × 80 = 600 gallons
3. A tank has a volume of 1,250 cubic centimeters.	
a. What is the volume in liters?	1,250 ÷ 1,000 = 1.25 liters
b. What weight of water would it hold in kilograms?	1.25 liters of water × 1 = 1.25 kilograms

4. A tank has a capacity of 6.75 cubic meters.

a. What is the volume in liters? $6.75 \times 1{,}000 = 6{,}750$ liters

b. What is the weight in kilograms? $6{,}750 \times 1 = 6{,}750$ kilograms

c. What is the weight in metric tons? $6{,}750 \div 1{,}000 = 6.75$ metric tons

Exercise

For each volume, find (a) the gallons and (b) the weight of water that could be held.

	Volume	Gallons	Weight in Pounds
1.	1,600 cu in		
2.	5,000 cu in		
3.	18 cu ft		
4.	275 cu ft		
5.	115.75 cu ft		

For each volume, find (a) the liters and (b) the weight of water that could be held.

	Volume	Liters	Weight in Kilograms
6.	3,450 cm^3		
7.	362 cm^3		

Unit 48

Rectangular Solids

Objectives

- Do additional volume problems.
- Learn how earth and concrete volumes are expressed.

Volume is always determined from three dimensions multiplied together. The rectangular shape may be a tank, a swimming pool, a ditch, or other trade application.

In order to obtain an answer in cubic inches, multiply inches × inches × inches. For cubic feet multiply feet × feet × feet. Earth and concrete volumes are usually expressed in cubic yards. They are calculated as cubic feet and changed to cubic yards by dividing by 27.

Sample Problems

1. A flush tank is 21" × 6½" and has water to a depth of 13". How many cubic inches of water does it contain? How many gallons does it contain?

   V = LWH

   V = 21" × 6.5" × 13" = 1,774.5 cubic inches

   V = 1,774.5 cu in ÷ 231 = 7.68 gallons

2. A flat roof 25' × 46' has water 2 inches deep on it. How many cubic feet of water does it contain? How much does the water weigh?

   V = LWH

   $V = 25' \times 46' \times \frac{2'}{12'} = 191\frac{2}{3}$ cubic feet

   Weight = 191.67 cu ft × 62.5 = 11,979.375 or 11,979⅜ pounds

3. A flat roof 7.9 m × 15 m is 5 cm deep with water. How many liters of water does it hold? What is the weight of the water in kilograms?

   V = LWH

   V = 790 cm × 1,500 cm × 5 cm = 5,925,000 cm^3

   V = 5,925,000 ÷ 1,000 = 5,925 liters

   Weight = 1 × 5,925 = 5,925 kilograms

4. A pipe trench is 60' long, 2' 6" wide and has an average depth of 4' 9". How many cubic yards were removed?

   V = LWH

   V = 60' × 2.5' × 4.75' = 712.50 cubic feet

   V = 712.50 cu ft ÷ 27 = 26.39 cubic yards

Exercise

1. A flush tank is 19" × 10½" and holds water to a depth of 11". How many cubic inches of water does it hold? How many gallons?
2. A swimming pool is 22' × 30' with an average depth of 3' 3". How many cubic feet of water can it contain? How many gallons?
3. In digging a trench 50' 0" long, 2' 3" wide and 3' 6" deep, how many cubic feet of earth were removed? How many cubic yards?
4. A flat roof is 14' × 23' is 4" deep with water. How many cubic feet of water are on the roof? How many gallons? How many pounds?
5. A wading pool is 4.7 m long × 4 m wide × 25 cm deep. How many liters of water does it take to fill the pool? What is the weight of the water?
6. A septic tank holding 840 gallons is 7 feet long and 4 feet deep, inside measurements. What is its capacity in cubic feet? How wide should this tank be?

Unit 49

Cylinders

Objectives

- Calculate volumes of cylindrical tanks.
- Determine gallons held by different pipe sizes.
- Learn a shortcut to solving for volumes in cylinders.

The basic formula for the volume of a cylinder is $V = \pi r^2 h$, or $V = 0.7854d^2h$. Notice that there are three dimensions as in any volume. The constant adjusts the size of the answer for the shape of the cylinder. Use 3.14 as π with the radius or 0.7854 with the diameter.

A shortcut to determine volumes of cylinders in gallons is to use the formula $V = 0.0408d^2h$ if *d* is measured in inches and *h* is measured in feet. This formula works because a cylinder 1 inch in diameter and 1 foot long holds 0.0408 gallon.

For standard weight pipe, the number of gallons that are held in one foot of a particular size of pipe can be found on a data sheet (see Data #3, Appendix). The number of gallons held in one foot is multiplied by the number of feet of pipe to find how much the length of pipe will hold. In draining or filling a pipe line it is sometimes desirable to know how much water is held in the pipe.

On occasion, the plumber or pipefitter may find a need to insulate or even construct a cylindrical tank. In this case it would be helpful to know how to calculate the area of the tank's shell. As has been previously noted, the worker must be able to visualize geometric shapes in complex patterns. This is one of those cases. A cylinder is a shape with circular ends of the same size, and a rectangle formed into a tube shape (see Figure 49–1). To calculate the surface we would have to know only the diameter and the length of the cylinder. From the diameter we could derive the area of both ends and the width of the rectangle.

- The diameter × 3.14159 (pi or π) would be equal to the width of the rectangle. The perimeter or circumference of a circle is equal to pi × diameter.
- The radius of the end circles would be ½ × diameter, and therefore the area of the end circle would be 3.14159 × radius^2. The area of a circle is equal to pi times the radius squared.

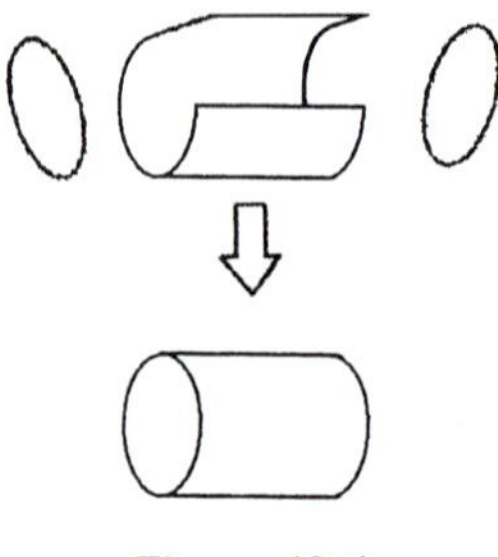

Figure 49–1

Sample Problems

1. A circular pool is 38 inches in diameter and 15 inches deep. What is its volume in gallons?

   $V = \pi r^2 h$

   $V = 3.14 \times (19")^2 \times 15" = 17{,}003.1$ cubic inches

   $V = 17{,}003.1$ cu in $\div 231 = 73.6$ gallons

2. How many gallons will a range boiler 13 inches in diameter and 5 feet high hold?

   $V = 0.0408d^2h$

   $V = 0.0408 \times 13" \times 13" \times 5'$

   $V = 34.476$ gallons

3. How many gallons of water can a 3-inch standard weight pipe hold if it is 18 feet long? (See Data #3, Appendix.)

   0.384 gallon per linear foot × 18 feet = 6.912 gallons

4. How many square inches of insulation would I need to cover a pressure tank 3 feet long and 20 inches in diameter?

   Insulation area = area of rectangle + (2 × area of end circle)

   Area of rectangle = Circumference × length of tank

   Circumference = 3.14159 × 20" = 62.8318"

   Length of tank = 3' = 36"

   Area of rectangle = 62.832" × 36" = 2261.952 square inches

   2 × area of circle = 2 × radius^2 × π

   Radius = diameter/2 = 10"

   2 × area of circle = 2 × 100 × 3.14159 = 628.318 square inches

   Insulation area = 2261.952 + 628.318 = 2890.27 square inches

Note: *Practically speaking, there would be no need to be so precise if this were a real job, but this is about math, not insulation.*

Exercise

1. A water tank is 8' 0" in diameter and 12' 0" high. What is its capacity in cubic feet and in gallons?
2. A gasoline storage tank is 42 inches in diameter and 40 inches high. What is its volume in cubic inches and in gallons?
3. A range boiler is 16" in diameter and 4' 0" high. Find its volume in gallons.
4. Determine the capacity, in gallons, of each of the following sizes and lengths of standard weight pipe. (Use Data #3.)
   a. 2" pipe, 45' 0" long
   b. 4" pipe, 28' 0" long
   c. 10" pipe, 30' 0" long
   d. ¾" pipe, 120' 0" long
5. What is the weight, in pounds, of each standard weight pipe listed? (Use Data #3.)
   a. 2" pipe, 45' 0" long
   b. 4" pipe, 28' 0" long
   c. 10" pipe, 30' 0" long
   d. ¾" pipe, 120' 0" long
6. A cylindrical tank has a diameter of 50 cm and a height of 120 cm. Compute the volume in liters.
7. How many square feet of ½-inch steel plate would be required to build a 20 foot high, 15 foot diameter, cylindrical holding tank?
8. How many square miles of material would be required to enclose the earth in a cylindrical tank? (Assume that the earth is 8,000 miles in diameter and perfectly spherical.)

Unit 50 Spheres

Objectives

- Characteristics of the sphere shape.
- Calculate volumes of spheres.
- Calculate volumes of half-spheres.

A sphere or half-sphere has nearly equal forces pushing in all directions when it is filled with water. Thus, the shape adds to the strength of the structure. Some tanks are built as spheres but more often a spherical bottom is used on a cylindrical tank.

The formula for the volume of a sphere may be seen in several slightly different forms. $V = 0.5236d^3$ is a convenient way of writing the formula. For a half-sphere, use one-half the volume of a whole sphere, or $0.2618d^3$.

Sample Problem

A sphere has a diameter of 9 inches. Determine its volume in cubic inches.

$$V = 0.5236d^3 = 0.5236 \times 9'' \times 9'' \times 9'' = 381.7044 \text{ cubic inches}$$

What is the volume of a half-sphere 9 inches in diameter?

$$V = \frac{381.70 \text{ cu in}}{2} = 190.85 \text{ cubic inches}$$

Exercise

1. A marble top lavatory has a semispherical porcelain bowl with an inside radius of 7". How much water, in cubic inches, will the bowl hold?

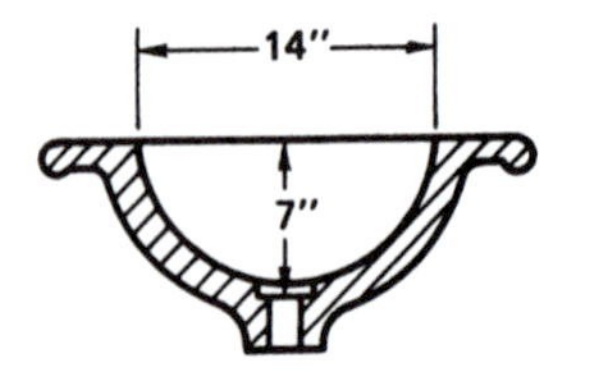

2. How many gallons of water does the illustrated circular water tower contain?

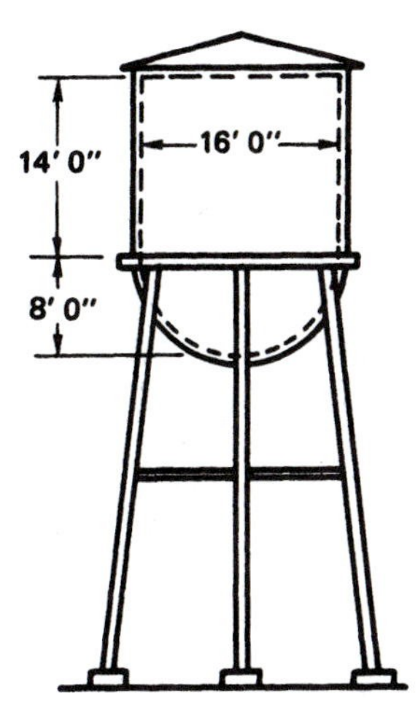

3. What is the pressure per square inch at the base of a standpipe terminating 30 feet below the tank bottom if the illustrated tank is filled? (Pressure is 0.434 pounds per square inch per foot of height.)

4. A spherical gas tank has an inside diameter of 14 feet. Determine the number of cubic feet of gas in the tank when it is full.

5. If a spherical gas tank has an inside radius of 6' 6" and is half-full, how many cubic feet of gas does it contain?

6. Which of the following has the greater capacity and how much greater: a spherical gas chamber with an inside diameter of 11 feet or a cylindrical-shaped chamber with a 10-foot inside diameter and a 10-foot height?

7. A spherical float is 42 cm in diameter. What is its volume in liters?

Unit 51 Segments

Objectives

- State the formula for partly filled tanks.
- Calculate the volumes of partly filled tanks.

Some cylindrical tanks are installed on the side instead of standing on end. When such a tank is partly filled there is a segment volume calculation. The volume is equal to the area of the end times the length of the tank.

The simplest formula for finding the approximate area of a segment is given.

$$A = \frac{4h^2}{3}\sqrt{\frac{d}{h} - 0.608} \text{ when}$$

A = area

d = diameter

h = height of segment

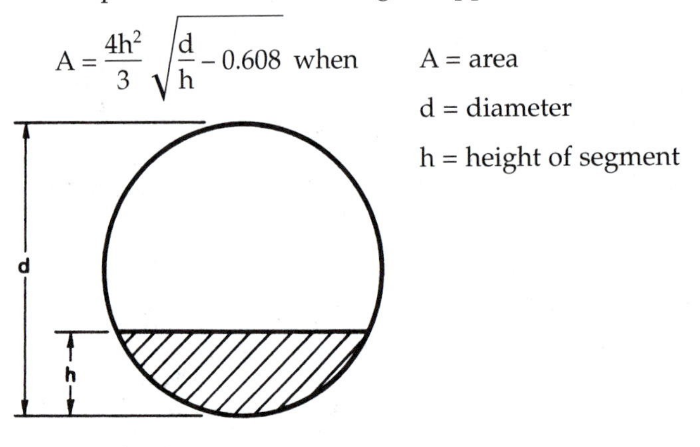

Sample Problem

a. Determine the cubic inch volume of the illustrated tank.

b. Convert this volume to gallons.

Solution

$$A = \frac{4h^2}{3}\sqrt{\frac{d}{h} - 0.608}$$

$$A = \frac{4}{3} \times 144\sqrt{\frac{18}{12} - 0.608}$$

$$A = \frac{576}{3}\sqrt{1.5 - 0.608}$$

$$A = 192\sqrt{0.892}$$

$A = 192 \times 0.94$

$A = 180.48$ sq in

$V = AL$

$V = 180.48$ sq in $\times$ 22 in

$V = 3{,}970.56$ cu in

$\text{Gallons} = \frac{3970.56}{231} = 17.18 \text{ gal}$

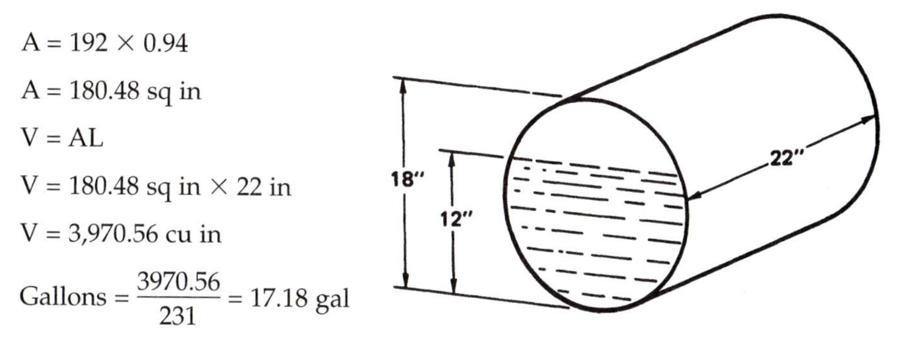

Exercise

1. What is the area of a segment of a circle if the diameter is 42 inches and the height of the segment is 14 inches?

2. Find the number of cubic feet of steam space in the gas tube boiler, as shown in the sketch. The boiler is 18 feet long between heads and has an inside diameter of 66 inches. The mean water line is 20 inches from the top.

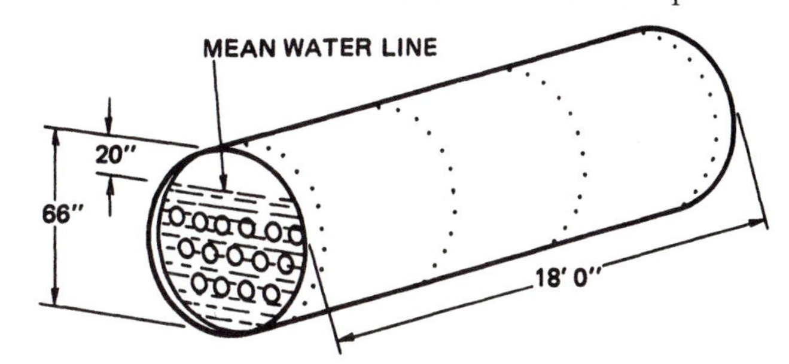

3. A pneumatic water system has a tank that is 25 feet long with a 4½-foot inside diameter and is set in a horizontal position. If this tank is filled to a level equal to ⅔ of its diameter, how many gallons of water does it contain?

4. An oil tank which is 12 feet long and 4 feet in diameter, inside measurements, is placed in a horizontal position. Using the illustration, make up a table showing the number of gallons of oil there will be in the tank for each of the levels shown.

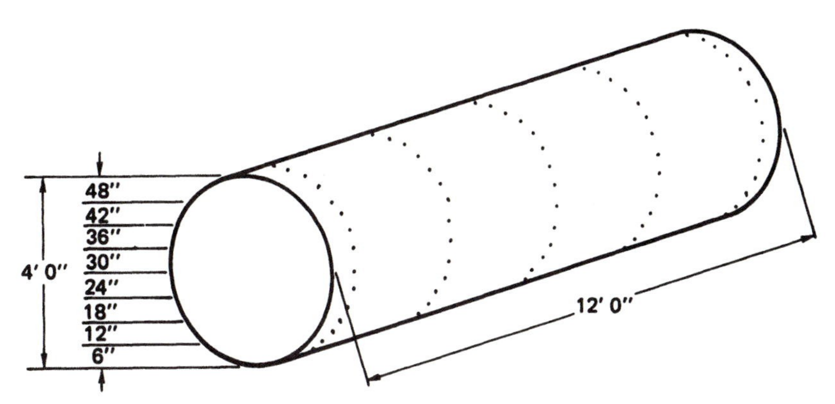

Unit 52 Partially Filled Containers of Varying Shapes

Objectives

- Identify simple elements in compound shapes.
- Develop attack strategies for calculating partial volumes.
- Calculate partial volumes and weights.

Swimming pools, linked tanks, heat exchangers, and natural water reservoirs are all examples of containers that may be inconsistent in shape and also only partially filled. The craftsperson may be called upon to treat the liquid remaining in a partially filled tank, to render the contents potable, to prevent metal etching and scaling, or perhaps to calculate the cost/time of raising the tank level.

As the complexity of the tank's shape increases, the probability of calculation error grows. Stationary engineers make charts plotting time, or water level in a sight glass, against volume for filling heat exchangers, boiler water, and other applications when the tanks have internal piping that take up inconsistent amounts of space at different levels. One way this is done is by testing the quantity of water supplied by a given fill line per minute and then filling the empty tank and making marks beside the sight glass as the water raises. (The sight glass is a tube of glass mounted into a tank that reveals the level of the liquid inside.)

Occasionally calculations have to be made, and the simplicity of the tank's shape, positioning, and inner contents will affect the confidence to be placed in the resulting calculation. Liquids seek their own levels, meaning that the surface of the liquid assumes a configuration parallel to the earth's surface. This complicates matters when we are calculating the amount of propane remaining in a truck that has run off the road into a ditch, for instance.

Some useful equivalents:

You have	You want	Multiply by
Cubic feet	Gallons	7.48
Cubic inches	Gallons	0.00433
Gallons of water	Pounds	8.33
Cubic feet of water	Pounds	62.31
Cubic inches	Cubic feet	0.0005787

Sample Problem

Figure 52–1 is a steel water tank. How many gallons will be needed to fill it?

Solution

1. Since the dimensions are in linear feet, find the total number of cubic feet and change the answer to gallons. Make sure that you are using dimensions of the same kind (feet × feet or inches × inches).

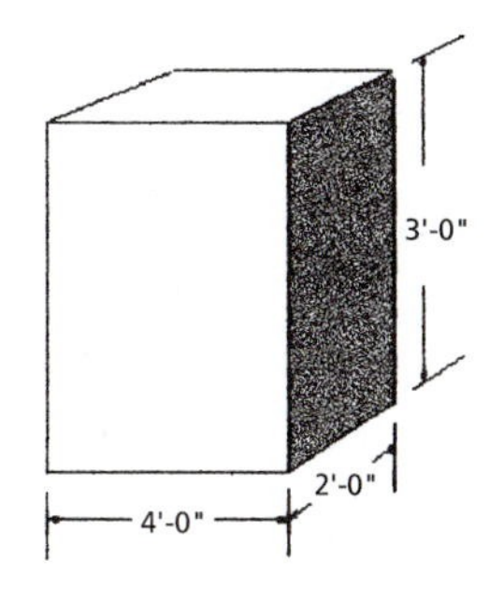

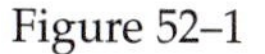
Figure 52–1

2. Volume of a rectangular prism = Length × Width × Height. Therefore, volume is equal to

   $4 \times 2 \times 3 = 24$ cubic feet

3. Convert the volume in cubic feet to gallons using the conversion table above.

   $24 \times 7.48 = 179.52$ gallons

Figure 52–2 shows the same steel tank as above, but now it already has 1 foot of water in it. How many gallons will it take to fill it? This is a simple problem of shapes within shapes. See the one major prism with two others inside. Derive the dimensions of the empty one and solve as we did above. Remember that in volumes of parallel-sided shapes, we usually multiply three dimensions that lie at right angles to each other together. Another way of looking at it is to say we find the area of the end and multiply this by the height. But suppose one end has an area that is different from the other end and the sides are not parallel?

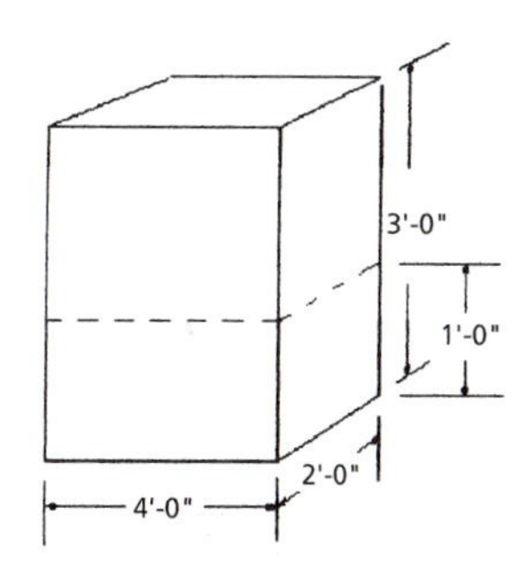

Figure 52–2

For calculating the volume of both frustums, we would find the areas of the top and bottom of the shapes and add them together and set that aside. Then we would multiply the areas together and find the square root of our answer. This last answer we would add to the first and multiply the result by the height divided by 3. The pyramid frustum:

$[(20 + 80) + (\sqrt{20} \times 80)] \times 7/3 =$ 326.666 cubic inches

For the cone, we simply multiply the area of the base by the height and divide the answer by 3.

$5 \times 5 \times 3.14 \times 12/3 = 314$ cubic inches

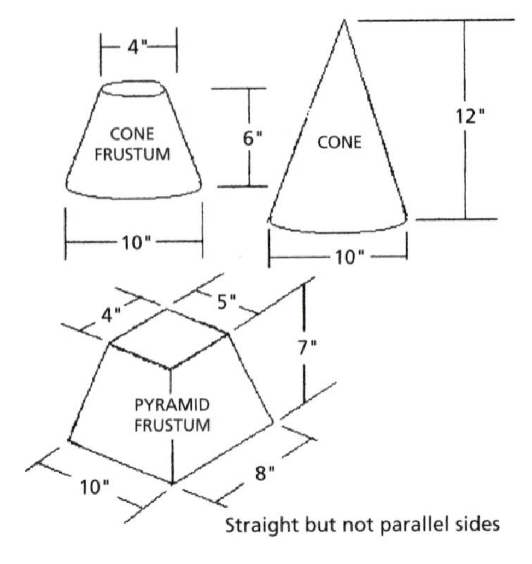

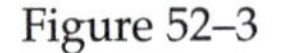
Figure 52–3

Suppose that we encountered a shape like that in Figure 52–4? In this case, if we took the area of the top or bottom and multiplied by the height, we would fail to get a correct answer because when visualized that way, the sides between the top and the bottom are not parallel. You probably see an additional trapezoidal prism attached to the side of the rectangular prism and recognize that with sufficient measurements, the shape could be easily divided into two solid figures and the answer derived. Or perhaps you have mentally rotated the figure so that the large sides are now the top and the bottom, and you now recognize that all you have to do is calculate the area of the top, now a complex figure, and multiply that times the new height, which has parallel sides, and which used to be the thickness or width. When looking for a way to solve these kinds of problems, use your ability to turn things around in your mind.

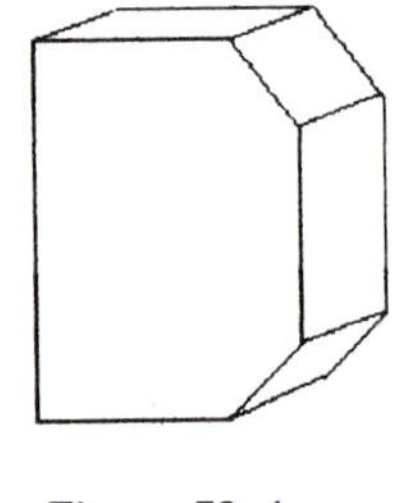

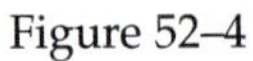
Figure 52–4

Suppose a cylinder lies at a 15-degree angle (see Figure 52–5). Could we call that a parallel-sided figure?

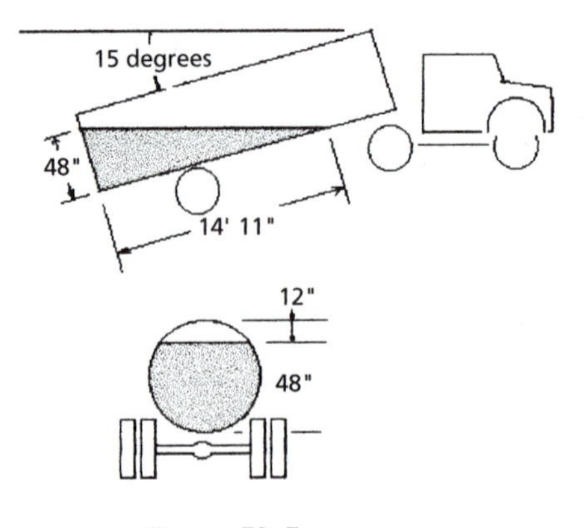

Figure 52–5

Sample Problem

How many pounds of water will it take to fill the tank illustrated by Figure 52–6?

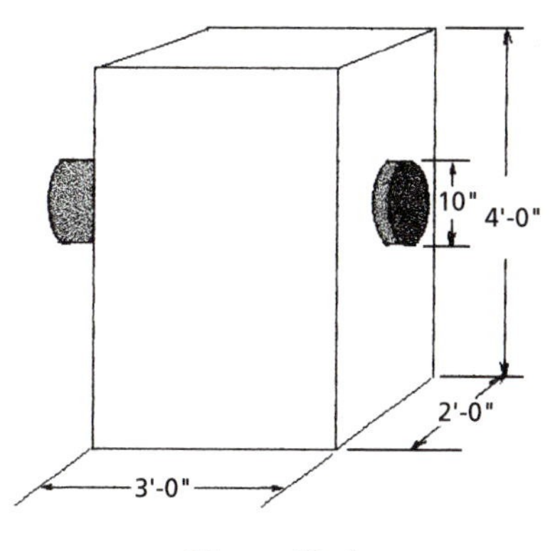

Figure 52–6

Solution

1. We know that the rectangular tank measures 4 feet by 2 feet by 3 feet and is penetrated by a 10-inch diameter cylinder which for our purposes is only 36 inches long. The pipe or cylinder will reduce the capacity of the rectangular tank.
2. If the tank were completely full of water then it would have $3 \times 2 = 6$ (the area of the end or top or bottom, the name doesn't matter) times the height: $6 \times 4 = 24$ cubic feet of water in it. Now we must discover how much capacity the pipe subtracts from 24.
3. The 10-inch pipe is a parallel-sided figure with an effective length of 3 feet or 36 inches. So find the area of the end of it. The formula for the area of a circle is Area = $3.14159 \times \text{radius}^2$. Since a pipe size (10") describes the diameter of a pipe and we need the radius, we divide the diameter by 2 and get 5 inches, the radius.

   $5 \times 5 \times 3.14159 = 78.54$ square inches (the area of the end, ignoring the pipe wall thickness)

   That times the length (78.54×36) equals 2827.44 cubic inches. Since we have been working with cubic feet up until now, you must change your answer in cubic inches to cubic feet

   $2827.44 \times 0.0005787 = 1.636$ cubic feet

4. The capacity of the tank in cubic feet of water less the volume of the pipe is

   $24 - 1.636$, or 22.364 cubic feet

   The answer is required to be in pounds, so convert cubic feet to pounds of water by multiplying cubic feet by 62.31 (see chart on page 185).

   $22.364 \times 62.31 = 1393.5$ pounds

If a container has an irregular shape then the possibility of accurately computing its volume becomes increasingly difficult. The engineer may have to estimate if the container is a lake, for instance, or use water displacement techniques, or filling with a known volume of liquid, if the container is small.

Exercise

1. Using the sample problem above, suppose that the water inside the tank was filled only to the center of the penetrating pipe and the center of the pipe was 18 inches down from the top of the tank. How many pounds would the water now weigh? ____________ pounds.
2. The minimum size for a septic tank in this particular location is 1,500 gallons. This tank is 10 feet long, 4 feet 2 inches wide, and 5 feet deep at the exiting invert. If 12 inches of hard sludge accumulates in the bottom, what is the tank's new effectiveness expressed as a percentage of capacity? ____________%

3. A manufacturer of fiberglass septic tanks wishes to apply a 6-inch radius to the tank's edges so that the molds will separate more easily (see Figure 52–7). What will be the new capacity of the tank, and will it then be within specifications? The local sewer authority requires a 1,500-gallon septic tank. It will accept a tank 10 feet long, 4 feet wide, and 5 feet high to the exit pipe's invert, but the capacity must be within 99% of code specifications. Your answer should be expressed as a percentage of 1,496 gallons. (Notice that this tank in new condition, and with squared corners, will retain only 1,496 gallons. Use gross lineal feet, ignoring the small amounts lost at the corner joints. Note also that only the wetted inside surface is significant.) ______________% Will the finished product still be within 99% of code specifications? ______________ yes/no

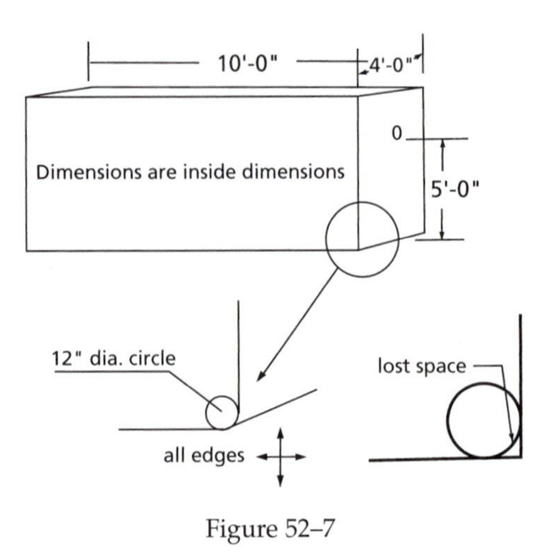

Figure 52–7

4. A rectangular crude oil preheater measures 19 feet long by 10 feet high by 5 feet thick. At the 5-foot by 10-foot ends, we can count 72, 1½" steam pipe penetrations that we are told run the length of the tank's inside. How many gallons of crude oil will the tank hold? *Note: Use the outside dimension for 1½" steel pipe, 1.900 inches.* ______________ gallons

5. A man needed to know how much space his outboard motor took up, so he put it in a plastic bag and then lowered it into his 5-foot diameter hot tub and took note that the water level in the hot tub increased by 2 inches. Give the volume occupied by his outboard motor in cubic feet to three decimal places. ______________ cubic feet

Unit 53 Water Pressure, Head, and Force

Objectives

- Relate the term "head" to pressure.
- Learn about force.
- Calculate pressure from height of water.

In plumbing practice, water pressure is measured by a gauge, incremented in pounds per square inch (psi) or the *head,* the height, in linear feet, of the standing water from the surface down to the point at which the pressure reading is significant. If we wished to know what the head is at the bottom of the tank, then the entire height of the standing water is significant.

We know that a cubic foot of water weighs 62.5 pounds. We know that a cubic foot of water could be a block of ice with 6 sides each measuring 12" by 12". We could use a band saw to cut a 1" × 1" corner out of that block (see Figure 53–1). Since each surface of the block is 12" × 12" we know there are 144 of these 1" × 1" slices 1 foot long in that block. Therefore, each 1" × 1" × 1' piece should weigh 62.5 pounds divided by 144 or 0.434 pounds. Since it would also weigh 0.434 pounds standing on one of its 1-square-inch ends and the column of frozen water is 1 foot high, we can also say that for every foot of height in a column of water the pounds over 1 square inch at the base would be 0.434 pounds. Pressure gauges are calibrated to give measurements in PSI. So couldn't we now predict that if we filled a pipe of any diameter to a height of 10 feet with water that the pressure gauge inserted into the bottom would show a pressure of 4.34 psi? (10 × 0.434)

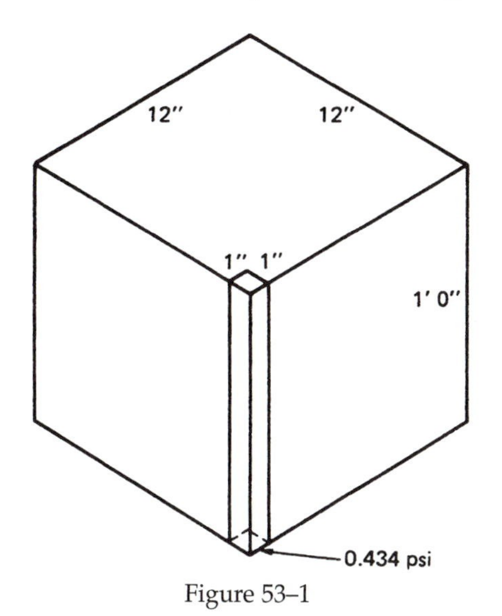

Figure 53–1

To determine how high a given pressure can lift water, it is necessary to convert psi into feet of head. Previously, we learned that, knowing the height of a column of water in feet, we can discover what the pressure (psi) will be at the bottom by multiplying the height of the column by 0.434. Turning that around, if we knew the pressure at the base we could predict the height by *dividing* the pressure by 0.434. Furthermore, if we preferred to multiply instead of divide we could take the reciprocal of 0.434 (2.304) and multiply by that to get the same answer.

$$\frac{1}{0.434} = 2.304 \text{ ft, the height necessary to produce 1 psi}$$

psi = 0.434 × H, where psi = pounds of pressure per square inch and H = height of water in feet.

- Each and every foot of head (or height) produces 0.434 pound pressure per square inch.
- Each and every pound per square inch of pressure requires a head of 2.304 feet.
- The shape or slope of the sides of a tank do not affect the pressure on the bottom of a tank.

Sample Problems

1. A stack 42 feet high above a test plug is filled with water. Determine the psi on the test plug.

   0.434 × 42' = 18.228 psi

2. How high will 25 psi lift water?

   2.304' × 25 = 57.600 feet

3. Determine the total force on a 6-inch diameter test plug with a water head of 15 feet.

   0.434 × 15' = 6.5100 psi

   $A = \pi r^2$

   A = 3.14 × 3" × 3" = 28.26 square inches

   28.26 sq in × 6.51 = 183.9726 pounds

Exercise

1. What will be the pressure in pounds per square inch on the base of a 4-inch soil stack 45 feet high that is filled with water?
2. What would be the height in feet of a column of water necessary to develop 13.02 pounds pressure per square inch?
3. What head of water will produce a 65-pound pressure per square inch? What head will produce 72 pounds per square inch? What head will produce 88 pounds per square inch?
4. What pressure is produced by a 28-foot head? A 49½-foot head? A 67½-foot head?
5. Determine the pounds of force on a 4-inch diameter test plug in a stack with a head of 20 feet.
6. Determine the pounds of force on a 3-inch diameter test plug in a stack with a head of 48 feet.

Unit 54

Ratio of Pipe Capacities

Objectives

- Learn to compare pipe-carrying capacities.
- Define the comparison of cross-sectional area method.
- Use a refinement of the cross-sectional area method.

The information given here is intended to help the plumber in case a flow problem occurs. However, most pipe is sized by the architect or engineer.

Pressure, friction, and cross-sectional area of the pipes control the amount of flow in piping. Water pressure is beyond the control of the plumber in most installations. Friction is somewhat reduced by proper design of a piping system to run as directly as possible. The material used may also affect friction. Copper tubing and PVC pipe have smoother walls than steel or wrought iron pipe. The smoother surface has less frictional resistance.

The plumber can install a larger pipe size to provide adequate flow to each faucet. Only in a fire sprinkler system must all outlets have full flow at the same time. In plumbing a percent of full use is expected. The architect sizes pipe for the expected use, using small size pipe whenever possible for the sake of economy.

There are two ways of computing the ratio of pipe capacities. The first, $D^2 \div d^2$, is a simplification of AREA ÷ area and makes no allowance for friction. This method is simple to use and gives a practical answer when the large diameter is not more than twice the small diameter. Also, pipe lengths should be short, which means 10 feet or less for pipe sizes 1" and smaller and up to 20 feet for larger sizes of pipe.

A more accurate comparison is obtained by $\sqrt{\left(\frac{D}{5}\right)^5}$ which does allow for friction.

Friction is greater in smaller pipes because a greater percentage of the total water drags against the pipe wall.

Sample Problems

1. How many 1" pipes can be supplied by a 2" pipe?

$R = \frac{D^2}{d^2}$ where R = number of smaller pipes

D = diameter of larger pipe

d = diameter of smaller pipes

$R = \frac{4}{1} = 4$ smaller pipes

2. How many ¾" pipes can be supplied by a 1" pipe?

$$R = \frac{D^2}{d^2}$$

$$R = \frac{1 \times 1}{0.75 \times 0.75} = \frac{1}{0.5625}$$

R = 1.8 or about 2 smaller pipes

3. How many 2" pipes will a 4" pipe supply, allowing for friction?

$$R \text{ varies as } \sqrt{\left(\frac{D}{d}\right)^5}$$

R = number of smaller pipes the larger one will supply

D = diameter of larger pipe

d = diameter of smaller pipes

$$R = \sqrt{\left(\frac{4}{2}\right)^5} = \sqrt{(2)^5} = \sqrt{32} = 5.657 \text{ or about 6 smaller pipes}$$

Exercise

Solve each problem using both formulas.

1. How many 2" pipes will a 3" pipe supply?
2. How many ¾" pipes will a 2" pipe supply?
3. How many 1¼" pipes will a 3½" pipe supply?
4. How many ½" pipes will a 1¼" pipe supply?
5. How many 1" pipes will a 1½" pipe supply?

Unit 55 Pipe Sizing

Objectives

- Learn the "units of flow" method for pipe sizing.
- Calculate a practical example with the "units of flow" method.

The selection of a size of pipe to use for water service is made by use of a table that has been found to work. It is based on the rate of flow from a ½-inch size faucet. Each supply to a fixture is rated as one or more units of flow. The units of flow are matched to a satisfactory pipe size.

UNITS OF FLOW FOR SELECTED FIXTURES			
Water Closet, Tank	1	Slop Sink	1
W.C. Flush Valve	16	Urinal, Flush Valve	4
Lavatory	1	Drinking Fountain	1
Sink	1	Hose Bibb	1
Shower	1	Hose Bibb ¾"	4

UNITS OF FLOW FOR EACH PIPE SIZE										
6"	5"	4"	3"	2½"	2"	1½"	1¼"	1"	¾"	½"
1	2	4	8	16	32	64	128	256	1024	4096
	1	2	4	8	16	32	64	128	512	2048
		1	2	4	8	16	32	64	256	1024
			1	2	4	8	16	32	128	512
				1	2	4	8	16	64	256
					1	2	4	8	32	128
						1	2	4	16	64
							1	2	8	32
								1	4	16
									1	4
										1

Sample Problem 1

How many 1-inch taps will a 5-inch water main serve?

Find 5" in the top row of the table.

Follow down to the 1 unit.

Follow across until under 1" in the top row.

Read 128 under the 1".

Number of services = 128.

Sample Problem 2

A water supply serves a building with 2 flush valve water closets, a urinal, 3 lavatories, and a tank W.C. How many units of flow are needed for these fixtures? What size water service is needed?

W.C. Flush Valve	$2 \times 16 = 32$
Urinal Flush Valve	$1 \times 4 = 4$
Lavatories, cold & hot	$3 \times 2 = 6$
W.C. Tank	$1 \times 1 = 1$
Total Units of Flow	43

To find the pipe size from the table,

Start with the number 1 under the ½" size.

Follow the column up to numbers 32 and 64.

Use 64 because 32 is smaller than 43.

Follow across from the 64 to the 1.

Look at the size at the top of this column.

Read 1½".

Water service needed = 1½".

Exercise

1. Using the pipe diagram shown, determine the number of units of flow and the pipe size at each of the locations indicated by a number in a circle.

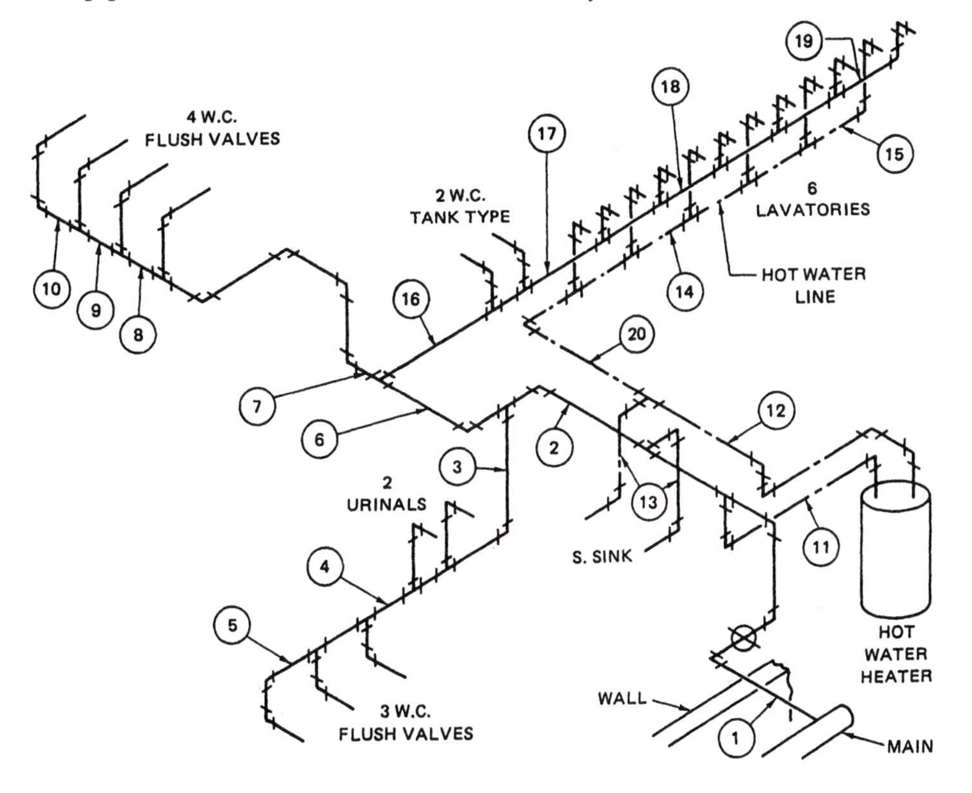

2. How many ¾-inch water service lines will a 4-inch water main serve?

Section Five
PHYSICS

Unit 56 Mechanical Advantage

Objectives

- Calculate mechanical advantages.
- Solve problems in hydraulics.
- Apply the wheel and axle method.
- Calculate equilibrium.
- Understand inclined planes.
- Overcome friction.

"Give me a lever long enough, and a place upon which to rest it, and I will move the earth."

—Archimides 287 B.C.–212 B.C.

The Inclined Plane and Mechanical Advantage

From the physicist's point of view, no work is accomplished when weights are moved horizontally, only when weights are lifted.

Mechanical advantage (M.A.) is what we use to enable us to lift and move heavy objects with only our own strength. But there is no free lunch. The worker lifts a 500-pound boiler 3 feet by pushing it up a long inclined ramp or by pulling down 100 feet of feet of rope looped around pulleys and attached to the boiler, or by moving a 2-foot handle up and down 50 times. In other words, the craftsperson uses rigging tools to do the work necessary by lifting or pushing a lesser weight through a much longer distance than the 3 vertical feet necessary. Distance is traded for weight. When rigging tools are used, another factor enters the scene—*friction*. Friction within tools like chain falls, cable-jacks or come-alongs, pulleys, floor jacks, and ramps have an effect on our calculations. But friction between surfaces like ramps and floors often also needs to be reduced (see Figure 56–1). In the lower example of the drawing, we can see that friction between the heavy wooden crate and the concrete ramp would be considerable, while in the upper case the workers have slipped rollers (usually short lengths of steel pipe) beneath the crate and have made the task much easier by reducing friction. Theoretically speaking (meaning without friction being taken into account), if the crate weighed 500 pounds and the ramp was 10 feet long and 1 foot high at the high end, then the mechanical advantage would be 10:1. The worker would experience the lifting as if it were only 50 pounds instead of 500 pounds. But friction would play a major role here and would probably make the task too difficult if rollers were not utilized to reduce friction. Friction or drag depends on the weight and the qualities of the materials in direct contact. So the task—to lift a 500-pound crate up 1 foot above the ground—is accomplished when the worker slides the work up the 10-foot-long ramp which is 1 foot higher at one end. The trade off is that the crate must also be pushed 10 feet sideways to achieve the 10:1 M.A. The alert student may be asking; "What about raising the crate onto the rollers? Didn't the craftsperson have to lift the 500-pound crate with only his own strength there?" The answer is: Not necessarily. The worker could have used another inclined plane called a wedge or even a lever to lift the crate high enough to put rollers beneath the load.

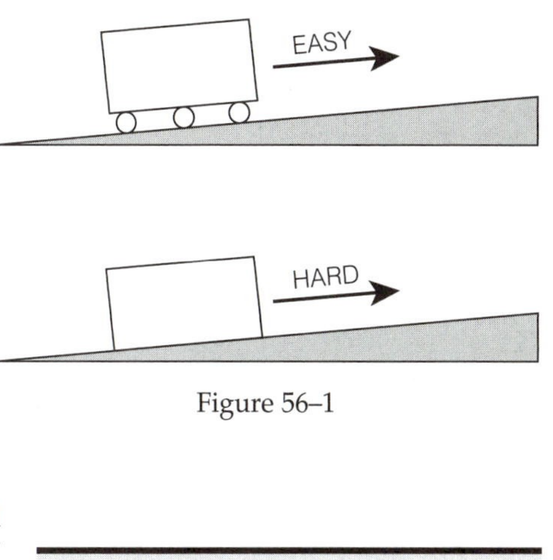

Figure 56–1

The calculations in this unit are for theoretical M.A.s. Friction always reduces the calculated M.A. and in some cases may even erase it.

Sample Problem

What is the M.A. for a wooden wedge 4 feet long and 3 inches thick at its thickest end?

Solution

4 feet = 48 inches and the wedge can lift the test weight 3 inches in that distance. Therefore the M.A. of the wooden wedge would be in the ratio of 48:3 and 48/3 = 16 which is the same as 16/1 therefore the M.A. of the wooden wedge is 16:1.

Levers and Leverage

Levers and leverage are also at the core of what we must know about lifting heavy objects with forces of less weight than the object to be lifted or turned. In order to understand this we must first examine equilibrium. Picture the lever as being a long straight bar of any strong material. To act as a lever this bar must utilize a fulcrum. A *fulcrum* is whatever supports the bar at one point anywhere on that bar. A bar lying on the floor has no fulcrum because the floor supports the bar at many points along the bar's length. Note that in Figure 56–2 in illustration A, the bar is supported at its exact middle point and the 10-pound weights are equidistant on both sides of the point of the fulcrum. The lever and the forces acting upon it, two 10-pound weights and the fulcrum are in a state of balance (*equilibrium*).

And in illustration B of the same figure they are also in balance because the total force vectors on the left are the same as those on the right side. But the weight on the right side of the fulcrum is 5 pounds less than the weight on the left side of the fulcrum. Consider the fact that a 5-pound weight on the right side is balancing a 10-pound weight on the left. This is because the weight on the right has a 2:1 M.A. because of its position along the bar. Notice that $10 \times 3.5 = 5 \times 7$.

It is helpful to think of the fulcrum as a rotational center for the forces that are acting upon it. The weights are attempting to rotate around the tip of the fulcrum.

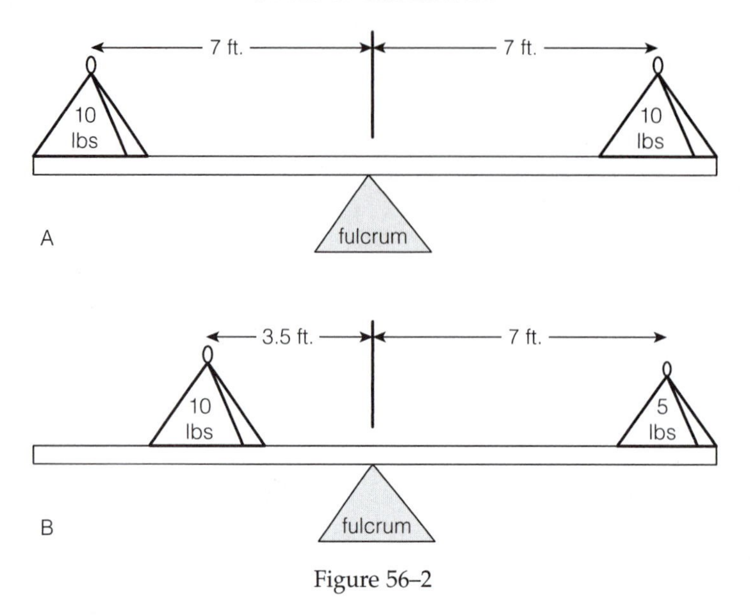

Figure 56–2

Sample Problem Where would you have to place the 10-pound weight in respect to the fulcrum if the 5-pound weight was 4 feet on its side of the fulcrum?

Solution

$10 \times \text{Distance} = 5 \times 4 \Rightarrow 10 \times \text{Distance} = 20 \Rightarrow \text{Distance} = 20/10 \Rightarrow \text{Distance} = 2$ *therefore*, the distance that the 10-pound weight would have to be placed from the fulcrum to balance the 5-pound weight on the other side would be 2 feet.

The Built-In Fulcrum

A lifting tool called a pinch-bar (Figure 56–3) is used to raise heavy objects just an inch or so at a time often to get enough room to slide rollers or planks beneath it. The pinch-bar carries its fulcrum with it in the design of the tool. The pinch-bar uses the lever to multiply *torque*, or rotational power. In this case, the balancing bar is the tool itself and as the long radius of the bar is pressed downward through a rotation of 24 inches, the short radius (the working end) moves upward perhaps 2 inches also in an arc. In this case we may calculate the M.A. by placing 24 over 2 and then reducing that fraction for an M.A. of 12:1.

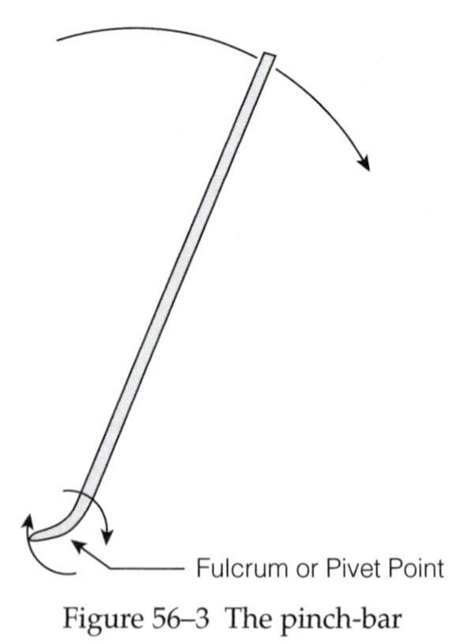

Figure 56–3 The pinch-bar

The Screw Jack

A screw jack uses an inclined plane also to provide a mechanical advantage (Figure 56–4). In this case the inclined plane is provided by the threads on the stem of the jack. To calculate M.A. of the screw jack provided by the threads, the diameter of the stem is measured at the midpoint of the thread depth, the circumference is calculated, and the distance between threads is measured. The ratio of the circumference to the distance between threads is the M.A. This and most jacks use a lever in conjunction with the inclined plane to accomplish their work. To operate the particular jack shown in the figure, the operator sticks a bar into one of the stem holes and rotates the stem in or out of the jack body. The threaded stem rises upward with the M.A. of the inclined plane of the threads.

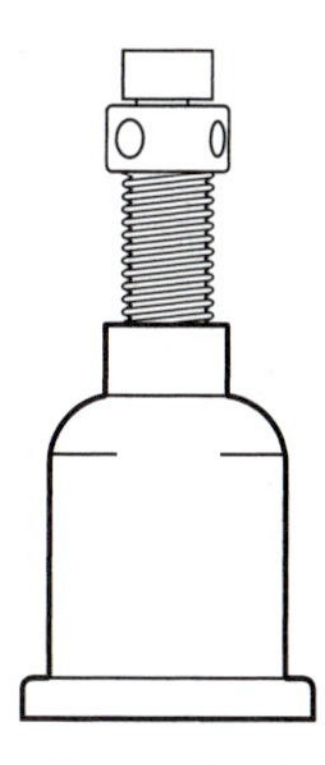

Figure 56–4

The Wheel and Axle

A spoked wheel is essentially a circle of levers. An automobile's steering wheel applies leverage to the steering column of the vehicle because the power is being applied at a distance perpendicular to the steering column. In the previous section, we saw how the mechanic could turn the stem screw on the screw jack by inserting a long bar into the holes in the stem head and applying rotational force to the stem. It is unlikely that the stem of the screw jack with weight pressing downward on it could have been turned by hand at all; yet with aid of the bar it can be turned easily. If the outside surface of the steering column bar is rotated 1 inch and the steering wheel is rotated through an arc 12 inches long, then the M.A. is 12:1. Everywhere that wheels with axles are working there are leverages at work. The block and tackle with its round pulleys, the oxen walking in endless circles yoked to a wheel and grinding grain, a stream's water turning a water wheel, even the wrench tightening a bolt—all of these are examples of the wheel and axle at work.

Sample Problem

If a wheel has a stem diameter of 1¼ inches and an outside diameter of 3 feet, what would be its calculated mechanical advantage without taking friction into account?

Solution

For comparing the distances traveled by the two arcs—the arc on the stem and the arc the wheel rim—we might just assume a complete circle and then all we would have to do is compute and compare the circumferences of the stem and the outer circle of the wheel.

Circumference is equal to the diameter times pi.

$C = d \times \pi$ and pi = 3.14159

Let SC = stem circumference and WC = wheel circumference.

> **Mechanical Advantage is always affected adversely by friction, so it would probably be wiser to round downward when arriving at whole numbers for our ratio.**

$SC = 1.25 \times 3.14159 \Rightarrow SC = 3.927$

$WC = 36 \times 3.14159 \Rightarrow WC = 113.097$

$MA = WC/SC \Rightarrow MA = 113.097/3.927 \Rightarrow MA = 28.8{:}1 \Rightarrow MA = 28{:}1$

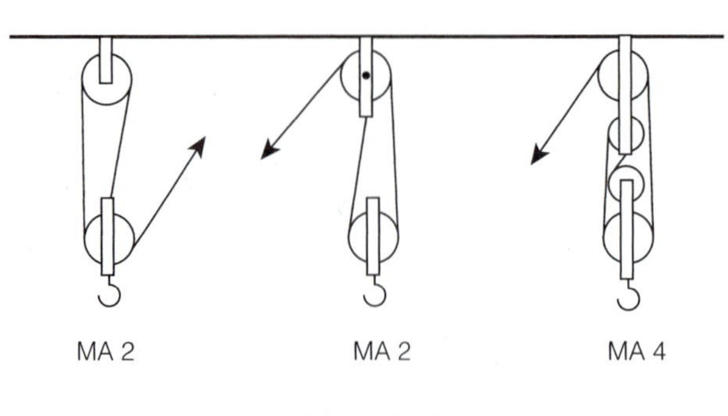

Figure 56–5

Extra Credit

What would happen if we simply compared diameters instead of circumferences? After you try it, explain what happened.

Pulleys

Pulleys are wheels but the M.A. attributable to pulleys (block and tackle) is gained through applying effort to a lifting task through the use of unattached pulleys. Imagine a pulley attached to an overhead beam. A rope is fed through the pulley and the end is dropped to the floor below. Now the rope forms a loop over the pulley and both ends of it are laying on the floor below.

A bucket of bolts is fastened to one end of the rope, and a student grasps the other end and begins pulling down on the rope. The bucket of bolts rises all the way up to the pulley above. What is the M.A. of this arrangement? If you answered none or one, you are correct. (*Note:* An M.A. equal to 1 is no advantage at all). For every foot to rope you pulled down, the bucket would have raised 1 foot, which is no advantage. As a matter of fact, the friction that is always present would have made the real-world M.A. less than 1. That's not to say that you wouldn't have had a physical advantage by being able to use your weight to help you haul the bucket up as opposed to being on a scaffold and having to draw the bucket up with your hands and arms.

Imagine a floating pulley with a hook on it to which a load could be fastened (see Figure 56–6). This arrangement would actually have an M.A. of 2. Notice there is one floating block.

In general, the M.A. will be: 1) the same as the number of vertical ropes excluding the open rope-end, 2) double the number of floating pulleys, or 3) the feet of rope pulled divided by the distance the bucket raised upward.

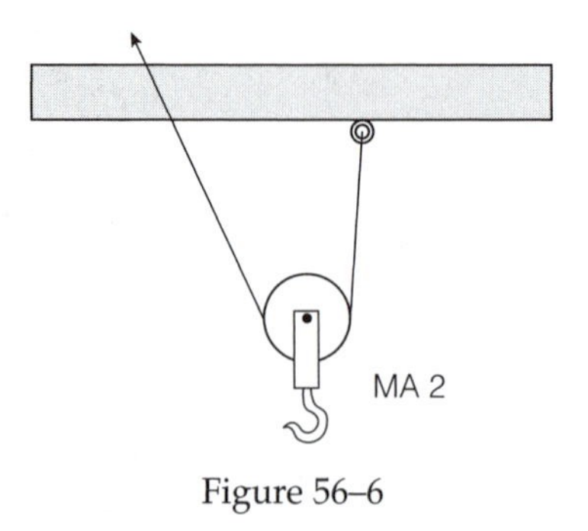

Figure 56–6

Hydraulics

Hydraulics have many applications but probably the most common hydraulically operated tool is the car jack. A handle is pumped up and down many times while the vehicle's tire rises slowly off of the ground. This is the (by now) familiar trade-off between a light weight being able to lift a much heavier weight when it possesses M.A. At the core of the hydraulically operated tool is an incompressible fluid-like oil. There is a strong tank that has two cylinders with pistons (Figure 56–7). The larger cylinder lifts the vehicle, and the smaller serves as a pump to increase the pressure on the fluid in the tank. The mechanical advantage is a comparison of the squares of the two piston diameters. $MA = LD^2/SD^2$ where LD is the diameter of the larger piston and SD is the diameter of the smaller piston. This formula is a simplification of $MA = ((LD/2)^2 \times \Pi)/((LS/2)^2 \times \Pi)$. This is a comparison of the two cylinder's areas, which provides a basis to discuss pounds per square inch (PSI).

> **The hydraulic jack has a reservoir to supply more oil when needed and a check valve to keep the oil in the main tank until the job is finished.**

Pascal's law states that when there is an increase in pressure (PSI) at any point in a confined fluid, there is an equal increase at every other point in the container. So if the smaller 1-inch diameter piston descends with a force of 20 PSI, the larger 3-inch piston will ascend (it must because oil is incompressible) with a force of 20 PSI. But since there are three times as many square inches at the top of the larger piston it will ascend with a power capable of overcoming a 60-pound weight.

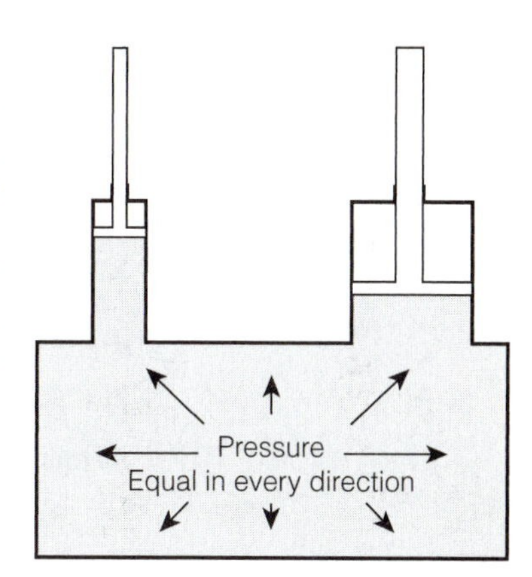

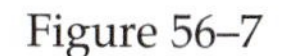
Figure 56–7

The pump piston is activated by a lever and pivot (fulcrum) arrangement and so, as in many cases, there are two power multiplying systems at work (Figure 56–8).

Sample Problem

What would be the theoretical M.A. of a hydraulic jack with a ½-inch pump piston and 2-inch output piston? Do not consider the handle lever in your answer.

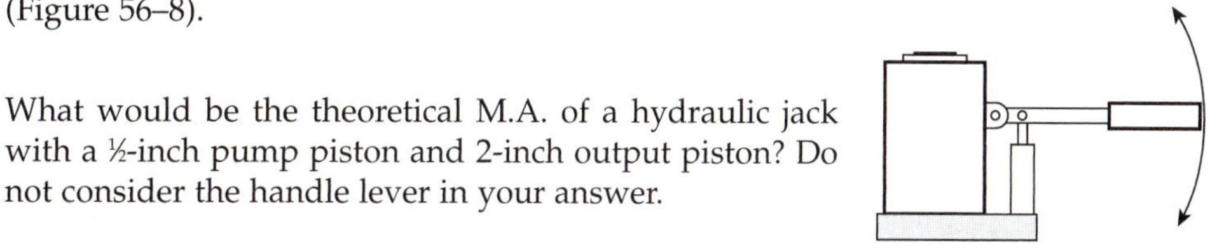
Figure 56–8

Solution

$MA = LD^2/SD^2$

$MA = 2^2/0.5^2 \Rightarrow MA = 4/0.25 \Rightarrow MA = 16{:}1$

Exercise

1. A 20-foot-long ramp leads up to a 4-foot-high loading dock. The workers place a 300-pound barrel of cooking oil on a dolly to reduce friction and push the barrel up the ramp and onto the loading dock. What is the M.A. of this arrangement?

   a. 75:1　　b. 5:1

   c. 3:2　　d. 1:1

2. In the previous problem, suppose that the workers unrolled a long carpet from the top to the bottom of the ramp and laid the barrel on its side on the lower end of the ramp and pulled the end of the rug over the top of the barrel and then continued to pull the end on the ramp until they came even with the top. Where would the barrel be?

   a. close to the top of the ramp

   b. close to the middle of the ramp

3. Was there any mechanical advantage in using the rug that way?

   a. Yes　　b. No

4. If an 8-pound weight was placed 3 feet to the left of the fulcrum, how far to the right of the fulcrum would you have to place a 5-pound weight to balance it?

   a. 4.8 feet
   b. 5 feet
   c. 8.2 feet
   d. 10.1 feet

5. A block and tackle assembly that has two floating blocks should have an M.A. of what?

   a. 1
   b. 2
   c. 3
   d. 4

6. If you have to pull 100 feet of rope to move a weight 25 feet into the air using a block and tackle, what is the M.A. of the arrangement?

   a. 12:1
   b. 8:1
   c. 4:1
   d. 2:1

7. What is the major mechanical advantage principle utilized by the screw jack?

   a. rack and pinion
   b. the lever
   c. inclined plane
   d. hydraulics

8. If a hydraulic jack had a ¾-inch pumping cylinder and a 3-inch main cylinder, what would be the M.A. for that particular jack (disregarding the leverage of the pumping handle)?

   a. 3:1
   b. 8:1
   c. 16:1
   d. 22:1

Unit 57

The States of Matter and Why It Matters

Objectives

- Learn about the states of matter.
- Discover the uses for latent heat.
- Calculate pressures and heat transfers pertaining to states of matter.
- Understand the scientific units for heat transfer.
- Change given temperatures between Fahrenheit and centigrade.

States of Matter

Perhaps it is most helpful to explain what "states" of matter are by simply describing the three states in which water and many materials exist: solid, liquid, and gas. In the special case of water that would be ice, water, and steam or water vapor. Not all materials come in three conditions or states. Wood, for instance, is either solid or it is gaseous; there is no liquid state. Iron, while solid at normal earth temperatures, can be melted into a liquid with the application of heat energy and also turned into a gas with the application of still more heat energy.

Water is the only substance that exists on earth in all of its states at normal earth temperatures.

At temperatures below 32° Fahrenheit (F) water is in its solid form. Between the temperatures of 32°F and 212°F it is a liquid, and above 212°F it becomes gaseous (water vapor or steam). These are the temperatures at which freezing or melting takes place. Pressure can affect the freezing, melting, or condensation temperatures of a substance. For instance, on a mountain top, water will boil (begin to turn gaseous) at a lower temperature than 212°F. To cooks this means that boiled foods must be cooked longer at higher altitudes. For this unit we will assume that we are at sea level with ambient pressure of 14.7 psi.

Water has the curious property of expanding when it freezes. There is almost no limit to the power it will exert outward as it does so. Pipes and other closed containers must either expand as the water freezes or they will rupture, causing all matter of havoc.

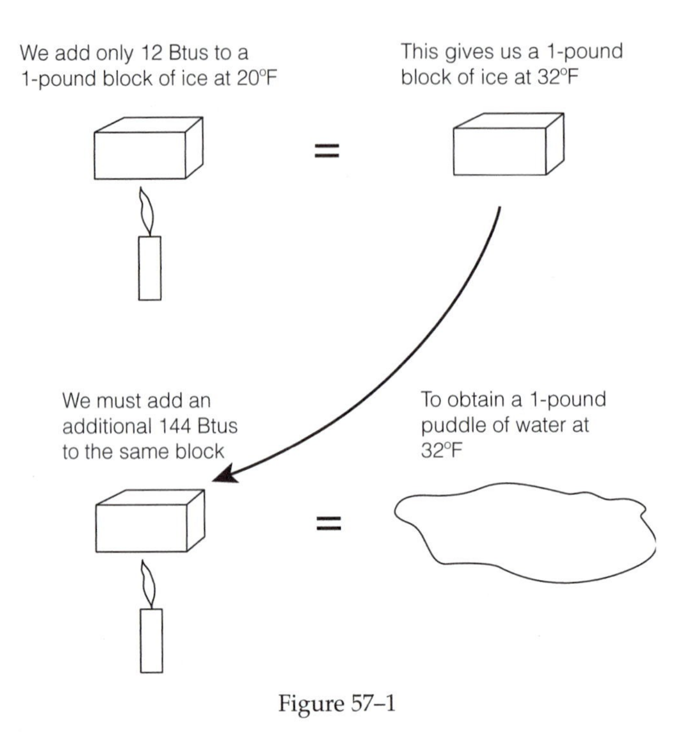

Figure 57–1

An interesting and sometimes valuable property of all materials is that they do not change temperature in an even fashion. Normally, one British thermal unit (Btu) will raise the temperature of one pound of water 1° Fahrenheit. But let's take a pound of ice in a bucket, put a thermometer through the side down low and into the ice, and begin applying heat to the bucket. We have a special insulated electric bucket that has a gauge telling us the number of Btus that are being applied. As the one pound of ice warms up from, let's say, 20°F to 32°F, it absorbs one Btu for each degree that it warms.

However, to simply change from ice at 32°F to water at 32°F it must absorb an additional 144 Btu. This is water's *latent heat of fusion* (Figure 57–1). Once it has changed to water and we keep supplying heat and the liquid water comes to 212°F and begins to boil, the temperature in the bucket will not raise another degree until all of the liquid water in the bucket has boiled away. This is called the *latent heat of vaporization*.

In the refrigeration and air conditioning cycle, a gas is repeatedly liquified and allowed to return to its gaseous state to take advantage of its latent heat of vaporization (see Figure 57–2).

Latent heat is the heat necessary to change the state of a material without changing its temperature.

It is important to understand that for steam at 212°F to turn to water at 212°F (condense), it *must* give off 970 Btus of its latent heat. For water at 212°F to turn to steam at 212°F, it *must* absorb 970 Btus of heat from some outside source. Water at 32°F *must* lose 144 Btus per pound before it can turn to ice at 32°F. Ice *must* gain 144 Btus per pound before it can turn to water at 32°F.

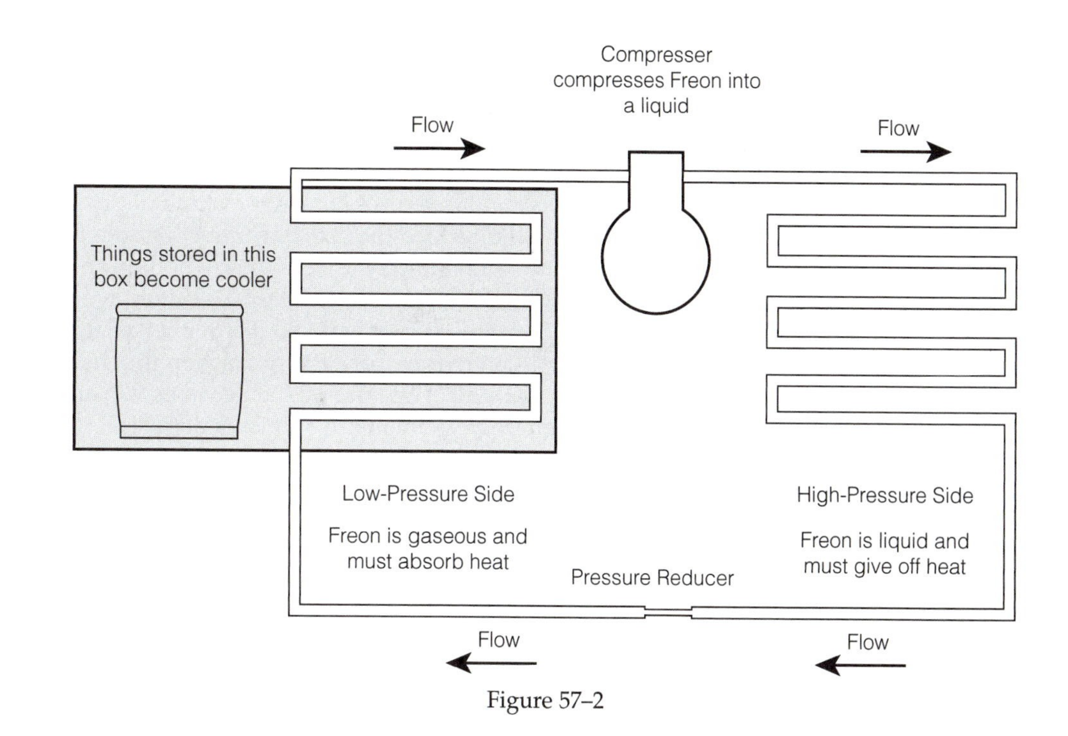

Figure 57–2

Sample Problem How many Btus of heat must I apply to 10 pounds of ice at 20°F to obtain 10 pounds of water for tea at 200°F?

Answer

Tea = Number of pounds of ice times the number of degrees that the water must be raised to get to the melting point of ice PLUS the number of pounds of ice times the latent heat that must be added per pound to change ice to water PLUS the number of Btus that must be applied to 10 pounds of water at 32°F to bring it up 200°F.

$$\text{Tea} = (10 \times 12) + (10 \times 144) + (10 \times 168)$$

$$\text{Tea} = 120 + 1440 + 1680 = 3240 \text{ Btus}$$

Scientific Heat Notation

It has been the trend to talk about heat transfer in metric terms. With global free trade, more heating and cooling equipment is being received with metric specifications. The pipe and heating trades student should become familiar with these terms.

You may recall that it takes 1 Btu to raise the temperature of 1 pound of water 1° F., and that water freezes at 32°F and it boils at 212°F. Another way of describing this increase is with calories, which is the unit of heat equal to 1/100 the quantity required to raise the temperature of 1 gram of water from 0°C to 100°C.

Converting Temperature Readings

There are 212 – 32 = 180 Fahrenheit degrees between freezing and boiling temperatures for water, but the metric degree centigrade scale has only 100 of its degrees between freezing and boiling. The freezing point of water is 0°C and boiling is 100°C. So we have a ratio here of 1.8:1 between the size of the Celsius (centigrade) and the Fahrenheit increment on the thermometer scale. The ratio 180:100 is no different from these ratios: 1800:1000 or 1.8:1. Multiplying or dividing both sides of a ratio by the same number results in an equivalent ratio, and dividing both sides of 180:100 by 100 results in a convenient equivalent ratio of 1.8:1. Having a 1 on one side of the ratio allows us to make "for each" comparisons, such as "There are 1.8°F for each degree C."

Ratios can also be expressed like a fraction: $\frac{3}{4} = 3:4$

or even 180/100, which could express that for every 180 of x there are 100 of y. As has been stated, we could reduce the size of numbers in the ratio by dividing both sides of the ratio by the same number. Let's divide both numbers of our ratio by 2: 180/2 = 90 and 100/2 = 50. Now we have 90:50. So now we have learned that for every 50 centigrade degrees there are 90 Fahrenheit degrees. Something else becomes apparent. We can reduce our ratio or fraction much more by dividing both the top and the bottom by 10. Then our 90/50 becomes 9/5 and now we know that for every 9 Fahrenheit degrees there are 5 Celsius degrees.

Perhaps we could say that C = 5/9 F and be able to convert to Celsius from Fahrenheit. If that's the case, we should be able to use 32°F (the freezing point of water Fahrenheit) and obtain 0°C (the freezing point of water on the centigrade scale).

This could be wrong. $C = 5/9 \times 32/1$ then $C = 160/9$ and then $C = 17.7$

The centigrade scale starts at freezing with 0°C for the freezing point of water, while Fahrenheit starts at an arbitrary 0°F point and has 32°F as the freezing point of water. This must be allowed for when doing conversions. We didn't allow for those 32 degrees that Fahrenheit water freezing starts with.

$$\mathbf{C = (F - 32) \times 5/9}$$

$$C = (32 - 32) \times 5/9$$

$$C = 0 \times 5/9$$

$$C = 0$$

Now it works, and the derived formula below should be okay for finding the Fahrenheit temperature. We know that water boils at 100°C and at 212°F. So suppose we test the formula below with 100 Celsius degrees and see if the answer is 212?

$$\mathbf{F = (C \times 9/5) + 32}$$

$$F = (100 \times 9/5) + 32$$

$$F = (100 \times 9/5) + 32$$

$$F = 180 + 32 = 212$$

Sample Problem What is the degree reading in centigrade for 0° Fahrenheit?

Answer

$$C = (F - 32) \times 5/9$$

$$C = (0 - 32) \times 5/9$$

$$C = -32 \times 5/9$$

$$C = -17.8°$$

Exercise

1. As the temperature goes down, at what temperature does water begin to turn to ice?

   a. 0° Fahrenheit
   b. 32° Fahrenheit
   c. 32° centigrade
   d. 212° centigrade

2. How many Btus will have to be applied to change one pound of ice at 0° to water at 100°F?

   1. 100 Btus
   2. 144 Btus
   3. 32 Btus
   4. 276 Btus

3. How many Btus are required to change one pound of water at 212°F to steam at 212°F?

   a. 212
   b. 970
   c. 144
   d. 244

4. How many Btus must be added to water at 32°F to get it to the boiling point?

   a. 180
   b. 212
   c. 244
   d. 1182

5. What percentage of the heat applied to a 1-pound block of ice at 0°F to turn it completely into steam is not used to increase the temperature?

   a. 4%
   b. 34%
   c. 64%
   d. 84%

In the following problems, change centigrade to Fahrenheit or Fahrenheit to centigrade.

6. 100°C ________________
7. 32°C ________________
8. 0°F ________________
9. 32°F ________________
10. 212°F ________________

Unit 58 Properties of Gases

Objectives

- Use Boyle's Law.
- Understand the properties of common gases found on the job.
- Compute propane cylinder gas quantities.
- Solve gas mixture problems.

Plumbers and pipefitters install and repair gas lines of all kinds. They also use compressed gases to apply heat to solder joints. Gases are also present in storage tanks and in pipelines for heating and medical purposes. Some gases, like natural gas, are lighter than air and will rise. Therefore, they are easy to remove from an enclosed area by simply opening windows and doors. But others, like propane, will flow downward filling cellars and trenches until perhaps they encounter a source of heat hot enough to ignite. Because some of these gases are toxic and flammable, the student should understand safe handling. The oxygen content in the air we breathe, for instance, is 21%. If that percentage should reach 25%, some common substances such as lubricating oil can spontaneously ignite. Nothing containing oil should be in close contact with oxygen supply. Oxygen in compressed form is found in medical care facilities and in garages and metalworking establishments.

Diesel fuel, gasolene, naphtha, and alcohols—while liquids at normal temperatures—continuously shed molecules into the air from their exposed surfaces. These molecule clouds are the gaseous form of these liquids.

Boyle's Law

Boyle's Law makes the assumption that the temperature remains constant at all times.

The law states that volume is inversely proportional to pressure, and this can be expressed by the following equation:

$$P\text{Initial} \times V\text{Initial} = P\text{Final} \times V\text{Final},$$

where *P*Initial is the initial pressure and *V*Initial is the initial volume and *P*Final is the final pressure and *V*Final is the resulting volume. And so, with a little rearranging, we can say that:

$$P\text{Final} = \frac{P\text{Initial} \times V\text{Initial}}{V\text{Final}} \quad \text{and} \quad V\text{Final} = \frac{P\text{Initial} \times V\text{Initial}}{P\text{Final}}$$

From this, we may deduce that if a hypothetical balloon with a certain amount of helium at one atmosphere of pressure (14.7 psi) was placed into a pressure chamber and the pressure on the balloon was doubled, the size or volume of the balloon would reduce by one-half.

Sample Problem

Let's assign an initial volume of 100 cubic inches to the balloon and test that idea. Use the situation in the previous paragraph.

Solution:

$V\text{Final} = P\text{Initial} \times V\text{Initial}/P\text{Final}$

$V\text{Final} = 14.7 \times 100/29.4$

$V\text{Final} = 1470/29.4$

$V\text{Final} = 50$ cubic inches

Note: Our hypothetical balloon has limitless elasticity; therefore, the pressure on the inside of the balloon and the pressure on the outside will always be the same. As the 29.4 psi is pumped into the air chamber, the molecules of the gas in the balloon press closer together and increase the internal pressure to exactly match the balloon's external pressure. The balloon delineating the new volume of the helium shrinks to half its former size.

Gases Under Pressure

Natural Gas

Natural gas is delivered through pipelines at pressures of up to 1000 psi and at speeds of 25 miles per hour. As it nears the point of use the pressure is reduced until, at the home or place of business, the pressure is only 0.25 psig. It is non-toxic, meaning that breathing it will not poison humans. However humans require oxygen, and natural gas contains no oxygen. A heavy concentration of natural gas is dangerous. In its natural state, it is also odorless. The familiar smells of both natural gas and propane (also known as LPG or just LP-gas) are introduced to make their presence easily detected.

> **Psig means *pounds per square inch gauge*. For example, there is still 14.7 psi gas product in the tank or pipeline at 0.0 psig.**

Natural gas will only ignite when there is an air-to-gas mixture of between 5 and 15 percent gas. Any air mixture with less than 5 percent or more than 15 percent natural gas will not ignite. At 48°F, air weighs 1.25 ounces per cubic foot while natural gas weighs 0.81 ounces.

Sample Problem

How many cubic feet of natural gas will make a 1200-cubic-foot room hazardous?

Solution:

The minimum percentage of natural gas needed to support combustion is 5 percent. The formula to solve this follows:

RCF = cubic feet in the room

GCF = cubic feet of contaminant gas

TP = ignition threshold percentage in the form: 5% = 0.05

GCF = RCF × TP

GCF = 1200 × 0.05

GCF = 60 cubic feet of natural gas is 5% by volume

Note that gases and some liquids combine by sharing space. In other words 1200 cubic feet of air plus 60 cubic feet of natural gas will not require 1260 cubic feet of space.

Sample Problem

At 0.25 psig, what is the absolute pressure?

Solution

At sea level with average barometric pressure, the absolute pressure is 14.95 psi; above the ocean of air around us, 0.25 psi.

Propane, LPG, and LP-gas

Propane, LPG, and LP-gas are the same fuel product. LPG stands for *liquified petroleum gas*. It is distributed under pressure in steel tanks. At normal temperatures but highly compressed, the gas becomes a liquid; its *state* is changed from a gas to a liquid. When the valve is opened, the pressure drops from tank pressure to atmospheric pressure. The liquid must absorb heat from the surrounding area. If the liquid touches human skin, frostbite may result. Fortunately many tanks are equipped with a ball check valve that will block the gas's passage if outlet pressure suddenly drops, such as if a hose ruptures. Propane is heavier than air. As a gas it weighs 1.90 ounces per cubic feet while air weighs 1.25 ounces per cubic foot. It is neither poisonous nor is it harmful to vegetation or water, but under certain circumstances it can explode and it can asphyxiate.

Propane is generally measured by weight. As was mentioned, it is supplied in a steel tank under pressure. When full, the propane inside the tank is nearly all liquid. As propane is used, the space over the top of the liquid in the tank is occupied with propane gas, but the pressure remains nearly the same until much of the propane has been removed.

The 20-pound propane tank is a common size, used for barbeque grills, RV fuel, and heating torches. The full propane tank will weigh about 42–44 pounds. The empty propane tank will weigh about 22 pounds; this is called the *tare weight*. You will find the tare weight of the tank stamped into the steel safety collar of the cylinder. Note that tare weight plus 20 pounds of propane equals 42 pounds.

Propane is also available by the gallon. A gallon of propane weighs 5.1 pounds and so a 20-pound propane tank should hold approximately 4 gallons.

Sample Problem

If my propane tank with a tare weight of 22 pounds weighs 34 pounds, how much fuel is left in the tank?

Solution

P = propane

Tare*W* = tare weight

Tank*W* = tank weight with propane in it

P = Tank*W* − Tare*W*

P = 34 − 22 = 12

P = 12 pounds of propane left

Sample Problem Using the data from the previous problem, how many gallons of propane would be left in the tank?

Solution

PG = PP/PW

PG = total gallons of propane

PP = pounds of propane = 12

PW = weight of one gallon of liquid propane = 5.1 pounds

PG = 12/5.1

PG = 2.35 gallons

Exercise

1. What is the percentage difference between the amount of oxygen in the air we normally breathe and the amount of oxygen in the air that can support spontaneous combustion?

   a. 21% b. 14%

   c. 9% d. 4%

2. We know that gasoline is a liquid. Can it also be a gas?

   a. Yes b. No

3. What is normal atmospheric pressure at sea level?

   a. 0.25 psi b. 5.12 psi

   c. 14.7 psi d. none of the choices

4. Boyle's Law makes what preliminary assumption?

   a. That communication satellites have no atmosphere.

   b. That the temperature remains constant.

   c. That balloons are hypothetical.

   e. That gas is lighter than air.

5. If a balloon with 200 cubic inches of helium in it was placed in a pressure vessel and the pressure on the outside of the balloon was reduced from 14.7 psi to 5.0 psi, the size of the balloon would

   a. expand to 588 cubic inch.

   b. reduce to 50 cubic inch.

   c. expand to 2940 cubic inch.

   d. stay precisely the same.

6. What pressure must be added to a standard pressure gauge reading to obtain absolute pressure at sea level?

   a. 21 psi b. 4 psi

   c. 0.25 psi d. 14.7 psi

7. If a diver took a 200-cubic-inch balloon down on a 60-foot dive, would the balloon get bigger or smaller?

   a. bigger b. smaller

8. If the unlit gas range was left on in a 3840-cubic-foot apartment and three hours later the apartment exploded, what was the probable hourly rate in cubic feet per hour of the natural gas leakage?

   a. 192
   b. 64
   c. 3
   d. none of the answers provided

9. Tare weight means

   a. the weight of the fuel to be added to the tank.
   b. the weight of the fuel already in the tank.
   c. the weight of the tank.
   d. the weight to be allowed for accidental seepage.

10. If a 22-pound propane tank has two gallons of propane in it, what should it weigh?

    a. 32.2 pounds
    b. 22 pounds
    c. 38 pounds
    d. 14.7 pounds

Section Six
HEATING

Unit 59
Heat Loss vs. Radiator Size

Objectives

- Describe heat loss theory.
- Define the British thermal unit and its use in heat loss.
- Compare steam heating to hot water heating.
- Use a heat loss chart to solve a problem.

Radiator and boiler size for heating are not difficult to compute. The principle is to balance the heat loss with the heat supply. While the principle and type of calculation are easily understood, the variations are many. The unit on heating presented here will illustrate the method based on a circulating hot water system using the baseboard-type radiation. A study such as this should provide basic information and a starting point for other radiation systems.

Manufacturers and distributors of heating equipment have detailed information for heat calculation. They will supply information and even the service of a heating expert for most installations. The calculations of radiation require specific information. To know how the calculations are made is an aid to supplying the necessary data about the building to be heated.

The unit of heat is the British thermal unit (Btu). It is the amount of heat that can raise the temperature of one pound of water one degree Fahrenheit.

Heat is lost from a room through any area exposed to a lower temperature. It is also lost by the heating of incoming air due to infiltration and to ventilation. Different materials and different types of construction vary the rate of heat loss.

The steam radiator will supply 240 Btu per hour for each square foot of radiator surface. The number of Btu per hour per square foot from radiator or baseboard radiation with circulating hot water heat depends on the design of the system. Boiler pressure, water temperature, and rate of flow are design factors. Hot water heating has radiation from 150 to 240 Btu per hour per square foot. The problems of this text will use 200 Btu per hour per square foot for baseboard-style radiation. A solar heat collector does not readily produce high temperatures. Thus a hot water system with solar heat might have radiator size based on no greater than 150 Btu per hour per square foot.

Boiler capacity is measured in Btu per hour. Thus a boiler size is the same for steam or hot water and for different radiator rates.

This table shows the heat loss per square foot per hour for one degree Fahrenheit of temperature difference.

HEAT LOSS PER SQUARE FOOT PER HOUR
Btu Heat Loss for Various Standard Constructions—Walls, Ceilings, and Windows

	Btu Per Degree Fahrenheit			
WALL CONSTRUCTION	No Insulation	½" Batts	3⅝" Batts	6" Batts
Frame—Siding or shingles, building paper, sheathing, plaster, or wall board	0.26	0.20	0.08	0.045 (2×6 studs)
Frame—Brick veneer, building paper, sheathing, plaster, or wall board	0.28	0.21	0.08	
8" brick—1" furring, lath, and plaster	0.32	0.22		
8" hollow tile or cinder block, 1" stucco finish	0.37			
4" brick and 8" tile or cinder block	0.34			
12" concrete, plain	0.57			
CEILING CONSTRUCTION				
Ceiling—½" plaster board Flooring on joists	0.25	0.10	0.07	0.05
Ceiling—lath and plastered Unfloored	0.70	0.26	0.10	0.05
WINDOWS AND DOORS				
Single glass—1.13 With storm windows—0.75 Double glass with sealed space between—0.65 Doors, with or without glass, same as windows Glass block, 3⅝" thick—0.49				

Sample Problem A single glass window is 3' 0" × 4' 0". The outside temperature is 0° F and the inside temperature is 70° F.

a. How much heat is lost through the window?

b. How much hot water radiation is needed to offset the heat loss?

Solution

a. Heat loss = constant × temperature difference (°F) × area (sq ft)

Heat loss = 1.13 × 70 × 12

Heat loss = 949.2 Btu

b. 949.2 Btu ÷ 200 Btu/sq ft = 4.746 = 4.75 sq ft of radiation

Exercise

For each of the following, compute the heat loss and the square feet of hot water radiation to supply heat equal to the loss.

1. A single glass window 3' 6" × 4′ 0″ with temperature 75° inside and 0° outside.
2. A double glass (storm window) 2' 9" × 4' 6" with 70° inside and 0° outside.
3. A ceiling, no insulation, 20' 0" × 12' 6" with 70° temperature difference.
4. A ceiling with 3⅝" of rock wool, 32' 0" × 14' 9", with 70° temperature difference.
5. A 12" thick plain concrete wall, no openings, 18' 0" × 8' 0", with 70° inside and 10° outside.

Unit 60

Radiation Sizing for Total Heat Loss of a Room

Objectives

- Learn the principles for "heat lossing" a space.
- Define infiltration heat loss.
- Calculate radiator size for a floor of rooms.

A single radiator is large enough to equal several heat losses in a room. The usual heat losses are through walls, through windows and doors, and in heating cold air due to infiltration and for ventilation. The ceiling is another source of heat loss unless there are heated rooms above. Heat loss through the floor may be omitted when there is a full basement with a boiler located in it.

Since most houses have all the rooms heated, it is necessary to include only outside walls in the total wall area. The sample problem shows a method for grouping heat losses in a room.

The table shows infiltration heat loss per cubic foot for each degree Fahrenheit temperature difference.

HEAT LOSS PER CUBIC FOOT PER HOUR

Btu Heat Loss to Incoming Air by Infiltration for Various Window Construction and Walls Exposed

	WEATHER STRIP OR STORM SASH	
EXPOSURE	**No**	**Yes**
An outside wall of room	0.018	0.009
Two walls	0.027	0.014
Three walls	0.036	0.018
Entrance halls	0.036	0.018
Sun rooms	0.054	0.027

Sample Problem

Determine the heat loss and the hot water radiator size for a dining room in a one-floor house. The room is 11' 2" × 11' 2" × 8' 0". There are two outside walls, each 11' 2" × 8' 0", and two windows, each 1' 6" × 4' 6". The windows are double glass with sealed space between. The walls are framed with wood siding and 3⅝" batts. The ceiling has 6" batt insulation. The temperature difference is 70° F.

Solution

Dining Room Room dimensions 11' 2" × 11' 2" × 8' 0"

	Calculation	Btu Loss
Volume cubic feet	998.16 × 0.014 × 70	978
Total wall area	178.72	
Less: window and door areas	22.5 × 0.65 × 70	1024
Net wall area	156.22 × 0.08 × 70	875
Square feet of ceiling	124.77 × 0.05 × 70	437
Total		3314
Square feet of radiation	3314 ÷ 200 = 16.57	

Exercise

For each room listed, follow the procedure as in the sample problem. Compute the heat loss and the hot water radiator size for each of the following locations in the one-floor house. The walls are standard frame construction with brick veneer. There is 3⅝" batt wall insulation and 6" batt ceiling insulation. The windows are double glass with sealed space. The temperature difference is 65°F.

1. Bedroom
2. Bathroom
3. Entry space
4. Living room including hall space
5. Kitchen-dining room including the side entrance space

The stairwell heat comes from the basement.

DOOR AND WINDOW SCHEDULE			
A	3' 0" × 3' 3"	D	7' 0" × 5' 4"
B	6' 0" × 6' 8"	E	2' 0" × 3' 3"
C	3' 0" × 6' 8"	F	5' 0" × 3' 3"

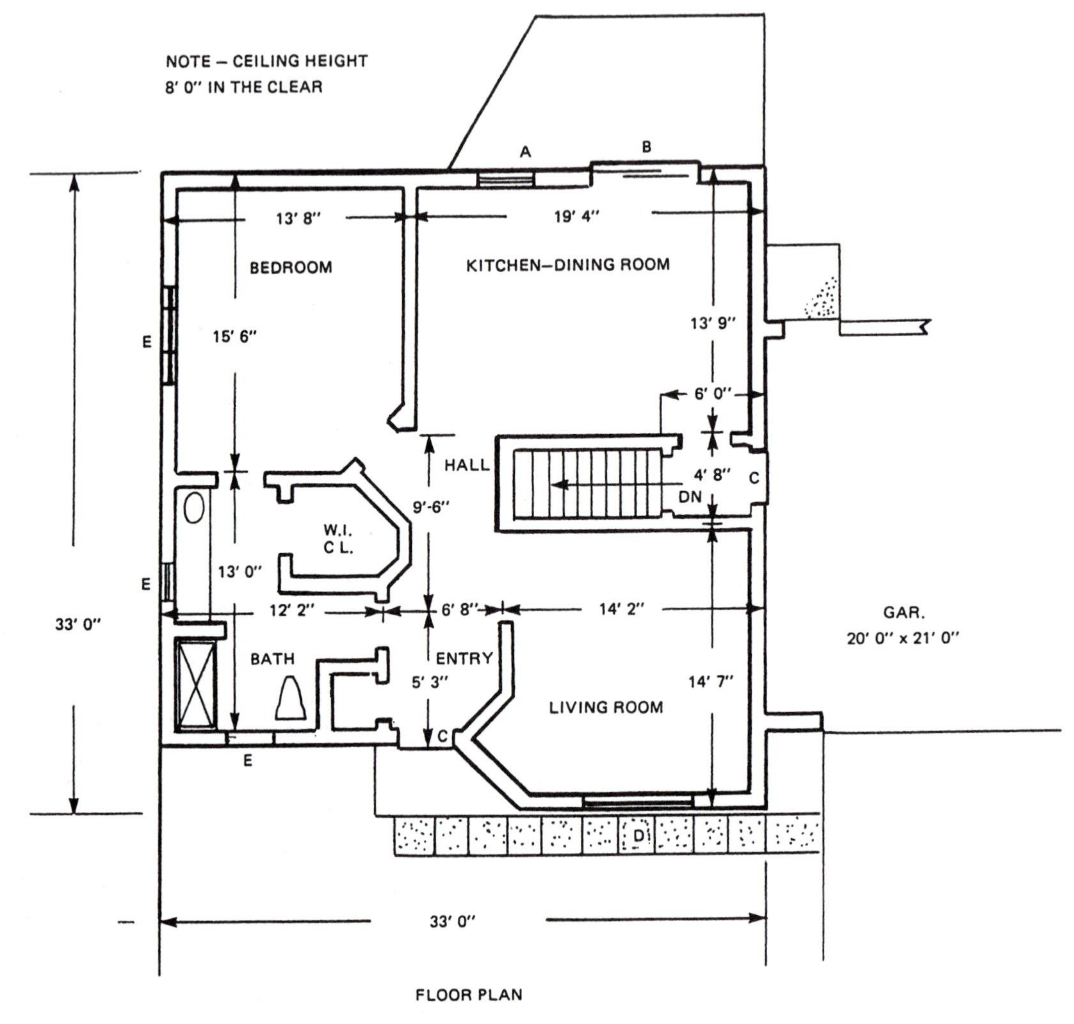

FLOOR PLAN

SINGLE-FLOOR HOUSE

SCALE 1/8" = 1' 0"

Unit 61

Estimating Size of Piping

Objectives

- Allow for restriction of flow by fittings.
- Use the flow method to size pipe for heating.
- Use pipe charts to allow for "flow need."

The one-pipe circulating hot water system is one of several piping systems and is used for home heating. The one pipe starts at the outlet tap on the boiler, goes around the basement near the ceiling and close to the walls, and then returns to the boiler at the inlet tap. There is a supply and a return connection for each radiator from and to the one-pipe system. The tee fittings have a partial restriction to aid hot-water flow through the radiator. The smaller supply and return lines are designed and installed with short runs and few fittings. An air cushion tank holds the closed system to about 12 psi so that the water can be heated to above 210°F without boiling. The circulating pump gives a difference of head within the heating system to cause a circulation flow. One or more thermostats control the water temperature and the circulation within the system.

The steel pipe or copper tube size is found from three factors of the one-pipe system. These factors are used with a table (Table 61–1) to give a pipe size for the main circuit and for each radiator. The factors are considered as follows:

a. The Circulation Head

   The pump head, in feet of water, is kept low to reduce the power needed. There are conditions where a higher head permits a smaller pipe size, which reduces installation cost. The problems of this text use six feet of head of water for circulation.

b. The Rate of Flow in Gallons Per Minute (GPM)

   The planned temperature drop is often 20°F. This is a good choice for a house of small size. For larger installations a temperature drop of 30°F and even 40°F can permit a smaller pipe size.

$$\text{Flow (GPM)} = \frac{\text{Btu/hr}}{\underset{(\text{Btu/lb})}{20} \times \underset{(\text{lb/gal})}{8.33} \times \underset{(\text{min/hr})}{60}}$$

Sample Problem

$$\text{Flow} = \frac{\text{Btu/hr}}{10{,}000}$$

What flow rate is needed to supply a hot water radiator with 659 Btu/hr?

Solution

$$\text{Flow} = \frac{659}{10{,}000} = .0659 \text{ GPM}$$

c. Length of pipe

The length of pipe is best found by making a pipe diagram. However, for the full basement one-pipe system, the length of pipe will almost equal the perimeter of the building plus 12 feet for each radiator.

Sample Problem Determine the length for a one-pipe hot water heating system for a house 40' 0" × 24' 0" that is to have 7 radiators.

Solution

$$\text{Length} = 2\,(l + w) + 12n$$

$$= 2\,(40 + 24) + 12 \times 7$$

$$= 128 + 84 = 212 \text{ ft}$$

The circulation head, pipe length, and flow rate are used in Table 61–1 to determine the pipe or tube size for the circuit. The sizes of supply and return branches for each.

TABLE 60-1
PIPE SIZING TABLE FOR ONE-PIPE SYSTEM

Available Head in Ft. of Water	TOTAL LENGTH OF CIRCUIT Measured length of circuit (not including branch piping) plus an allowance of 12 feet for each heat-distributing unit connected to circuit.															
4	35	45	50	60	65	70	75	80	90	100	110	130	150	180	220	290
5	45	60	65	70	80	90	95	100	120	130	140	160	190	230	290	360
6	55	70	80	90	100	110	120	130	140	160	180	200	240	290	350	450
7	65	90	100	110	120	130	140	150	170	190	210	240	290	340	420	540
8	75	100	110	130	140	150	160	180	200	220	250	290	330	400	490	620
9	85	110	130	150	160	170	190	200	230	250	290	330	380	450	560	710
10	100	130	140	170	180	190	210	230	260	290	320	370	430	510	620	790
11	110	140	160	190	200	220	240	260	290	320	360	410	480	570	690	880
12	120	160	180	200	220	240	260	290	320	350	400	450	530	620	760	960
14	150	190	210	250	260	290	310	340	380	420	470	540	620	730	900	1120

Pipe or Tubing Sizes: Trunk or Circuit	Branch	GALLONS PER MINUTE CAPACITIES OF TRUNK OR CIRCUIT AND BRANCHES Pipe or Copper Tubing with One-Pipe Fittings (except ³/₈" branches)															
¾"		5.0	4.3	4.1	3.8	3.7	3.6	3.4	3.2	3.1	2.9	2.8	2.6	2.4	2.2	2.0	1.8
	⅜"*	0.9	0.8	0.8	0.7	0.7	0.7	0.7	0.7	0.7	0.4	0.4	0.4	0.4	0.4	0.4	0.3
	½"	1.8	1.5	1.5	1.2	1.2	1.2	1.2	1.2	1.2	0.8	0.8	0.8	0.8	0.8	0.8	0.6
1"		9.6	8.3	7.7	7.3	7.0	6.8	6.5	6.3	5.9	5.7	5.5	5.0	4.6	4.3	3.8	3.4
	⅜"*	0.7	0.7	0.6	0.6	0.6	0.5	0.5	0.5	0.4	0.4	0.4	0.4	0.3	0.3	0.2	0.2
	½"	1.4	1.3	1.1	1.1	1.1	1.0	1.0	1.0	0.8	0.8	0.8	0.8	0.7	0.7	0.5	0.5
	¾"	2.4	2.1	1.9	1.9	1.9	1.6	1.6	1.6	1.3	1.3	1.3	1.3	1.1	1.1	0.9	0.9
1¼"		..	18	17	16	15	15	14	14	13	12	11	11	9.7	9.0	8.3	7.3
	⅜"*	..	0.9	0.9	0.8	0.8	0.8	0.7	0.7	0.7	0.7	0.6	0.6	0.4	0.4	0.4	0.3
	½"	..	1.8	1.8	1.6	1.5	1.5	1.4	1.4	1.3	1.2	1.1	1.1	0.9	0.9	0.8	0.7
	¾"	..	2.9	2.9	2.6	2.4	2.4	2.3	2.3	2.1	2.0	1.8	1.8	1.5	1.5	1.3	1.2
1½"		..	..	25	24	23	22	21	20	19	18	17	16	15	13	12	11
	⅜"*	..	..	1.0	1.0	1.0	0.9	0.9	0.8	0.8	0.8	0.7	0.7	0.7	0.6	0.5	0.4
	½"	..	..	2.0	2.0	1.9	1.8	1.7	1.6	1.6	1.5	1.4	1.3	1.2	1.1	1.0	0.9

Sample Problem (continued)

radiator are determined from Table 61–1 with regard to trunk or circuit size. The use of Table 61–1 is shown in the solution to the sample problem.

A one-floor house is to have a one-pipe hot water heating system. The circulation pump head is six feet. The pipe length is 212 feet. The circulation flow is 1.7 GPM.

a. Determine the pipe or the tube size for the circuit.

b. What size supply for a radiator with a flow of .07 GPM?

c. What size supply branch for a radiator with a flow need of .47 GPM?

Solution

Using Table 61–1, start with the six-foot head, in upper left, follow across to 240 feet, since 210 feet is less than the length of 212 feet. Follow down from 240 feet to the lower part of Table 61–1. A flow rate of 2.6 GPM will satisfy the 1.7 GPM needed. Follow to the left to find the circuit pipe or tube size.

a. From the left column find a ¾" circuit size.

b. For .07 GPM the ⅜" tube would be used. Note that the ⅜" tube has a flow capacity of .4 GPM in relation to the ¾" circuit size.

c. For .47 GPM a ½" branch is needed.

Exercise

Item	Heat Loss (Btu)	Flow (GPM)	Head (Ft Water)	Pipe Length (Feet)	Pipe Size	
					Circuit	Branch
Total			6			X
Bedroom			6	X	X	
Bathroom			6	X	X	
Entry			6	X	X	
Living Room			6	X	X	
Kitchen			6	X	X	

Unit 62 Sizing Ventilation

Objectives

- Size ventilation for commercial buildings.
- Calculate pulley sizes to adjust blower outputs.
- Learn circulating and fresh air requirements for health.

Ventilation Principles

Ventilation is a very important factor in human health. From six to nine complete air changes per hour in a living or working space are necessary to avoid health problems. Some of this air flow should be fresh and re-oxygenated outside air. An amount equal to one-tenth to one-third of the total flow of air should be fresh outside air. This principle is often overlooked in homes and workplaces for the sake of economy. Costly heating and cooling energy must be used on this outside air to bring its temperature in line with building thermostat settings.

Domestic (home) space heating often relies on crackage (small fitting gaps around doors and windows) to re-supply fresh air to the living spaces. Boilers and furnaces that do not have an outside air supply consume living-space air to maintain the combustion. This creates a small negative pressure within the building. The outside air pressure, now being higher, forces outside air into the building through fitting gaps throughout the building.

Buildings with free public access or a large number of employees often have access doors being more or less continually opened and closed. This will admit outside air and this usage should be taken into account. How much or how little to allow is often a matter of experienced judgment.

This air should be filtered in cities to remove suspended particles. It should also be possible to shut off the incoming fresh air for short periods of time if particularly bad odors or noxious fumes are present on the outside.

Mathematical Factors in Ventilation

Blowers and ventilation fans (air handlers) are rated in cubic feet per minute (CFM).

Humans require at least six air changes per hour.

Heavy smoking and other air pollutants quintuple (multiply by 5) the fresh air requirement.

The volume of the building or space to be ventilated must be calculated in cubic feet.

Air Changes Per Hour

A rule of thumb is that six air changes per hour should be provided for comfortable and healthy human working conditions. All of the air in the building will pass through a filter at each designed air change, or six times per hour. A ventilating air handler will be responsible for moving all of the air through that filter the required number of times per hour. This is the air handler that we will be sizing in the following paragraphs.

Keep in mind that another air handler will probably be supplying one air change per hour of outside air to the same space. This air will be added to the building's circulating air. Outside air coming in may be preheated or cooled, thereby saving energy, by a careful arrangement of the outgoing stale air duct with respect to the incoming air duct.

Sample Problem

One of the preliminary problems that the designer might look at is the minimum required size of the air handler for a small building.

For a two-story building with a 50' × 25' floor plan and 8-foot ceilings,

> Calculate overall floor plan area. 50' × 25' × 2 = 2500 square feet. The 2 is used because this is a two-story building.
>
> To get the volume needed, multiply the total floor area by the ceiling height. 2500 × 8 = 20,000 cubic feet.
>
> Since the air must be circulated six times every hour, the handler must move six times this quantity every hour, or 20,000 × 6 = 120,000 cubic feet per hour.
>
> Air handlers are rated in cubic feet per minute, however; so divide the volume required, 120,000, by 60 minutes per hour. 120,000/60 = 2,000 cubic feet per minute (CFM). The air handler's rating must be 2,000 CFM.

Note: In larger buildings, allowance must be made for "friction loss," the drag on the flow of air induced by duct-work and fittings. For domestic use (use in the home) the air handler usually does not run continuously. It is expected that windows will be opened when the air feels "stuffy." Smaller commercial buildings may rely on overhead vent openings and exposed fans to circulate fresh air.

Larger buildings must have a continuously operating air handling system to supply air-conditioned (air that is cleaned, heated and/or cooled) air to the rooms because most rooms may not have access to the outside air.

Within reasonable limits, the CFM of most blowers or air handlers may be adjusted by changing the drive pulley sizes in the case of belt-driven blowers, or by connecting the electrical supply to alternate terminals in the case of direct motor-driven blowers. With directly connected blowers the drive shaft of the motor is directly connected to the shaft of the air handler.

The Eureka Blower Company

Air Volume (cfm)	Blower Model Number
500	MBL-05BD
1000	MBL-10BD
1200	MBL-12BD
1400	MBL-14BD
1600	MBL-16BD
1800	MBL-18BD
2000	MBL-20BD
2200	MBL-22BD
2400	MBL-24BD
2600	MBL-26BD

A way to fine-tune the fresh air per person ratio, especially in smaller work environments, is to have a fresh air inlet blower that supplies air at the rates shown in the following table for the smoking and the non-smoking environments. See also a sample table from the Eureka Blower Company.

Fresh Air Inlet Table

Smoking	CFM per person
light or none	5–10
medium	10–25
heavy	25–40

Adjusting the CFM

Adjusting the output of a directly connected blower consists of changing the wire connections according to the manufacturer's specification. This will result in only a limited number of CFM options.

When the handler is belt-driven (Figure 62–1), the speed can be adjusted over a much wider range by changing the pulleys on the driving motor, the blower shaft, or both.

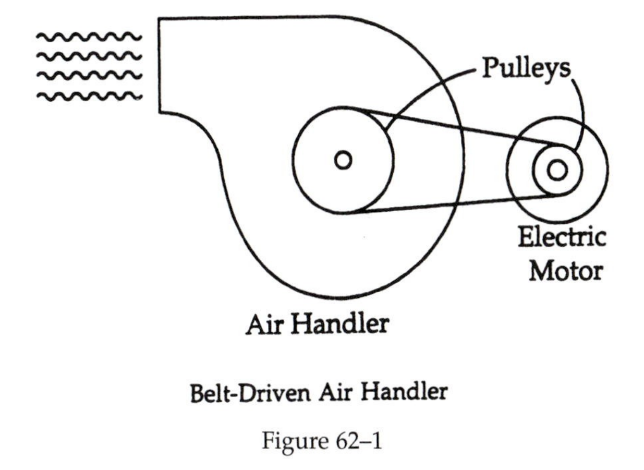

Belt-Driven Air Handler

Figure 62–1

The drive motor will have an RPM (revolutions per minute) rating on a tag which is usually riveted to the motor casing. The equation relating the speeds of the blower and the motor to the respective pulley diameters is as follows:

Blower RPM × Blower Pulley Diameter = Motor RPM × Motor Pulley Diameter

By various re-arrangements of the above equation, three useful formulas can be derived, where:

BRPM = Blower RPM

MRPM = Motor RPM (this will be constant as per name plate)

MPD = Motor pulley diameter

BPD = Blower pulley diameter

$$BPD = \frac{MRPM}{BRPM} \times MPD$$

$$MPD = \frac{BRPM}{MRPM} \times BPD$$

The formula for blower rpm is as follows:

$$BRPM = MRPM \times \frac{MPD}{BPD}$$

Sample Problem

What size pulleys should be used to produce a blower RPM of 2000?

If a BRPM of 2000 is desired and the motor speed is 1750 RPM then:

$$2000 = 1750 \times \frac{MPD}{BPD}$$

Can you see that the value represented by MPD divided by BPD would have to be greater than one? Would the top or the bottom of the fraction have to be larger? If the bottom of the fraction were 1.0, what would the top have to be?

Answers: **yes, the top,** and 2000/1750 = **1.143** and

$$2000 = 1750 \times \frac{1.143}{1}$$

The motor pulley diameter *in this case* should be 1.143 times the diameter of the blower pulley.

Note: As a practical matter, the blower usually runs at a lower RPM than the motor; therefore, the blower pulley diameter is usually larger.

Blower Output (CFM)

The base equation is

$$\frac{CFM_1}{CFM_2} = \frac{RPM_1}{RPM_2}$$

CFM_1 (say "CFM sub one") means *the original CFM* and CFM_2 means *the final CFM.* This equation can be re-arranged into the formula:

$$CFM_2 = CFM_1 \times \frac{RPM_2}{RPM_1}$$

To put this formula in plain English we could say that, "If I double the RPM of the blower, I double the CFM; if I cut the RPM in half, I cut the CFM in half."

A combination and derivative of the two equation sets enables us to directly calculate the blower pulley diameter for a given blower output if we know the old blower output and the old blower pulley diameter.

Changing the blower pulley diameter

$$\text{New BPD} = \frac{\text{Old CFM}}{\text{New CFM}} \times \text{Old BPD}$$

Another more important derivative, because the blower speed is almost always changed by changing the motor pulley size, is the following formula:

Changing the motor pulley diameter

$$\text{New MPD} = \text{New}\,\frac{\text{CFM}}{\text{Old CFM}} \times \text{Old MPD}$$

Note: The blower output cannot be increased without limit. The electric motor will support outputs only up to the point that its horsepower rating will allow.

Review

Note: If the calculated blower size falls between sizes, pick the nearest size in the table. Use the model number in the answer. Don't consider ducting and friction loss here.

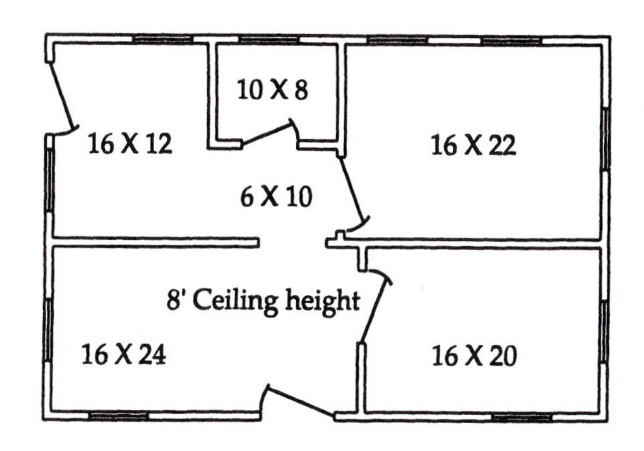

1. Select a blower from the Eureka Blower Company's chart (shown on page 225) for the small office shown for an air exchange rate of 5 times per hour.

   Ans. ______________________

2. Select a blower from the Eureka Blower Company's chart for the small office shown above for an air exchange rate of 9 times per hour.

   Ans. ______________________

3. Select a blower from the Eureka Blower Company's chart for the small office shown above for an air exchange rate of 1 time per hour.

   Ans. ______________________

4. Select a blower from the Eureka Blower Company's chart for the small office shown above for an air exchange rate of 1 time per hour. Include the maximum CFM from the Fresh Air Inlet Table for ten people who are medium smokers.

   Ans. ______________________________

For the following problems fill in the chart, reducing the CFM of the Eureka Company's blowers to provide a half-step between model sizes. The drive motor RPM is always 1750 and the drive motor pulley diameter is always 6 inches. Assume that you can order motor pulleys in decimal sizes, e.g., 2.114".

The Eureka Blower Company

Question	Air volume (CFM)	New air volume (CFM)	New motor pulley size	Blower model number
5	500	250		MBL-05BD
6	1000	750		MBL-10BD
7 & 8	1200			MBL-12BD
9 & 10	1400			MBL-14BD
11 & 12	1600			MBL-16BD
13 & 14	1800			MBL-18BD
15 & 16	2000			MBL-20BD
17 & 18	2200			MBL-22BD
19 & 20	2400			MBL-24BD
21 & 22	2600			MBL-26BD

Unit 63 Heating Problems for a Two-Floor House

Objectives

- Explain a procedure for heat-lossing a whole house.
- Use skills learned so far in a practical application.

In new home construction, the contractor may use a shell-load computation to size the gross heating needs. That is to say, a house is treated as if there were no rooms inside, and only the "shell" of the house—the walls, ceilings, and floors—exposed to outside temperatures is considered. Since the TD or temperature difference between the rooms is equal to zero, there is no heat loss to consider. When the amount of radiation for each room is considered, or if an existing structure is being added to or modified in a substantial way, or the design anticipates varying supply sizes to different rooms, then the contractor will use the room-by-room heat-lossing method.

DOOR AND WINDOW SCHEDULE			
A	3' 0" × 6' 8"	E	5' 0" × 3' 0"
B	9' 0" × 5' 0"	F	6' 0" × 6' 8"
C	7' 6" × 6' 8"	G	3' 0" × 4' 3"
D	1' 6" × 2' 0"		

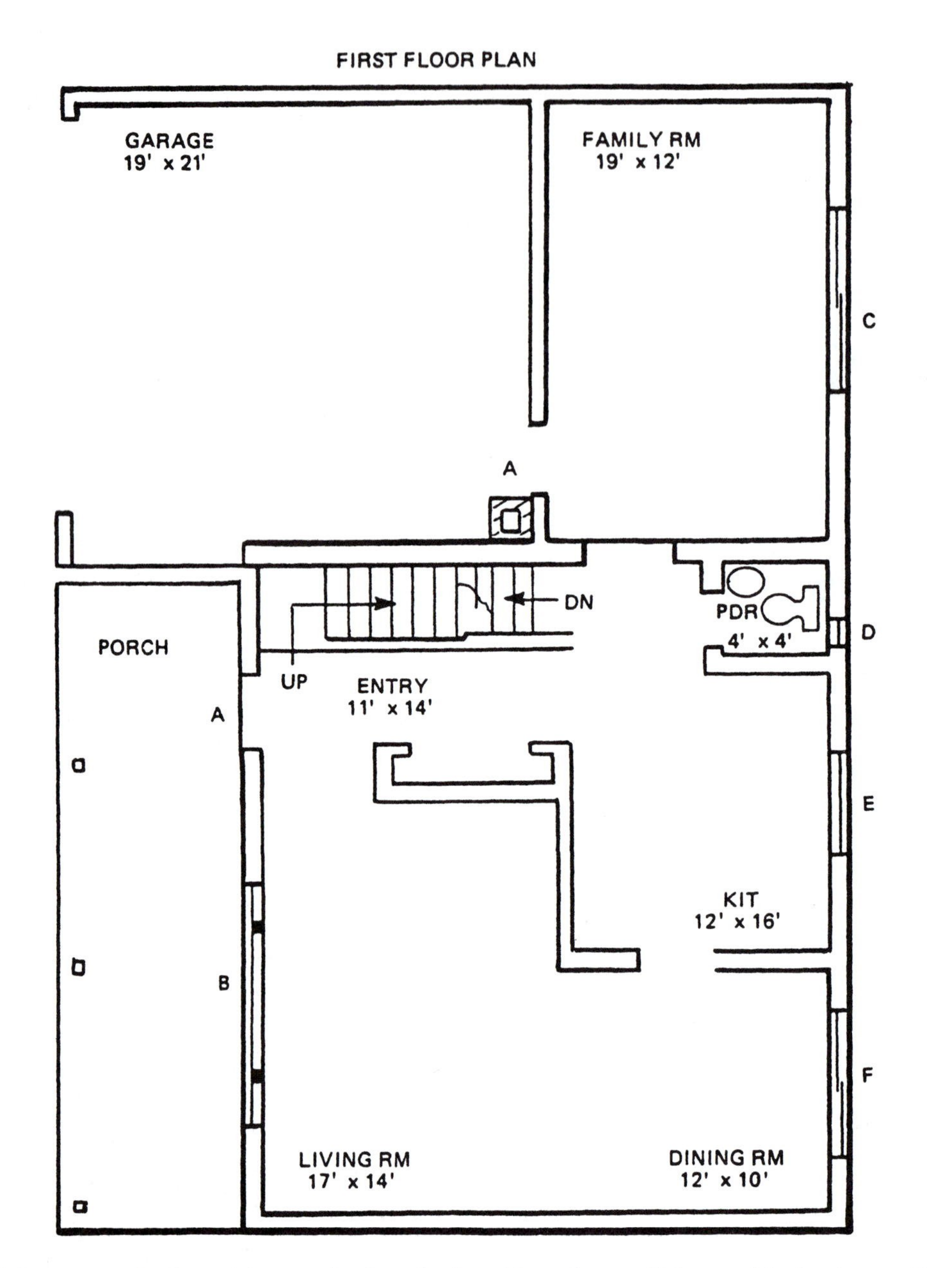

A furnace or boiler with a perfectly calculated heat loss will be too big for its application a great percentage of the time because maximum TD occurs rarely. This is reflected in the heater shutting down at regular intervals—less when the outside temperature is low and more as it rises. In practice, the calculations will result in a heat loss that is in between boiler sizes, and the next bigger boiler size will be selected for installation. This will provide a margin of excess capacity.

With warm water baseboard radiation, the entire home may be serviced with distribution loops all of the same size. The larger the home or place of business, the more complex the sizing problems will be.

Wherever there is a temperature difference between the living spaces and the other side of the walls, floors, or ceilings, there will be a heat loss. The living space design temperature will be 70° to 72° Fahrenheit. The outside design temperature must vary for different geographic localities. In the northern coastal states of the United States, the lowest temperatures expected are around 0° Fahrenheit, and so in these places the design TD is 70°. TD may be thought of as a heat pressure between the inside of the wall and its outside surface. The higher the TD, the faster heat will be lost to the outside environment.

On a warm winter afternoon of 70° and light winds, there will be no heat pressure from the inside to the outside and the heater or boiler may never come on. There might even be a overall heat gain because of solar radiation. (The same principle heats the inside of a car in a parking lot with the windows rolled up.)

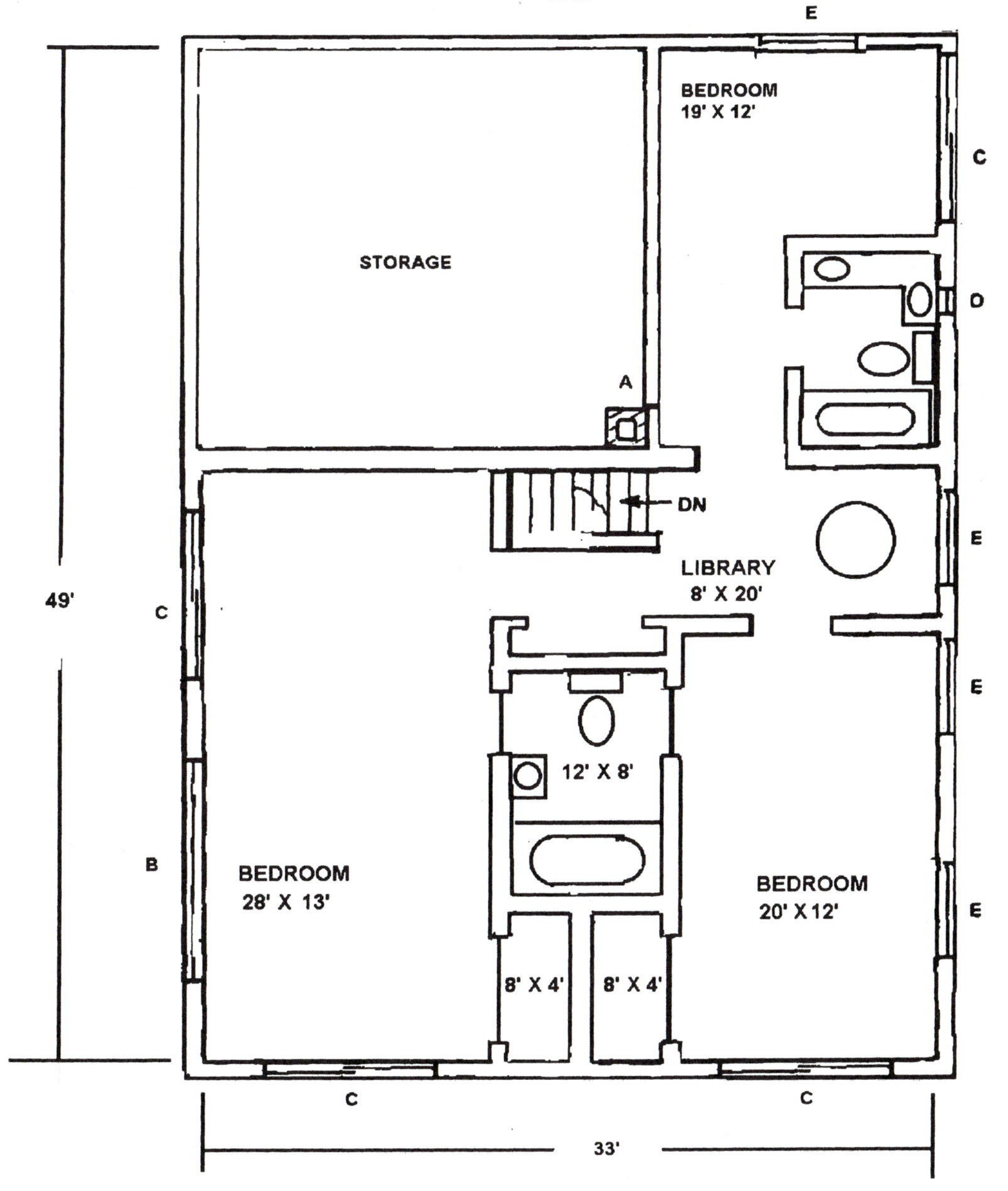

There is another kind of heat loss, sometimes called *crackage*. This is the heat that is lost by air transference around windows and doors. There are different ways of calculating this. In one method, the length of the openable perimeters of all windows and doors are calculated and then allowed for. In another there are general ratings for windows and doors that include allowances for doors and windows opened occasionally by humans. In some countries the crackage is included in a "number of air changes" calculation (three complete air changes in a room per hour is considered a safe allowance). In this book we will use the "infiltration" calculations on the first page of Unit 60, Radiation Sizing for Total Heat Loss of a Room. Note that air changes, infiltration, and crackage require heat applied to cubic feet of air rather than square feet of construction materials.

In order to calculate the heat loss for a two-floor house, the plans for both floors are examined. Second-floor rooms will have a heat loss through the ceilings. First-floor rooms without heated rooms above will have a ceiling heat loss. There are one-story sections on some two-story houses. There may be storage space that is unheated, as well as one-floor sections.

A partition between a room and an unheated space should be calculated as an exposed (outside) wall. Here, in practice, we may have to adjust the temperature difference (TD) to allow for special circumstances. Even though the lowest expected outside temperature in winter may be 0° Fahrenheit, yielding a TD of 70 (70° interior minus 0° exterior), at that same location the temperature in the unheated basement may never be less than 50°, yielding a TD of 20 between the basement and the floor above. This happens because the soil's temperature below the frost line is stable at all seasons.

Closets should be included in the heat loss of the room into which they open.

Some houses with finished basements require the use of basement radiation. A house with an unfinished basement may depend on heat from the boiler and the pipe of the heating circuit for heating the basement and the basement stairwell.

A stair space to the second floor must receive heat from adjacent rooms, as it usually has no radiator of its own. The room sizes shown on the floor plans of this unit make allowance for some adjacent space.

Exercise

The two-floor house shown in the plans (Figure 63–1 and Figure 63–2) is frame construction with wood shingle siding. The insulation is 3⅝-inch batts in the walls and 6-inch batts in the ceilings exposed to a lower temperature. Room height is 8 feet except for the family room, which has a wall height of 9 feet 6 inches. Use a temperature difference of 70°F. The circulation head within the one-pipe hot water heating system is 6 feet. Windows and doors have double-pane glass with a space between panes. All bathrooms and powder rooms have supplemental heat provided by heat lamps in the ceilings, so you may include the volume of the bathrooms within the volumes of the rooms, which include them, i.e., split the upper-floor bathroom between the bedrooms on either side.

Follow the chart and procedures of the Sample Problems in Unit 60, and the chart in Unit 59, to calculate each item. If a heated area has no dimensions, include it within an area that does. The family room on the first floor is completed (see following page) for an additional guideline. Make an answer chart including these items:

a. Heat loss for each room and total heat loss for the purpose of determining the size needed for a warm water boiler which will *not* be required to supply domestic hot water in addition to space heating.

b. Note that the longest upstairs bedroom overhangs the front porch by 9 feet. This portion of the bedroom floor is exposed to outside temperatures, and is insulated with 6-inch batt.

c. There is a basement beneath the living quarters, which maintains an ambient temperature of 55°F, but the floor above the basement has 6-inch insulation between the joists.

d. Use data for storm sash for infiltration values.

Sample Problem

sample room problem calculation

		Area square feet	Inside Temp	Outside Temp	Temp. diff. (TD)	Loss Value	Loss in BtuH
Family room	Ceiling	NA					0
	North wall	114	70	0	70	0.08	638

Section Seven
THE BUILDER'S LEVEL

Unit 64
Words Used in Leveling

Objectives

- Describe the builder's level.
- Define the special terms used in site leveling.
- Use the leveling terms in a practical application.

The builder's level is frequently used in the plumbing trade to determine elevations for pipe location. The instrument is a telescope supported on a tripod. The builder's level, when properly adjusted, gives the operator a level line of sight through the telescope. The telescope can be rotated to any direction. The instrument operator sights against a rod, which is a pole with measurements on it, held by another worker at chosen locations. The builder's level is a moderately expensive tool, but it "levels" in less time and more accurately than other methods. To work with this tool requires an understanding of terms and procedure.

1. The *Line of Sight* is an imaginary line through the telescope, centered on the crosshairs showing in the lens and directed toward a measuring rod.
2. The *Instrument Location* is the place where the tripod is set up, usually about midpoint between stations.
3. A *Station* is the point where the worker holds the rod. It is chosen with a purpose.
4. A *Bench mark* (B.M.) is a station of known elevation often established by others. The U.S. Coast and Geodetic Survey establishes bench marks with elevations above sea level. A construction site may have a bench mark located near the job, perhaps on the curb. For convenience the building bench mark may be listed at 100.00 feet elevation.
5. A *Turning Point* (T.P.) is a station located between instrument locations. It is used when a building or other obstacle prevents a direct line of sight.
6. The *Height of Instrument* (H.I.) is the elevation of the line of sight. It is the sum of the station elevation plus the back sight.
7. A *Back Sight* (B.S.) is the measurement on the rod at the line of sight when the rod is held on a station of known elevation, such as the bench mark. The back sight measurement is always added to the elevation to compute the height of the instrument.
8. A *Fore Sight* (F.S.) is the measurement on the rod at the line of sight when the rod is held on a station of unknown elevation. The station elevation is computed by subtracting the fore sight from the height of instrument.

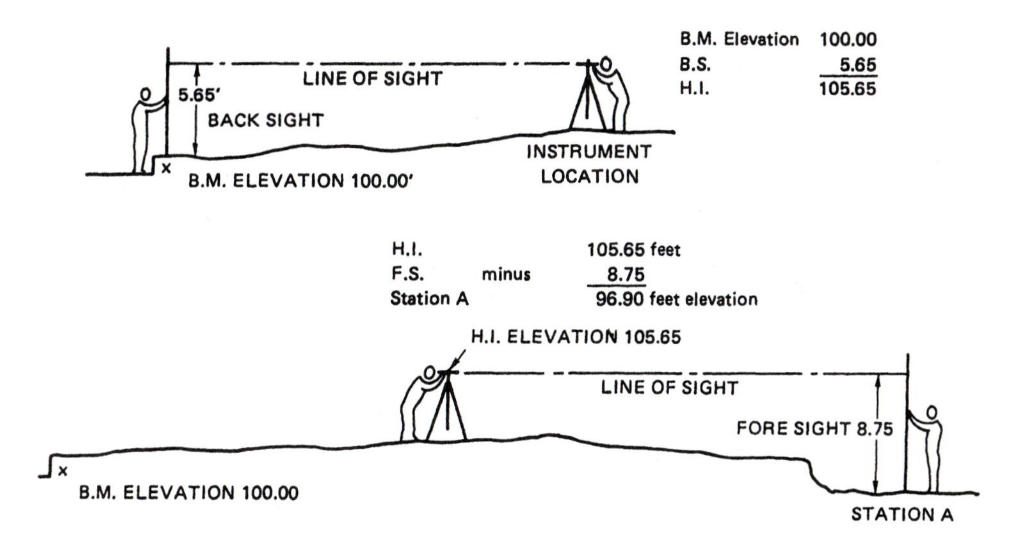

Exercise

The profile and plot plan shows a leveling job to locate two top of manhole elevation stakes. The letters are to be matched to the listed leveling terms. For example, u and v are lines of sight and $p + a$ is height of instrument.

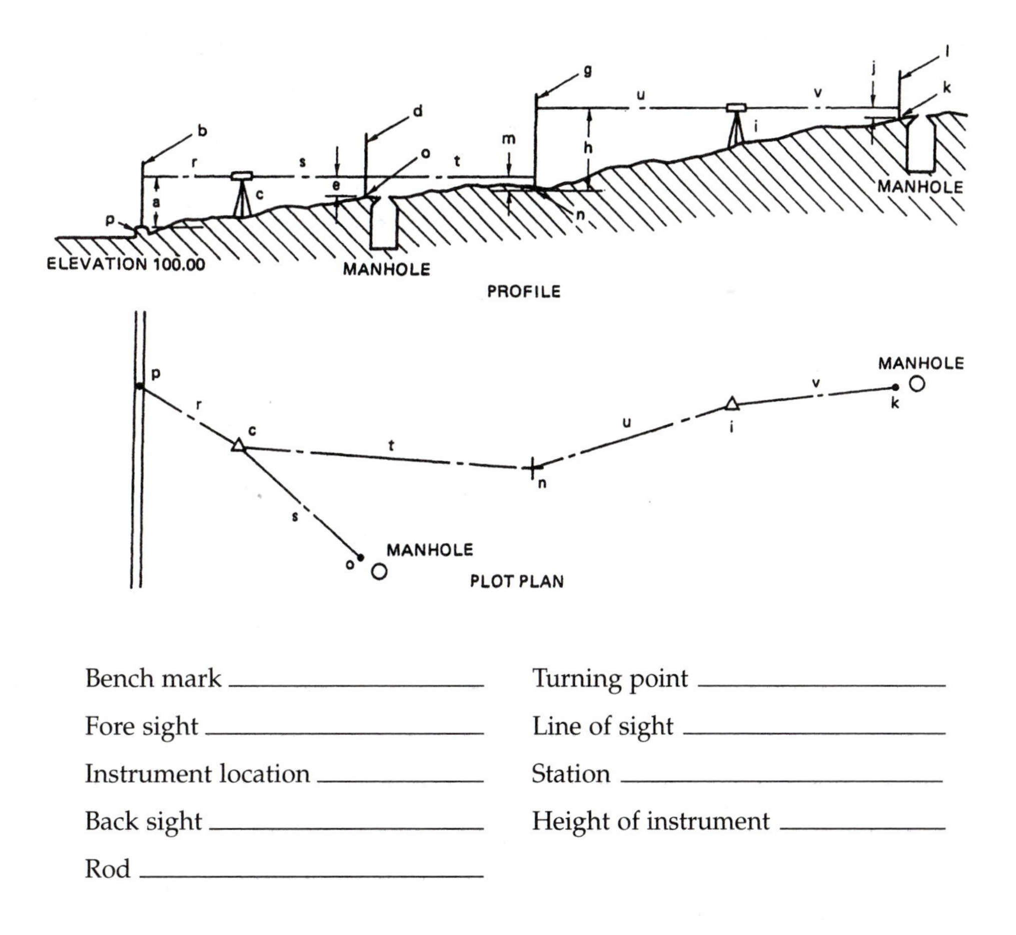

Bench mark ________________

Fore sight ________________

Instrument location ________________

Back sight ________________

Rod ________________

Turning point ________________

Line of sight ________________

Station ________________

Height of instrument ________________

Unit 65 Math Aids in Leveling

Objectives

- Explain the conventions used in site leveling.
- Practice keeping field notes.

Stations are located and numbered to indicate positions along a pipe line, such as a sewer. The major spacing distance of stations is 100 feet. Plumbing requirements often use 50-foot spacing in leveling. Any distance can be used for a station as needed. The low elevation or other starting point is station 0 + 00. Station 0 + 50 is fifty feet away and station 1 + 100 is 100 feet along the pipe line from station 0 + 00. Station 1 + 27 is 27 feet past station 1 + 00 and 127 feet from station 0 + 00. Numbering the stations in this way is good practice for long distances.

Field notes keep a record of the leveling work and make it possible to check for errors. The field note form and procedure shown in this unit are the result of the experience of many people. Checking for arithmetic errors or procedure errors is readily done to assure accuracy of the results. The use of two lines of notes per station avoids crowding and aids in checking elevations.

Sample Problems

1. If station 0 + 00 has an elevation of 137.56 feet, determine the elevation of station 0 + 50 when the grade is 0.01 ft/ft (⅛"/ft).
2. If the instrument height is 143.00 feet, what is the fore sight for station 0 + 50 in problem 1?

Solutions

1. 0.01 × 50' = 0.50'
   137.56'
   + .50'
   138.06' elevation

2. 143.00'
   −138.06'
   4.94' fore sight

Exercise

Complete the field notes based on the sketch and items that are completed. The station 0 + 00 stake has been established. The field notes have a check mark for each space in which information items need to be completed. The back sights and fore sights are shown in the sketch. The instrument location is moved to a new position by use of a turning point. The station stakes are firmly driven and each station elevation is 7 feet above the invert of the sewer to be installed. The station number is marked on the stake. The elevation is shown on the stake by a line mark with a nail driven at the correct height.

NAMES DATE

JOB SEWER FOR JK CO.

DESCRIPTION OF WORK SET GRADE STAKES 7 FT ABOVE INVERT OF SEWER.

ELEVATION STAKE AT MAIN SEWER BY CITY SURVEYOR

STATION	B.S. ( )	H.I.	F.S. (-)	ELEVATION	REMARKS
0 + 00				*137.56*	*On Stake - Set by City Surveyor*
	5.44	*143.00*			*137.56 + 5.44*
0 + 50				*138.06*	*137.56 + 0.50*
			4.94		*143.00 - 138.06*
✓		*143.00*		*138.56*	
✓		*143.00*		✓	
			✓		
T.P.		*143.00*	*5.27*	✓	*At Paint Mark on Boulder*
	5.71	✓			
✓				✓	
			✓		
✓				✓	
			✓		
✓				✓	
			✓		

Sketch and Description and Property Location

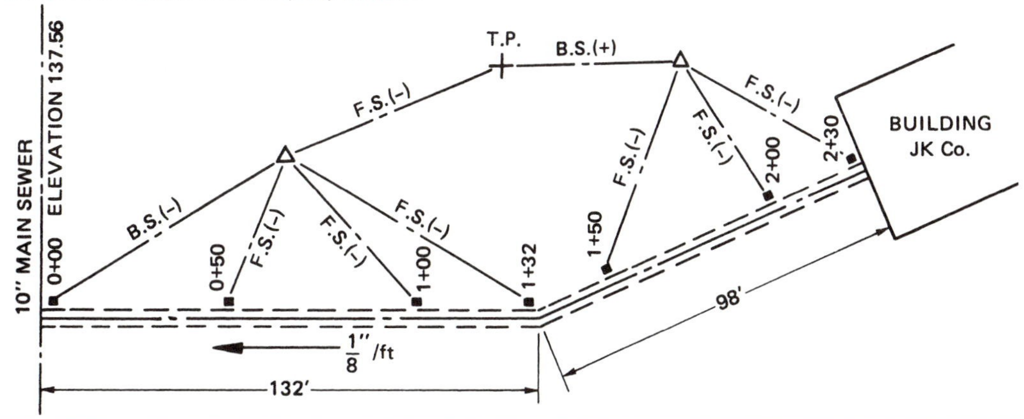

APPENDIX A

Note: The Square Root Tables A through D, pages 239–242, give the square roots of numbers from 1.00 to 99.9. By using the same tables, the square roots of numbers from 100 to 1000 may also be found by proper placement of the decimal point.

EXAMPLE 1:

Find the square root of 105.
The table shows the square root of 1.05 to be 1.025.
By moving the decimal point one place to the right, we find 10.25 to be the square root of 105.

EXAMPLE 2:

Find the square root of 327.
The table shows the square root of 3.27 to be 1.808.
Moving the decimal point one place to the right, we find 18.08 to be the square root of 327.

TABLE A

SQUARE ROOT OF NUMBERS 1.00–5.49

Number	0	1	2	3	4	5	6	7	8	9
1.0	1.000	1.005	1.010	1.015	1.020	1.025	1.030	1.034	1.039	1.044
1.1	1.049	1.054	1.058	1.063	1.068	1.072	1.077	1.082	1.086	1.091
1.2	1.095	1.100	1.105	1.109	1.114	1.118	1.122	1.127	1.131	1.136
1.3	1.140	1.145	1.149	1.153	1.158	1.162	1.166	1.170	1.175	1.179
1.4	1.183	1.187	1.192	1.196	1.200	1.204	1.208	1.212	1.217	1.221
1.5	1.225	1.229	1.233	1.237	1.241	1.245	1.249	1.253	1.257	1.261
1.6	1.265	1.269	1.273	1.277	1.281	1.285	1.288	1.292	1.296	1.300
1.7	1.304	1.308	1.311	1.315	1.319	1.323	1.327	1.330	1.334	1.338
1.8	1.342	1.345	1.349	1.353	1.356	1.360	1.364	1.367	1.371	1.375
1.9	1.378	1.382	1.386	1.389	1.393	1.396	1.400	1.404	1.407	1.411
2.0	1.414	1.418	1.421	1.425	1.428	1.432	1.435	1.439	1.442	1.446
2.1	1.449	1.453	1.456	1.459	1.463	1.466	1.470	1.473	1.476	1.480
2.2	1.483	1.487	1.490	1.493	1.497	1.500	1.503	1.507	1.510	1.513
2.3	1.517	1.520	1.523	1.526	1.530	1.533	1.536	1.539	1.543	1.546
2.4	1.549	1.552	1.556	1.559	1.562	1.565	1.568	1.572	1.575	1.578
2.5	1.581	1.584	1.587	1.591	1.594	1.597	1.600	1.603	1.606	1.609
2.6	1.612	1.616	1.619	1.622	1.625	1.628	1.631	1.634	1.637	1.640
2.7	1.643	1.646	1.649	1.652	1.655	1.658	1.661	1.664	1.667	1.670
2.8	1.673	1.676	1.679	1.682	1.685	1.688	1.691	1.694	1.697	1.700
2.9	1.703	1.706	1.709	1.712	1.715	1.718	1.720	1.723	1.726	1.729
3.0	1.732	1.735	1.738	1.741	1.744	1.746	1.749	1.752	1.755	1.758
3.1	1.761	1.764	1.766	1.769	1.772	1.775	1.778	1.780	1.783	1.786
3.2	1.789	1.792	1.794	1.797	1.800	1.803	1.806	1.808	1.811	1.814
3.3	1.817	1.819	1.822	1.825	1.828	1.830	1.833	1.836	1.838	1.841
3.4	1.844	1.847	1.849	1.852	1.855	1.857	1.860	1.863	1.865	1.868
3.5	1.871	1.873	1.876	1.879	1.881	1.884	1.887	1.889	1.892	1.895
3.6	1.897	1.900	1.903	1.905	1.908	1.910	1.913	1.916	1.918	1.921
3.7	1.924	1.926	1.929	1.931	1.934	1.936	1.939	1.942	1.944	1.947
3.8	1.949	1.952	1.954	1.957	1.960	1.962	1.965	1.967	1.970	1.972
3.9	1.975	1.977	1.980	1.982	1.985	1.087	1.990	1.992	1.995	1.997
4.0	2.000	2.002	2.005	2.007	3.010	3.912	2.015	2.017	2.020	2.022
4.1	2.025	2.027	2.030	2.032	2.035	2.037	2.040	2.042	2.045	2.047
4.2	2.049	2.052	2.054	2.057	2.059	2.062	2.064	2.066	2.069	2.071
4.3	2.074	2.076	2.078	2.081	2.083	2.086	2.088	2.090	2.093	2.095
4.4	2.098	2.100	2.102	2.105	2.107	2.110	2.112	2.114	2.117	2.119
4.5	2.121	2.124	2.126	2.128	2.131	2.133	2.135	2.138	2.140	2.142
4.6	2.145	2.147	2.149	2.152	2.154	2.156	2.159	2.161	2.163	2.166
4.7	2.168	2.170	2.173	2.175	2.177	2.179	2.182	2.184	2.186	2.189
4.8	2.191	2.193	2.195	2.198	2.200	2.202	2.205	2.207	2.209	2.211
4.9	2.214	2.216	2.218	2.220	2.223	2.225	2.227	2.229	2.232	2.234
5.0	2.236	2.238	2.241	2.243	2.245	2.247	2.249	2.252	2.254	2.256
5.1	2.258	2.261	2.263	2.265	2.267	2.269	2.272	2.274	2.276	2.278
5.2	2.280	2.283	2.285	2.287	2.289	2.291	2.293	2.296	2.298	2.300
5.3	2.302	2.304	2.307	2.309	2.311	2.313	2.315	2.317	2.319	2.322
5.4	2.324	2.326	3.328	2.330	2.332	2.335	2.337	2.339	2.341	2.343
	0	1	2	3	4	5	6	7	8	9

TABLE B

SQUARE ROOT OF NUMBERS 5.50–9.99

Number	0	1	2	3	4	5	6	7	8	9
5.5	2.345	2.347	2.349	2.352	2.354	2.356	2.358	2.360	2.362	2.364
5.6	2.366	2.369	2.371	2.373	2.375	2.377	2.379	2.381	2.383	2.385
5.7	2.387	2.390	2.392	2.394	2.396	2.398	2.400	2.402	2.404	2.406
5.8	2.408	2.410	2.412	2.415	2.417	2.419	2.421	2.423	2.425	2.427
5.9	2.429	2.431	2.433	2.435	2.437	2.439	2.441	2.443	2.445	2.447
6.0	2.449	2.452	2.454	2.456	2.458	2.460	2.462	2.464	2.466	2.468
6.1	2.470	2.472	2.474	2.476	2.478	2.480	2.482	2.484	2.486	2.488
6.2	2.490	2.492	2.494	2.496	2.498	2.500	2.502	2.504	2.506	2.508
6.3	2.510	2.512	2.514	2.516	2.518	2.520	2.522	2.524	2.526	2.528
6.4	2.530	2.532	2.534	2.536	2.538	2.540	2.542	2.544	2.546	2.548
6.5	2.550	2.551	2.553	2.555	2.557	2.559	2.561	2.563	2.565	2.567
6.6	2.569	2.571	2.573	2.575	2.577	2.579	2.581	2.583	2.585	2.587
6.7	2.588	2.590	2.592	2.594	2.596	2.598	2.600	2.602	2.604	2.606
6.8	2.608	2.610	2.612	2.613	2.615	2.617	2.619	2.621	2.623	2.625
6.9	2.627	2.629	2.631	2.632	2.634	2.636	2.638	2.640	2.642	2.644
7.0	2.646	2.648	2.650	2.651	2.653	2.655	2.657	2.659	2.661	2.663
7.1	2.665	2.666	2.668	2.670	2.672	2.674	2.676	2.678	2.680	2.681
7.2	2.683	2.685	2.687	2.689	2.691	2.693	2.694	2.696	2.698	2.700
7.3	2.702	2.704	2.706	2.707	2.709	2.711	2.713	2.715	2.717	2.718
7.4	2.720	2.722	2.724	2.726	2.728	2.729	2.731	2.733	2.735	2.737
7.5	2.739	2.740	2.742	2.744	2.746	2.748	2.750	2.751	2.753	2.755
7.6	2.757	2.759	2.760	2.762	2.764	2.766	2.768	2.769	2.771	2773
7.7	2.775	2.777	2.778	2.780	2.782	2.784	2.786	2.787	2.789	2.791
7.8	2.793	2.795	2.796	2.798	2.800	2.802	2.804	2.805	2.807	2.809
7.9	2.811	2.812	2.814	2.816	2.818	2.820	2.821	2.823	2.825	2.827
8.0	2.828	2.830	2.832	2.834	2.835	2.837	2.839	2.841	2.843	2.844
8.1	2.846	2.848	2.850	2.851	2.853	2.855	2.857	2.858	2.860	2.862
8.2	2.864	2.865	2.867	2.869	2.871	2.872	2.874	2.876	2.877	2.879
8.3	2.881	2.883	2.884	2.886	2.888	2.890	2.891	2.893	2.895	2.897
8.4	2.898	2.900	2.902	2.903	2.905	2.907	2.909	2.910	2.912	2.914
8.5	2.915	2.917	2.919	2.921	2.922	2.924	2.926	2.927	2.929	2.931
8.6	2.933	2.934	2.936	2.938	2.939	2.941	2.943	2.944	2.946	2.948
8.7	2.950	2.951	2.953	2.955	2.956	2.958	2.960	2.961	2.963	2.965
8.8	2.966	2.968	2.970	2.972	2.973	2.975	2.977	2.978	2.980	2.982
8.9	2.983	2.985	2.987	2.988	2.990	2.992	2.993	2.995	2.997	2.998
9.0	3.000	3.002	3.003	3.005	3.007	3.008	3.010	3.012	3.013	3.015
9.1	3.017	3.018	3.020	3.022	3.023	3.025	3.027	3.028	3.030	3.032
9.2	3.033	3.035	3.036	3.038	3.040	3.041	3.043	3.045	3.046	3.048
9.3	3.050	3.051	3.053	3.055	3.056	3.058	3.059	3.061	3.063	3.064
9.4	3.066	3.068	3.069	3.071	3.072	3.074	3.076	3.077	3.079	3.081
9.5	3.082	3.084	3.085	3.087	3.089	3.090	3.092	3.094	3.095	3.097
9.6	3.098	3.100	3.102	3.103	3.105	3.106	3.108	3.110	3.111	3.113
9.7	3.114	3.116	3.118	3.119	3.121	3.122	3.124	3.126	3.127	3.129
9.8	3.130	3.132	3.134	3.135	3.137	3.138	3.140	3.142	3.143	3.145
9.9	3.146	3.148	3.150	3.151	3 153	3.154	3.156	3.158	3.159	3.161
	0	1	2	3	4	5	6	7	8	9

TABLE C

SQUARE ROOT OF NUMBERS 10.0–54.9

Number	0	1	2	3	4	5	6	7	8	9
10.	3.162	3.178	3.194	3.209	3.225	3.240	3.256	3.271	3.286	3.302
11.	3.317	3.332	3.347	3.362	3.376	3.391	3.406	3.421	3.435	3.450
12.	3.464	3.479	3.493	3.507	3.521	3.536	3.550	3.564	3.578	3.592
13.	3.606	3.619	3.633	3.647	3.661	3.674	3.688	3.701	3.715	3.728
14.	3.742	3.755	3.768	3.782	3.795	3.808	3.821	3.834	3.847	3.860
15.	3.873	3.886	3.899	3.912	3.924	3.937	3.950	3.962	3.975	3.987
16.	4.000	4.012	4.025	4.037	4.050	4.062	4.074	4.087	4.099	4.111
17.	4.123	4.135	4.147	4.159	4.171	4.183	4.195	4.207	4.219	4.231
18.	4.243	4.254	4.266	4.278	4.290	4.301	4.313	4.324	4.336	4.347
19.	4.359	4.370	4.382	4.393	4.405	4.416	4.427	4.438	4.450	4.461
20.	4.472	4.483	4.494	4.506	4.517	4.528	4.539	4.550	4.561	4.572
21.	4.583	4.593	4.604	4.615	4.626	4.637	4.648	4.658	4.669	4.680
22.	4.690	4.701	4.712	4.722	4.733	4.743	4.754	4.764	4.775	4.785
23.	4.796	4.806	4.817	4.827	4.837	4.848	4.858	4.868	4.879	4.889
24.	4.899	4.909	4.919	4.930	4.940	4.950	4.960	4.970	4.980	4.990
25.	5.000	5.010	5.020	5.030	5.040	5.050	5.060	5.070	5.079	5.089
26.	5.099	5.109	5.119	5.128	5.138	5.148	5.158	5.167	5.177	5.187
27.	5.196	5.206	5.215	5.225	5.235	5.244	5.254	5.263	5.273	5.282
28.	5.292	5.301	5.310	5.320	5.329	5.339	5.348	5.357	5.367	5.376
29.	5.385	5.394	5.404	5.413	5.422	5.431	5.441	5.450	5.459	5.468
30.	5.477	5.486	5.495	5.505	5.514	5.523	5.532	5.541	5.550	5.559
31.	5.568	5.577	5.586	5.595	5.604	5.612	5.621	5.630	5.639	5.648
32.	5.657	5.666	5.675	5.683	5.692	5.701	5.710	5.718	5.727	5.736
33.	5.745	5.753	5.762	5.771	5.779	5.788	5.797	5.805	5.814	5.822
34.	5.831	5.840	5.848	5.857	5.865	5.874	5.882	5.891	5.899	5.908
35.	5.916	5.925	5.933	5.941	5.950	5.958	5.967	5.975	5.983	5.992
36.	6.000	6.008	6.017	6.025	6.033	6.042	6.050	6.058	6.066	6.075
37.	6.083	6.091	6.099	6.107	6.116	6.124	6.132	6.140	6.148	6.156
38.	6.164	6.173	6.181	6.189	6.197	6.205	6.213	6.221	6.229	6.237
39.	6.245	6.253	6.261	6.269	6.277	6.285	6.293	6.301	6.309	6.317
40.	6.325	6.332	6.340	6.348	6.356	6.364	6.372	6.380	6.387	6.395
41.	6.403	6.411	6.419	6.427	6.434	6.442	6.450	6.458	6.465	6.473
42.	6.481	6.488	6.496	6.504	6.512	6.519	6.527	6.535	6.542	6.550
43.	6.557	6.565	6.573	6.580	6.588	6.595	6.603	6.611	6.618	6.626
44.	6.633	6.641	6.648	6.656	6.663	6.671	6.678	6.686	6.693	6.701
45.	6.708	6.716	6.723	6.731	6.738	6.745	6.753	6.760	6.768	6.775
46.	6.782	6.790	6.797	6.804	6.812	6.819	6.826	6.834	6.841	6.848
47.	6.856	6.863	6.870	6.877	6.885	6.892	6.899	6.907	6.914	6.921
48.	6.928	6.935	6.943	6.950	6.957	6.964	6.971	6.979	6.986	6.993
49.	7.000	7.007	7.014	7.021	7.029	7.036	7.043	7.050	7.057	7.064
50.	7.071	7.078	7.085	7.092	7.099	7.106	7.113	7.120	7.127	7.134
51.	7.141	7.148	7.155	7.162	7.169	7.176	7.183	7.190	7.197	7.204
52.	7.211	7.218	7.225	7.232	7.239	7.246	7.253	7.259	7.266	7.273
53.	7.280	7.287	7.294	7.301	7.308	7.314	7.321	7.328	7.335	7.342
54.	7.348	7.355	7.362	7.369	7.376	7.382	7.389	7.396	7.403	7.409
	0	1	2	3	4	5	6	7	8	9

TABLE D

SQUARE ROOT OF NUMBERS 55.0–99.9

Numbers	0	1	2	3	4	5	6	7	8	9
55.	7.416	7.423	7.430	7.436	7.443	7.450	7.457	7.463	7.470	7.477
56.	7.483	7.490	7.497	7.503	7.510	7.517	7.523	7.530	7.537	7.543
57.	7.550	7.556	7.563	7.570	7.576	7.583	7.589	7.596	7.603	7.609
58.	7.616	7.622	7.629	7.635	7.642	7.649	7.655	7.662	7.668	7.675
59.	7.681	7.688	7.694	7.701	7.707	7.714	7.720	7.727	7.733	7.740
60.	7.746	7.752	7.759	7.765	7.772	7.778	7.785	7.791	7.797	7.804
61.	7.810	7.817	7.823	7.829	7.836	7.842	7.849	7.855	7.861	7.868
62.	7.874	7.880	7.887	7.893	7.899	7.906	7.912	7.918	7.925	7.931
63.	7.937	7.944	7.950	7.956	7.962	7.969	7.975	7.981	7.987	7.994
64.	8.000	8.006	8.012	8.019	8.025	8.031	8.037	8.044	8.050	8.056
65.	8.062	8.068	8.075	8.081	8.087	8.093	8.099	8.106	8.112	8.118
66.	8.124	8.130	8.136	8.142	8.149	8.155	8.161	8.167	8.173	8.179
67.	8.185	8.191	8.198	8.204	8.210	8.216	8.222	8.228	8.234	8.240
68.	8.246	8.252	8.258	8.264	8.270	8.276	8.283	8.289	8.295	8.301
69.	8.307	8.313	8.319	8.325	8.331	8.337	8.343	8.349	8.355	8.361
70.	8.367	8.373	8.379	8.385	8.390	8.396	8.402	8.408	8.414	8.420
71.	8.426	8.432	8.438	8.444	8.450	8.456	8.462	8.468	8.473	8.479
72.	8.485	8.491	8.497	8.503	8.509	8.515	8.521	8.526	8.532	8.538
73.	8.544	8.550	8.556	8.562	8.567	8.573	8.579	8.585	8.591	8.597
74.	8.602	8.608	8.614	8.620	8.626	8.631	8.637	8.643	8.649	8.654
75.	8.660	8.666	8.672	8.678	8.683	8.689	8.695	8.701	8.706	8.712
76.	8.718	8.724	8.729	8.735	8.741	8.746	8.752	8.758	8.764	8.769
77.	8.775	8.781	8.786	8.792	8.798	8.803	8.809	8.815	8.820	8.826
78.	8.832	8.837	8.843	8.849	8.854	8.860	8.866	8.871	8.877	8.883
79.	8.888	8.894	8.899	8.905	8.911	8.916	8.922	8.927	8.933	8.939
80.	8.944	8.950	8.955	8.961	8.967	8.972	8.978	8.983	8.989	8.994
81.	9.000	9.006	9.011	9.017	9.022	9.028	9.033	9.039	9.044	9.050
82.	9.055	9.061	9.066	9.072	9.077	9.083	9.088	9.094	9.099	9.105
83.	9.110	9.116	9.121	9.127	9.132	9.138	9.143	9.149	9.154	9.160
84.	9.165	9.171	9.176	9.182	9.187	9.192	9.198	9.203	9.209	9.214
85.	9.220	9.225	9.230	9.236	9.241	9.247	9.252	9.257	9.263	9.268
86.	9.274	9.279	9.284	9.290	9.295	9.301	9.306	9.311	9.317	9.322
87.	9.327	9.333	9.338	9.343	9.349	9.354	9.359	9.365	9.370	9.375
88.	9.381	9.386	9.391	9.397	9.402	9.407	9.413	9.418	9.423	9.429
89.	9.434	9.439	9.445	9.450	9.455	9.460	9.466	9.471	9.476	9.482
90.	9.487	9.492	9.497	9.503	9.508	9.513	9.518	9.524	9.529	9.534
91.	9.539	9.545	9.550	9.555	9.560	9.566	9.571	9.576	9.581	9.586
92.	9.592	9.597	9.602	9.607	9.612	9.618	9.623	9.628	9.633	9.638
93.	9.644	9.649	9.654	9.659	9.664	9.670	9.675	9.680	9.685	9.690
94.	9.695	9.701	9.706	9.711	9.716	9.721	9.726	9.731	9.737	9.742
95.	9.747	9.752	9.757	9.762	9.767	9.772	9.778	9.783	9.788	9.793
96.	9.798	9.803	9.808	9.813	9.818	9.823	9.829	9.834	9.839	9.844
97.	9.849	9.854	9.859	9.864	9.869	9.874	9.879	9.884	9.889	9.894
98.	9.899	9.905	9.910	9.915	9.920	9.925	9.930	9.935	9.940	9.945
99.	9.950	9.955	9.960	9.965	9.970	9.975	9.980	9.985	9.990	9.995
	0	1	2	3	4	5	6	7	8	9

TABLE E
POWERS AND ROOTS OF NUMBERS 1–100

	Powers		Roots			Powers		Roots	
Number	Square	Cube	Square	Cube	Number	Square	Cube	Square	Cube
1	1	1	1.000	1.000	51	2 601	132 651	7.141	3.708
2	4	8	1.414	1.260	52	2 704	140 608	7.211	3.733
3	9	27	1.732	1.442	53	2 809	148 877	7.280	3.756
4	16	64	2.000	1.587	54	2 916	157 464	7.348	3.780
5	25	125	2.236	1.710	55	3 025	166 375	7.416	3.803
6	36	216	2.449	1.817	56	3 136	175 616	7.483	3.826
7	49	343	2.646	1.913	57	3 249	185 193	7.550	3.849
8	64	512	2.828	2.000	58	3 364	195 112	7.616	3.871
9	81	729	3.000	2.080	59	3 481	205 379	7.681	3.893
10	100	1 000	3.162	2.154	60	3 600	216 000	7.746	3.915
11	121	1 331	3.317	2.224	61	3 721	226 981	7.810	3.936
12	144	1 728	3.464	2.289	62	3 844	238 328	7.874	3.958
13	169	2 197	3.606	2.351	63	3 969	250 047	7.937	3.979
14	196	2 744	3.742	2.410	64	4 096	262 144	8.000	4.000
15	225	3 375	3.873	2.466	65	4 225	274 625	8.062	4.021
16	256	4 096	4.000	2.520	66	4 356	287 496	8.124	4.041
17	289	4 913	4.123	2.571	67	4 489	300 763	8.185	4.062
18	324	5 832	4.243	2.621	68	4 624	314 432	8.246	4.082
19	361	6 859	4.359	2.668	69	4 761	328 509	8.307	4.102
20	400	8 000	4.472	2.714	70	4 900	343 000	8.367	4.121
21	441	9 261	4.583	2.759	71	5 041	357 911	8.426	4.141
22	484	10 648	4.690	2.802	72	5 184	373 248	8.485	4.160
23	529	12 167	4.796	2.844	73	5 329	389 017	8.544	4.179
24	576	13 824	4.899	2.884	74	5 476	405 224	8.602	4.198
25	625	15 625	5.000	2.924	75	5 625	421 875	8.660	4.217
26	676	17 576	5.099	2.962	76	5 776	438 976	8.718	4.236
27	729	19 683	5.196	3.000	77	5 929	456 533	8.775	4.254
28	784	21 952	5.292	3.037	78	6 084	474 552	8.832	4.273
29	841	24 389	5.385	3.072	79	6 241	493 039	8.888	4.291
30	900	27 000	5.477	3.107	80	6 400	512 000	8.944	4.309
31	961	29 791	5.568	3.141	81	6 561	531 441	9.000	4.327
32	1 024	32 768	5.657	3.175	82	6 724	551 368	9.055	4.344
33	1 089	35 937	5.745	3.208	83	6 889	571 787	9.110	4.362
34	1 156	39 304	5.831	3.240	84	7 056	592 704	9.165	4.380
35	1 225	42 875	5.916	3.271	85	7 225	614 125	9.220	4.397
36	1 296	46 656	6.000	3.302	86	7 396	636 056	9.274	4.414
37	1 369	50 653	6.083	3.332	87	7 569	658 503	9.327	4.431
38	1 444	54 872	6.164	3.362	88	7 744	681 472	9.381	4.448
39	1 521	59 319	6.245	3.391	89	7 921	704 969	9.434	4.465
40	1 600	64 000	6.325	3.420	90	8 100	729 000	9.487	4.481
41	1 681	68 921	6.403	3.448	91	8 281	753 571	9.539	4.498
42	1 764	74 088	6.481	3.476	92	8 464	778 688	9.592	4.514
43	1 849	79 507	6.557	3.503	93	8 649	804 357	9.644	4.531
44	1 936	85 184	6.633	3.530	94	8 836	830 584	9.695	4.547
45	2 025	91 125	6.708	3.557	95	9 025	857 375	9.747	4.563
46	2 116	97 336	6.782	3.583	96	9 216	884 736	9.798	4.579
47	2 209	103 823	6.856	3.609	97	9 409	912 673	9.849	4.595
48	2 304	110 592	6.928	3.634	98	9 604	941 192	9.899	4.610
49	2 401	117 649	7.000	3.659	99	9 801	970 299	9.950	4.626
50	2 500	125 000	7.071	3.684	100	10 000	1 000 000	10.000	4.642

TABLE F

STANDARD TABLES OF ENGLISH MEASURE

Linear Measure	
12 in	= 1 foot (ft)
3 ft	= 1 yard (yd)
16 1/2 ft	= 1 rod (rd)
5 1/2 yd	= 1 rod
320 rd	= 1 mile
1 760 yd	= 1 mile
5 280 ft	= 1 mile

Surface Measure	
144 sq in	= 1 sq ft
9 sq ft	= 1 sq yd
30 1/4 sq yd	= 1 sq rd
160 sq rd	= 1 acre
640 acres	= 1 sq mi
43 560 sq ft	= 1 acre

Cubic Measure	
1 728 cu in	= 1 cu ft
27 cu ft	= 1 cu yd
128 cu ft	= 1 cord

TABLE G

STANDARD TABLES OF METRIC MEASURE

Linear Measure		
Unit	Value in metres	Symbol
Micrometre	0.000 001	μm
Millimetre	0.001	mm
Centimetre	0.01	cm
Decimetre	0.1	dm
Metre (unit)	1.0	m
Dekametre	10.0	dam
Hectometre	100.0	hm
Kilometre	1 000.00	km
Megameter	1 000 000.0	Mm

Surface Measure		
Unit	Value in square metres	Symbol
Square millimetre	0.000 001	mm^2
Square centimetre	0.000 1	cm^2
Square decimetre	0.01	dm^2
Square metre	1.0	m^2
Square dekametre	100.0	dam^2
Square hectometre	10 000.0	hm^2
Square kilometre	1 000 000.0	km^2

Cubic Measure		
Unit	Value in cubic metres	Symbol
Cubic millimetre	10^{-9}	mm^3
Cubic centimetre	10^{-6}	cm^3
Cubic decimetre	10^{-3}	dm^3
Cubic metre	1	m^3
Cubic dekametre	10^3	dam^3
Cubic hectometre	10^6	hm^3
Cubic kilometre	10^9	km^3

DATA 1
DECIMAL INCH AND FRACTIONAL INCH EQUIVALENTS

Decimal (inch)	Fractional (inch)	Low limit	High limit
0.00000	0	0	0.03124
0.06250	1/16	0.03125	0.09374
0.12500	1/8	0.09375	0.15624
0.18750	3/16	0.15625	0.21874
0.25000	1/4	0.21875	0.28124
0.31250	5/16	0.28125	0.34374
0.37500	3/8	0.34375	0.40624
0.43750	7/16	0.40625	0.46874
0.50000	1/2	0.46875	0.53124
0.56250	9/16	0.53125	0.59374
0.62500	5/8	0.59375	0.65624
0.68750	11/16	0.65625	0.71874
0.75000	3/4	0.71875	0.78124
0.81250	13/16	0.78125	0.84374
0.87500	7/8	0.84375	0.90624
0.93750	15/16	0.90625	0.96874
1.00000	1	0.96875	1.03124

DATA 2
STANDARD WEIGHT PIPE – DIAMETERS AND THREADS

Nominal Size (inches)	Diameters				Threads			
	O.D. (inches)	I.D. (inches)	Wood Auger for Pipe (inches)	Tap Drill (inches)	Number per inch	Length on Pipe (inches)	Threads on Pipe	Thread-in (inches)
1/8	0.405	0.269	1/2	21/64	27	3/8	10	1/4
1/4	0.540	0.364	5/8	29/64	18	9/16	10	3/8
3/8	0.675	0.493	11/16	19/32	18	9/16	10	3/8
1/2	0.840	0.622	15/16	23/32	14	3/4	10 1/2	1/2
3/4	1.050	0.824	1 1/8	15/16	14	3/4	10 1/2	1/2
1	1.315	1.049	1 7/16	1 3/16	11 1/2	1	11 1/2	1/2
1 1/4	1.660	1.380	1 3/4	1 15/16	11 1/2	1	11 1/2	1/2
1 1/2	1.900	1.610	2	1 23/32	11 1/2	1	11 1/2	1/2
2	2.375	2.067	2 1/2	2 3/16	11 1/2	1	11 1/2	1/2
2 1/2	2.875	2.469	3	2 5/8	8	1 1/2	12	3/4
3	3.500	3.068	3 11/16	3 1/4	8	1 1/2	12	1
3 1/2	4.000	3.548	4 1/4	3 3/4	8	1 5/8	13	1
4	4.500	4.025	4 11/16	4 1/4	8	1 5/8	13	1
5	5.563	5.047	5 3/4	5 5/16	8	1 3/4	14	1 1/4
6	6.625	6.065	6 7/8	6 5/16	8	1 3/4	14	1 1/4
8	8.625	7.981	8 7/8	8 3/8	8	2	16	1 1/4
10	10.75	10.02	11	10 5/8	8	2	16	1 1/2
12	12.75	12.00	13	12 5/8	8	2 1/2	20	1 5/8

DATA 3
STANDARD WEIGHT PIPE – DIAMETERS AND CAPACITIES

Nominal Dia (inches)	Actual I.D. (inches)	Actual O.D. (inches)	Outside Circum (inches)	Outside Circum (feet)	Inside Cross-Sectional Area (sq in)	Inside Cross-Sectional Area (sq ft)	Surface Area per Lin Ft (sq ft)	Capacity per Lin Ft (gal)	Weight of Water per Lin Ft (lb)	Weight of Pipe per Lin Ft (lb)
1/8	0.269	0.405	1.272	0.106	0.057	0.000 4	0.106	0.003	0.024	0.246
1/4	0.364	0.540	1.696	0.141	0.104	0.000 7	0.141	0.005	0.045	0.426
3/8	0.493	0.675	2.121	0.177	0.191	0.001 3	0.177	0.009	0.082	0.570
1/2	0.622	0.840	2.639	0.220	0.304	0.002 1	0.220	0.015	0.131	0.855
3/4	0.824	1.050	3.299	0.273	0.533	0.003 7	0.273	0.027	0.230	1.140
1	1.049	1.315	4.131	0.343	0.864	0.006	0.343	0.044	0.374	1.690
1 1/4	1.388	1.660	5.215	0.433	1.496	0.010 3	0.433	0.077	0.647	2.290
1 1/2	1.610	1.900	5.969	0.497	2.036	0.014 1	0.497	0.105	0.881	2.740
2	2.067	2.375	7.461	0.622	3.356	0.023	0.622	0.174	1.453	3.690
2 1/2	2.469	2.875	9.032	0.751	4.778	0.033	0.751	0.248	2.073	5.85
3	3.068	3.500	11.00	0.843	7.393	0.051	0.843	0.384	3.201	7.66
3 1/2	3.548	4.000	12.566	1.045	9.90	0.068	1.045	0.515	4.290	8.98
4	4.026	4.500	14.14	1.18	12.73	0.088	1.178	0.661	5.512	10.9
5	5.047	5.563	17.49	1.455	20.01	0.139	1.455	1.039	8.662	14.9
6	6.065	6.625	20.81	1.73	28.89	0.2	1.734	1.500	12.51	19.2
8	7.981	8.625	27.10	2.26	50.03	0.35	2.258	2.598	21.66	28.9
10	10.020	10.750	33.772	2.81	78.85	0.545	2.81	4.096	34.12	40.5
12	12.000	12.750	40.055	3.38	113.09	0.984	3.38	5.88	48.96	49.56

DATA 4
COPPER TUBES – DIAMETERS

Nominal Size (inches)	Outside Diameter	Inside Diameter			
	Types K, L, M, and DWV (inches)	Type K (inches)	Type L (inches)	Type M (inches)	Type DWV (inches)
1/4	0.375	0.305	0.315		
3/8	0.500	0.402	0.430		
1/2	0.625	0.527	0.545		
5/8	0.750	0.652	0.666		
3/4	0.875	0.745	0.785		
1	1.125	0.995	1.025		
1 1/4	1.375	1.245	1.265	1.291	1.295
1 1/2	1.625	1.481	1.505	1.527	1.511
2	2.125	1.959	1.985	2.009	2.041
2 1/2	2.625	2.435	2.465	2.495	
3	3.125	2.907	2.945	2.981	3.035
3 1/2	3.625	3.385	3.425	3.459	
4	4.125	3.857	3.905	3.935	4.009
5	5.125	4.805	4.875	4.907	4.981
6	6.125	5.741	5.845	5.881	5.959
8	8.125	7.583	7.725	7.785	
10	10.125	9.449	9.625	9.701	
12	12.125	11.315	11.565	11.617	

DATA 5

ELBOWS – 90° and 45° – THREADED STD, COPPER, PVC, DRAINAGE

A is Center-to-Face Measure
B is Thread-in Measure
G is Fitting-Allowance Measure

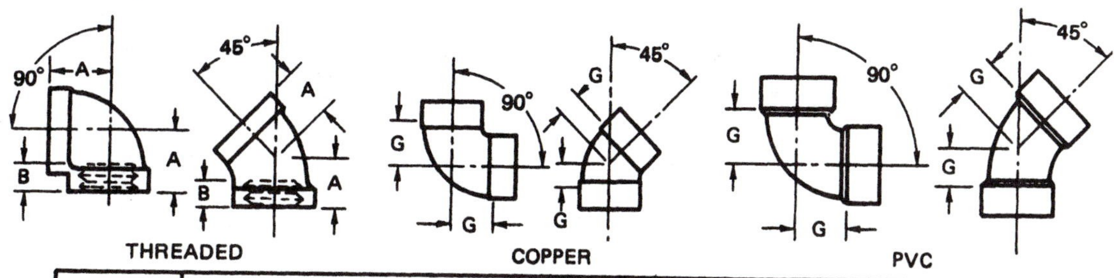

Nominal Pipe Size (inches)	Threaded			Copper		PVC	
	90°	45°		90°	45°	90°	45°
	A (inches)	A (inches)	B (inches)	G (inches)	G (inches)	G (inches)	G (inches)
3/8	1	3/4	3/8	5/16	3/16	3/8	1/4
1/2	1 1/8	3/4	1/2	3/8	3/16	1/2	1/4
3/4	1 3/8	1	1/2	1/2	1/4	9/16	5/16
1	1 1/2	1 1/8	1/2	3/4	5/16	11/16	5/16
1 1/4	1 3/4	1 5/16	1/2	1 1/8	7/16	1 9/16	1
1 1/2	1 15/16	1 7/16	1/2	1 5/16	9/16	1 3/4	1 1/8
2	2 1/4	1 11/16	1/2	1 7/8	3/4	2 5/16	1 1/2
2 1/2	2 11/16	2 1/16	3/4				
3	3 1/16	2 3/16	1	2 7/8	1 1/8	3 1/16	1 3/4
4	3 13/16	2 5/8	1	3 3/4	1 1/2	3 7/8	2 3/16

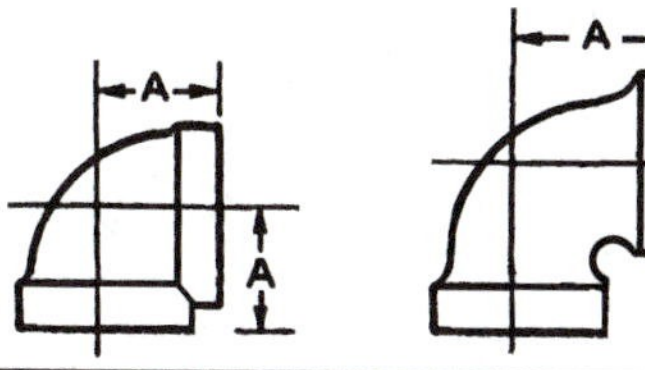

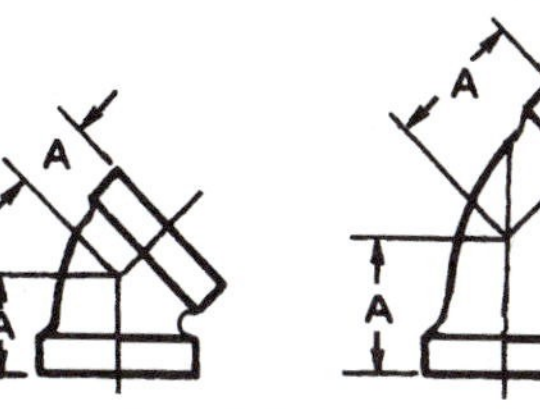

Nominal Pipe Size (inches)	90° Drainage Elbows			Nominal Pipe Size (inches)	45° Drainage Elbows		
	90° short turn	90° long turn			45° short turn	45° long turn	
	A (inches)	A (inches)	B Thread-in		A (inches)	A (inches)	B Thread-in
1 1/4	1 7/8	2 3/8	1/2	1 1/4	1 5/16	1 3/4	1/2
1 1/2	2	2 11/16	1/2	1 1/2	1 1/2	1 7/8	1/2
1 1/2 x 1 1/4	2	---	1/2 x 1/2	2	1 5/8	2 1/4	1/2
2	2 5/8	3 7/16	1/2	2 1/2	1 11/16	3 7/16	3/4
2 x 1 1/2	2 1/8	---	1/2 x 1/2	3	2 1/4	4 5/8	1
2 1/2	2 13/16	3 11/16	3/4	4	2 7/16	4 5/8	1
3	3 3/8	5 7/8	1				
4	3 15/16	6 3/4	1				

DATA 6
ELBOWS – 60°, 22 1/2°, and 11 1/4° – THREADED, COPPER, PVC

A is Center-to-Face Measure
B is Thread-in Measure
G is Fitting Allowance Measure

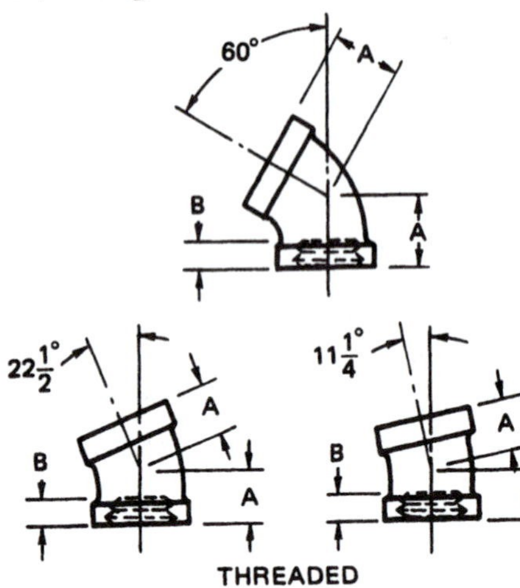

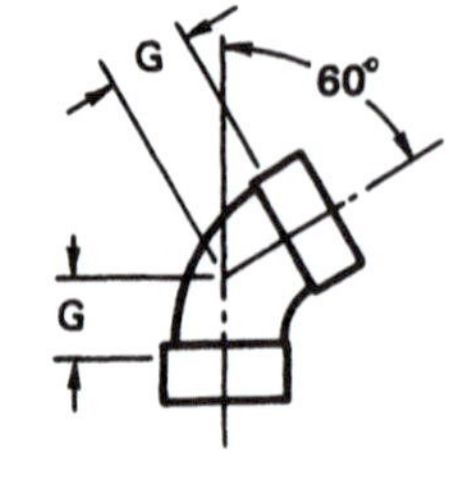

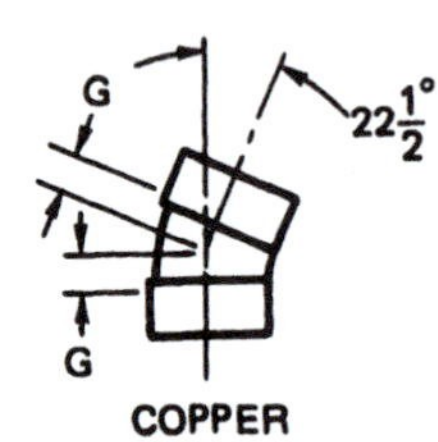

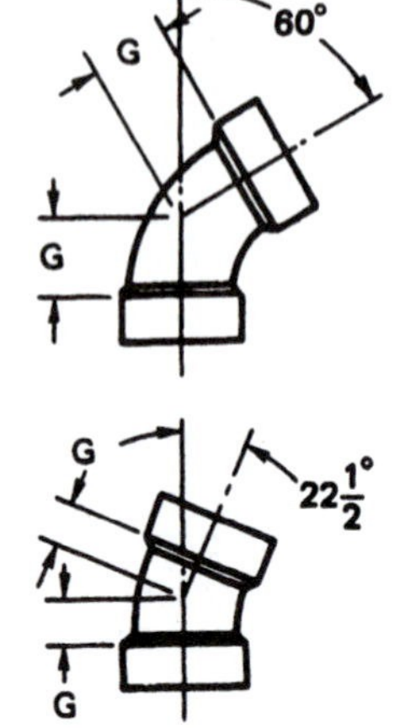

Nominal Pipe Size (inches)	Threaded 60° A (inches)	Threaded 22 1/2° A (inches)	Threaded 11 1/4° A (inches)	Threaded B (inches)	Copper 60° G (inches)	Copper 22 1/2° G (inches)	PVC 60° G (inches)	PVC 22 1/2° G (inches)
1 1/4	1 9/16	1 1/8	1 1/16	1/2	5/8	3/16		
1 1/2	1 3/4	1 1/4	1 1/4	1/2	13/16	1/4	1	1/2
2	2 3/16	1 1/2	1 3/8	1/2	1 1/16	5/16	1 5/16	11/16
2 1/2	2 1/2	2	1 5/8	3/4				
3	2 13/16	1 15/16	1 13/16	1	1 5/16	1/2	1 11/16	13/16
4	3 3/8	2 1/8	2 1/4	1	2 3/16	11/16	2 1/16	1

Dimensions for 22 1/2° and 11 1/4° DWV Elbows

22 1/2° Elbows Size (inches)	22 1/2° Elbows A (inches)	11 1/4° Elbows Thread-in (inches)	11 1/4° Elbows Size (inches)	11 1/4° Elbows A (inches)
1 1/4	1 1/16	1/2	1 1/4	1 1/16
1 1/2	1 1/4	1/2	1 1/2	1 1/8
2	1 7/16	1/2	2	1 3/8
2 1/2	1 1/2	3/4	2 1/2	1 7/16
3	1 7/8	1	3	1 5/8
4	1 7/8	1	4	1 11/16

DATA 7
TEES – THREADED, COPPER, PVC

C is Center-to-Face Measure
B is Thread-in Measure
H is Fitting-Allowance Measure

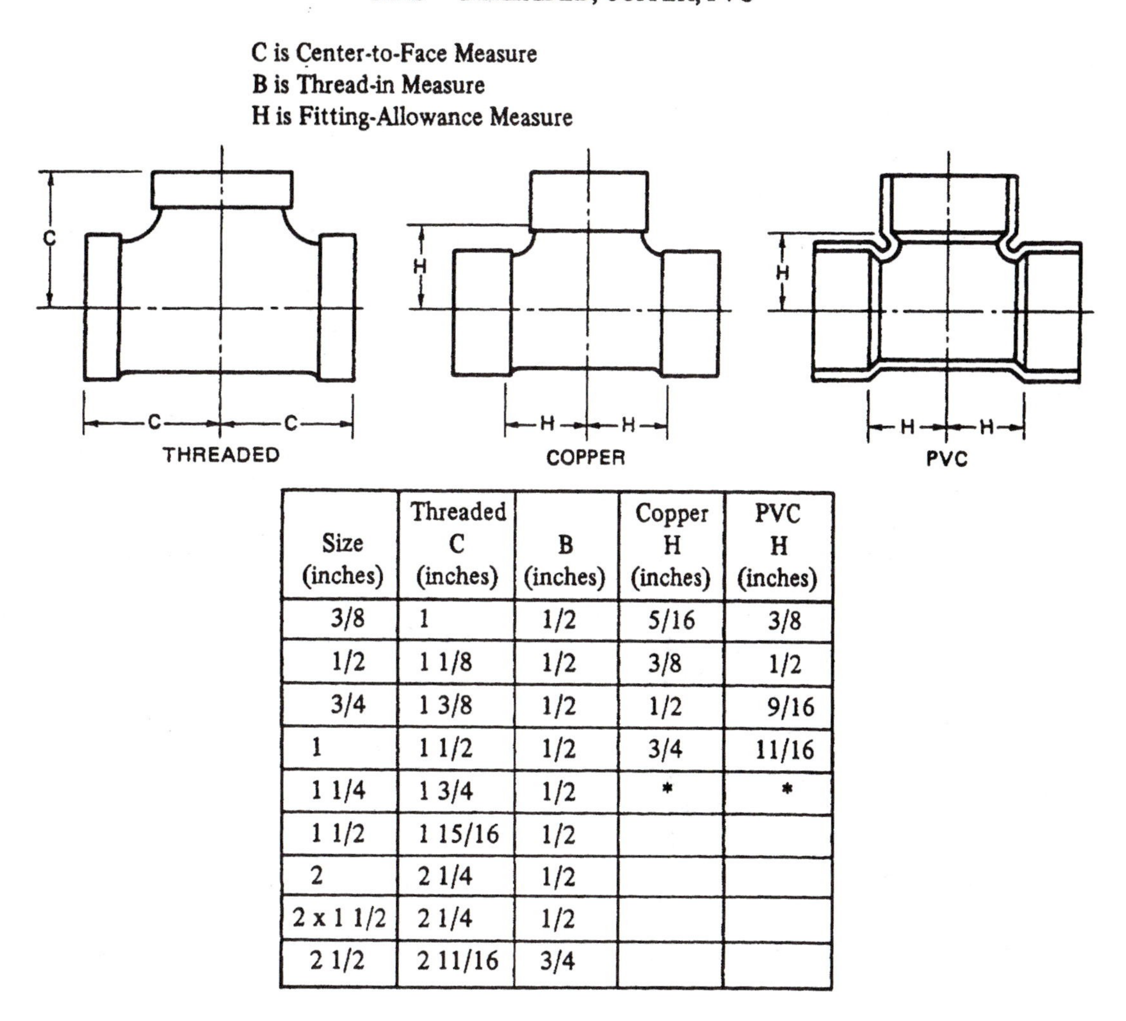

Size (inches)	Threaded C (inches)	B (inches)	Copper H (inches)	PVC H (inches)
3/8	1	1/2	5/16	3/8
1/2	1 1/8	1/2	3/8	1/2
3/4	1 3/8	1/2	1/2	9/16
1	1 1/2	1/2	3/4	11/16
1 1/4	1 3/4	1/2	*	*
1 1/2	1 15/16	1/2		
2	2 1/4	1/2		
2 x 1 1/2	2 1/4	1/2		
2 1/2	2 11/16	3/4		

**Copper and PVC use a tee wye on vent lines.*

DATA 8
REGULAR 45° WYE – THREADED, COPPER, PVC

C is Center-to-Face Measure (long run and branch)
D is Center-to-Face Measure (short run)
B is Thread-in Measure
H is Fitting-Allowance Measure (long run and branch)
I is Fitting-Allowance Measure (short run)

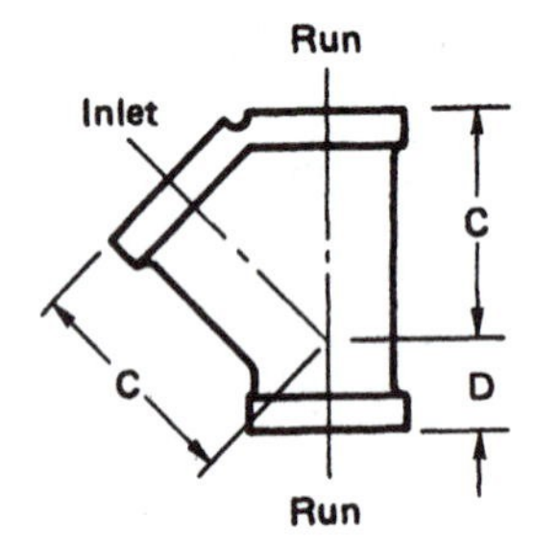

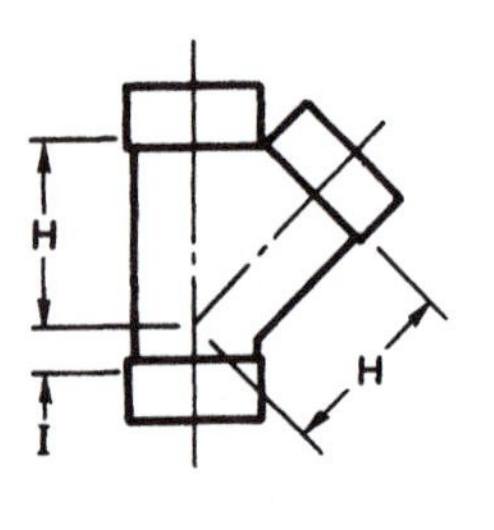

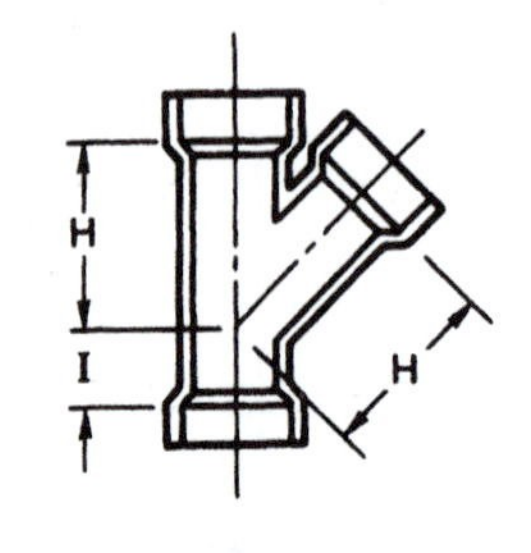

Nominal Pipe Size (inches)	Threaded			Copper		PVC	
	C (inches)	D (inches)	B (inches)	H (inches)	I (inches)	H (inches)	I (inches)
1 1/4	3 1/4	1 3/4	1/2	2	3/8	2 9/16	1 1/16
1 1/2	3 9/16	1 15/16	1/2	2 5/16	1/2	2 7/8	1 1/8
2	4 3/8	2 1/8	1/2	2 7/8	1/2	3 5/8	1 3/8
3	6 1/4	2 13/16	1	3 7/8	9/16	5	1 5/8
4	7 11/16	3 5/16	1	5 7/16	1 1/16	6 3/8	1 7/8

DRAINAGE 45° Y's – THREADED

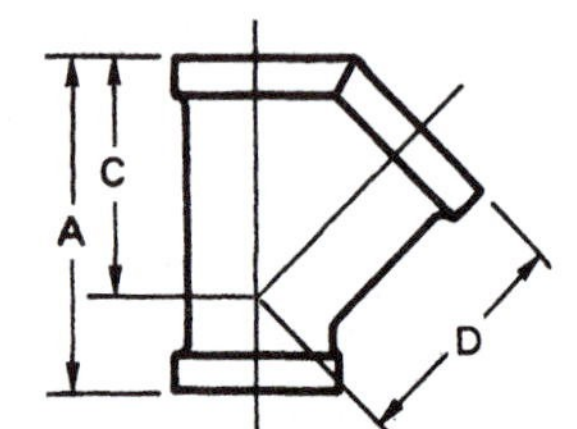

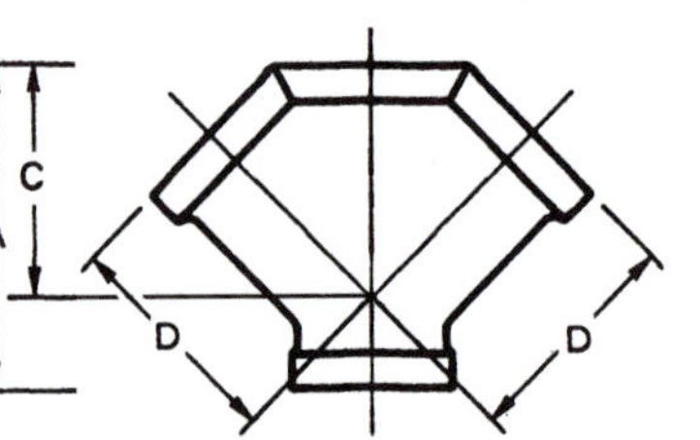

45° Y				Thread-in	DOUBLE 45° Y			
Size (inches)	A (inches)	C (inches)	D (inches)	Size TI	Size (inches)	A (inches)	C (inches)	D (inches)
1 1/4	5 3/16	3 7/16	3 7/16	1 1/4 = 1/2	1 1/4	5 3/16	3 7/16	3 7/16
1 1/2	5 1/4	3 7/16	3 7/16	1 1/2 = 1/2	1 1/2	5 1/4	3 7/16	3 7/16
2	6 1/4	4 1/8	4 1/8	2 = 1/2	2	6 1/4	4 1/8	4 1/8
2 1/2	8	5 1/4	5 1/4	2 1/2 = 3/4	2 1/2	8	5 1/4	5 1/4
3	8 3/8	6	6	3 = 1	3	8 3/8	6	6
4	9 5/8	7	7	4 = 1	4	9 5/8	7	7

DATA 9
REDUCED BRANCH 45° WYE – THREADED, COPPER, PVC

C is Center-to-Face Measure (long run)
D is Center-to-Face Measure (short run)
E is Center-to-Face Measure (branch)
B is Thread-in Measure

H is Fitting-Allowance Measure (long run)
I is Fitting-Allowance Measure (short run)
J is Fitting-Allowance Measure (branch)

Nominal Pipe Size (inches)	Threaded			
	C (inches)	D (inches)	E (inches)	B (inches)
2 x 1 1/2	4 1/16	1 13/16	4 1/8	1/2
3 x 1 1/2	4 15/16	1 11/16	5 1/16	1 and 1/2
3 x 2	5 5/16	2 1/16	5 3/8	1 and 1/2
4 x 2	6	1 11/16	6 5/16	1 and 1/2
4 x 3	6 7/8	2 3/8	7 3/16	1

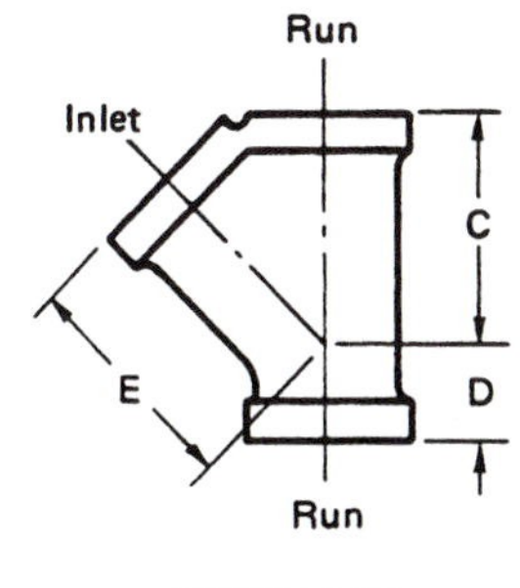

THREADED

Nominal Pipe Size (inches)	Copper		
	H (inches)	I (inches)	J (inches)
2 x 1 1/2	2 9/16	3/16	2 11/16
3 x 1 1/2	2 15/16	3/16	3 1/4
3 x 2	3 5/16	1/8	3 1/2
4 x 2	3 3/4	1/2	4 1/8
4 x 3	4 1/2	1/4	4 3/4

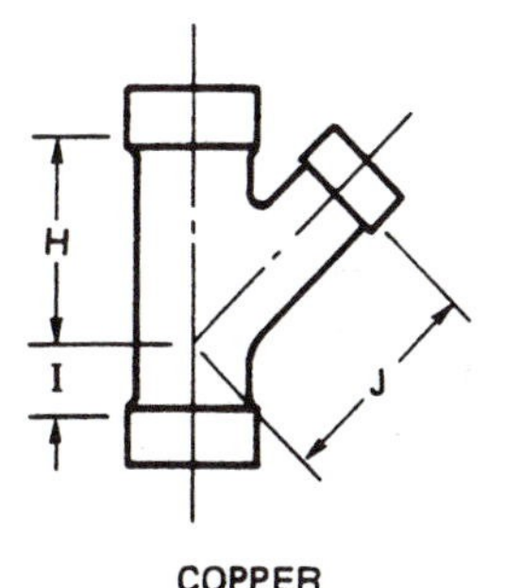

COPPER

Nominal Pipe Size (inches)	PVC		
	H (inches)	I (inches)	J (inches)
2 x 1 1/2	3 5/16	1 1/16	3 7/16
3 x 1 1/2	3 3/4	1/2	4 5/16
3 x 2	4 1/8	7/8	4 5/8
4 x 2	4 5/8	3/8	5 9/16
4 x 3	5 9/16	1 1/16	6

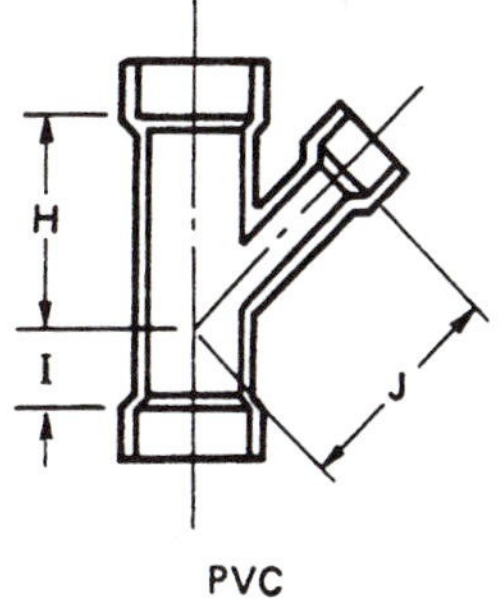

PVC

(Continued)

DATA 9 (continued)

DRAINAGE, REDUCING 45° Y's – THREADED

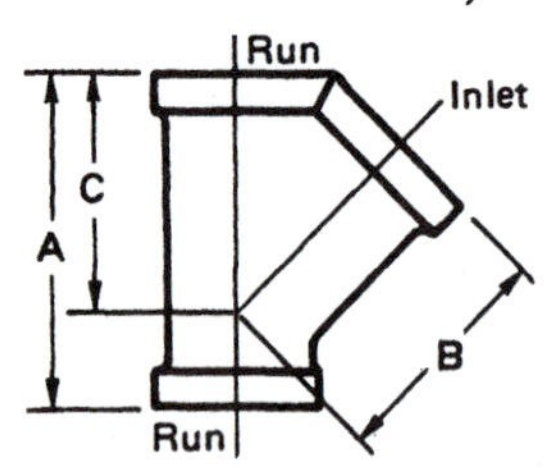

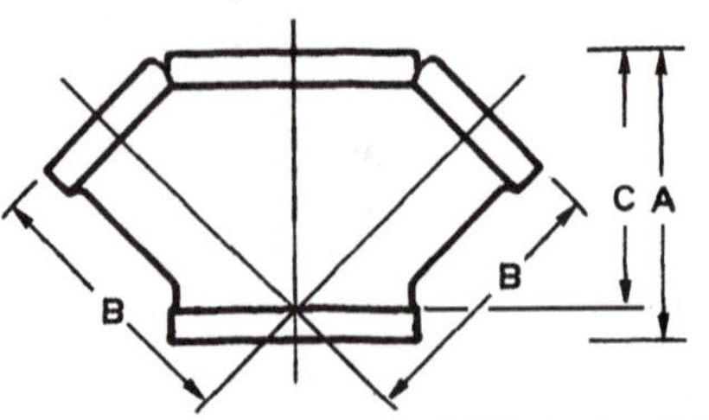

REDUCING 45° Y				Thread-in	REDUCING DOUBLE 45° Y			
Size (inches)	A (inches)	B (inches)	C (inches)	Size TI	Size (inches)	A (inches)	B (inches)	C (inches)
1 1/2 x 1 1/4	5 1/4	3 7/16	3 7/16	1 1/4 = 1/2	1 1/2 x 1 1/4	5 1/4	3 7/16	3 7/16
2 x 1 1/2	5 15/16	4 1/4	4 1/4	1 1/2 = 1/2	2 x 1 1/2	5 15/16	4 1/4	4 1/4
2 1/2 x 1 1/2	6 1/4	4 5/8	4 9/16	2 = 1/2	2 1/2 x 1 1/2	6 1/4	4 5/8	4 9/16
2 1/2 x 2	6 3/8	4 3/4	4 5/8	2 1/2 = 3/4	2 1/2 x 2	6 3/8	4 3/4	4 5/8
3 x 1 1/2	6 5/8	5 1/16	5 11/16	3 = 1	3 x 1 1/2	6 5/8	5 1/16	4 11/16
3 x 2	6 3/4	5 1/4	5 1/8	4 = 1	3 x 2	6 3/4	5 1/4	5 1/8
3 x 2 1/2	8	5 13/16	5 11/16					
4 x 1 1/2	7 5/8	5 5/8	5 7/16		4 x 2	6 5/8	5 7/8	5 5/8
4 x 2	6 5/8	5 7/8	5 5/8		4 x 3	8 1/2	6 5/8	6 3/8
4 x 2 1/2	7 1/4	5 7/8	5 3/4					
4 x 3	8 1/2	6 5/8	6 3/8					

DATA 10
REGULAR TEE WYE – THREADED, COPPER, PVC

C is Center-to-Face Measure (long run and branch)
D is Center-to-Face Measure (short run)
B is Thread-in Measure
H is Fitting-Allowance Measure (long run and branch)
I is Fitting allowance (short run)

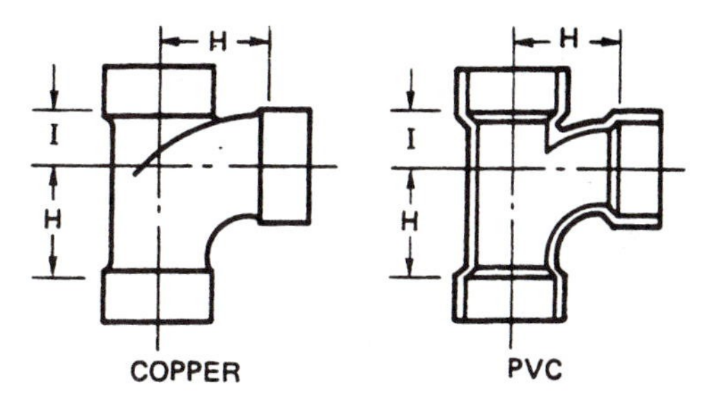

Nominal Pipe Size (inches)	Copper		PVC	
	H (inches)	I (inches)	H (inches)	I (inches)
1 1/4	1 1/8	13/16	1 9/16	11/16
1 1/2	1 3/8	7/8	1 3/4	1
2	1 7/8	1 1/8	2 5/16	1 3/8
3	2 7/8	1 9/16	3 1/16	1 13/16
4	3 13/16	2	3 1/8	2 1/4

DRAINAGE TY's – THREADED

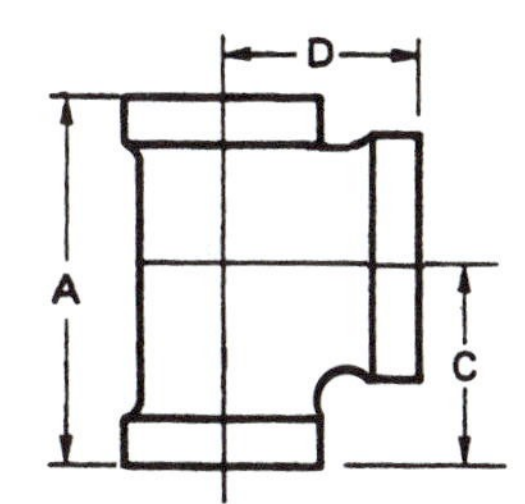

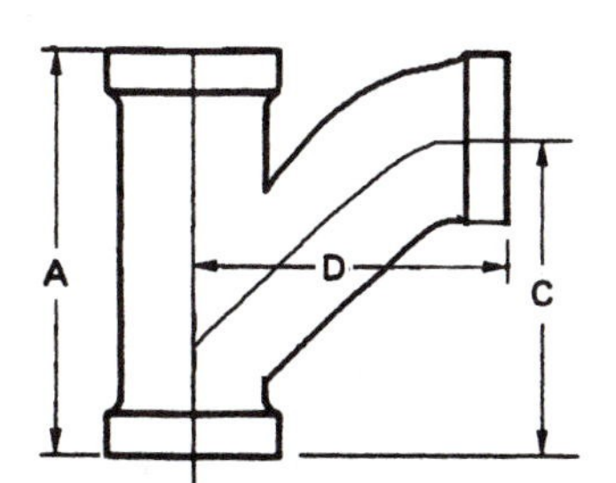

SHORT TY					LONG TY			
Size (inches)	A (inches)	C (inches)	D (inches)	Thread-in Size TI	Size (inches)	A (inches)	C (inches)	D (inches)
1 1/4	3 13/16	2 3/16	2 3/16	1 1/4 = 1/2	1 1/4	4 3/4	3 7/16	3 5/8
1 1/2	4 5/16	2 7/16	2 1/2	1 1/2 = 1/2	1 1/2	5 7/16	3 7/8	4 3/16
2	5 3/4	3 3/16	3 3/16	2 = 1/2	2	6 1/2	4 5/8	5 1/8
2 1/2	6 5/8	3 3/4	3 7/8	2 1/2 = 3/4	2 1/2	8 1/4	6 1/16	6 3/16
3	7 1/4	4 3/16	4 3/16	3 = 1	3	9	6 1/2	7 1/8
4	9 1/8	5 5/8	5 3/8	4 = 1	4	10 3/4	7 5/8	8 7/8

DATA 11
REDUCED BRANCH TEE WYE – THREADED, COPPER, PVC

C is Center-to-Face Measure (long run)
D is Center-to-Face Measure (short run)
E is Center-to-Face Measure (branch)
B is Thread-in Measure

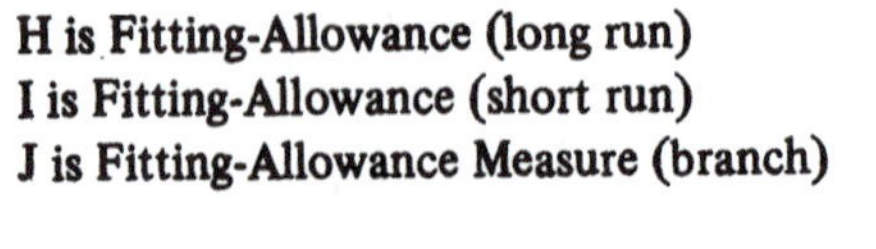
H is Fitting-Allowance (long run)
I is Fitting-Allowance (short run)
J is Fitting-Allowance Measure (branch)

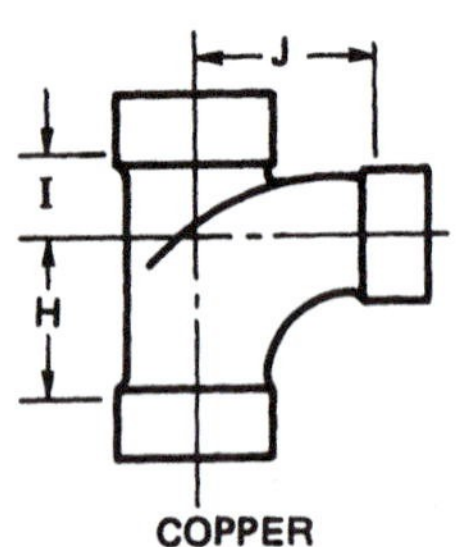

COPPER

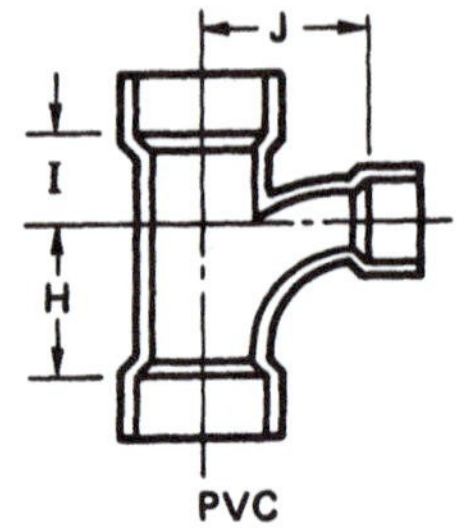

PVC

Nominal Pipe Size (inches)	Copper		
	H (inches)	I (inches)	J (inches)
2 x 1 1/2	1 3/8	15/16	1 5/8
3 x 1 1/2	1 5/16	3/4	2 1/16
3 x 2	1 7/8	1 1/8	2 3/8
4 x 2	1 13/16	1 5/16	2 7/8
4 x 3	2 7/8	1 9/16	3 5/16

Nominal Pipe Size (inches)	PVC		
	H (inches)	I (inches)	J (inches)
2 x 1 1/2	1 15/16	1 3/16	2 3/16
3 x 1 1/2	1 3/4	1 15/16	2 9/16
3 x 2	2 1/8	1 3/16	2 7/8
4 x 2	2 1/16	1 1/8	3 5/16
4 x 3	3	1 3/4	3 9/16

DRAINAGE, REDUCING 45° TY's – THREADED

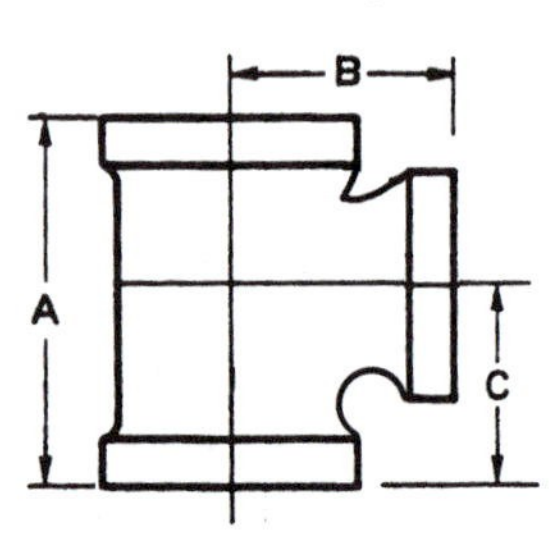

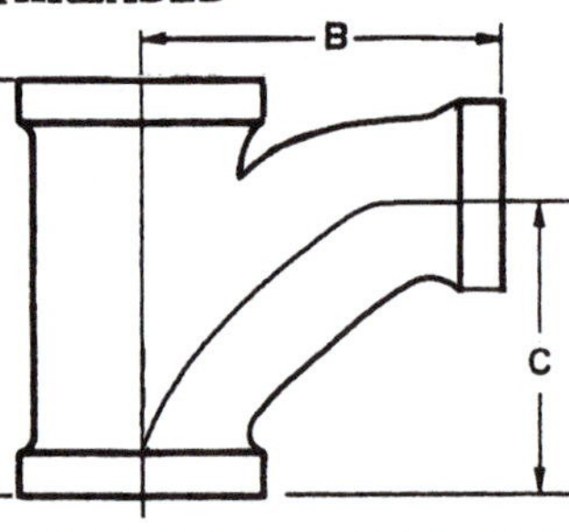

REDUCING SHORT TY				
Size (inches)	A (inches)	B (inches)	C (inches)	Thread-in Size TI
1 1/2 x 1 1/4	4 1/4	2 11/16	2 11/16	1 1/4 = 1/2
2 x 1 1/4	4 7/16	2 13/16	2 11/16	1 1/2 = 1/2
2 x 1 1/2	4 13/16	3	2 7/8	2 = 1/2
2 x 2 1/2	6 1/2	3 1/2	3 5/8	2 1/2 = 3/4
2 1/2 x 1 1/4	4 3/8	2 15/16	2 1/2	3 = 1
2 1/2 x 1 1/2	5 1/8	3 1/4	3	4 = 1
2 1/2 x 2	5 1/2	3 11/16	3 1/8	
3 x 1 1/2	5 1/16	3 5/16	2 15/16	
3 x 2	5 3/4	3 5/8	3 1/4	
4 x 1 1/2	5 1/4	3 13/16	3	
4 x 2	6	4 1/4	3 7/16	
4 x 2 1/2	6 5/8	4 7/16	3 13/16	
4 x 3	7 3/8	4 15/16	4 1/8	

REDUCING LONG TY			
Size (inches)	A (inches)	B (inches)	C (inches)
1 1/2 x 1 1/4	5 1/8	3 7/8	4 1/8
2 x 1 1/2	5 13/16	4 1/2	4 1/2
2 1/2 x 1 1/2	5 3/4	4 1/2	4 1/4
2 1/2 x 2	5 3/8	4 1/2	4 3/8
3 x 1 1/2	5 7/8	5 1/4	4 9/16
3 x 2	6 5/8	5 1/2	4 1/2
4 x 1 1/2	6 1/8	5 7/16	4 3/4
4 x 2	7	6 3/8	5 1/8
4 x 2 1/2	8 5/8	7 1/16	6 5/8
4 x 3	9 1/4	7 3/4	6 1/2

DATA 12
BENDS – 1/4 and 1/8 – HUB AND NO-HUB

D is Fitting-Allowance Measure (long run)
C is Fitting-Allowance Measure (short run)
X is Center-to-Face Measure

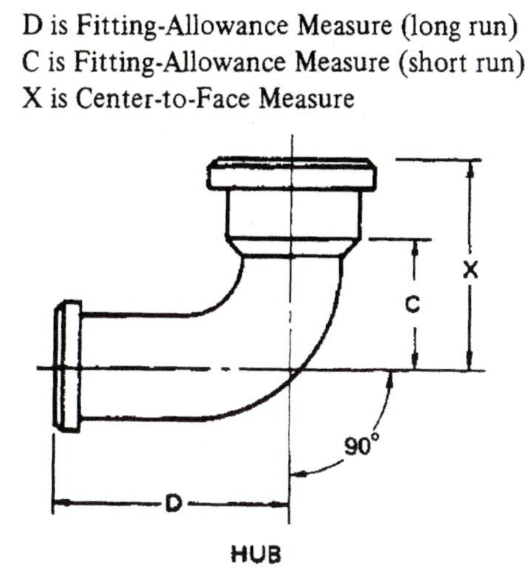

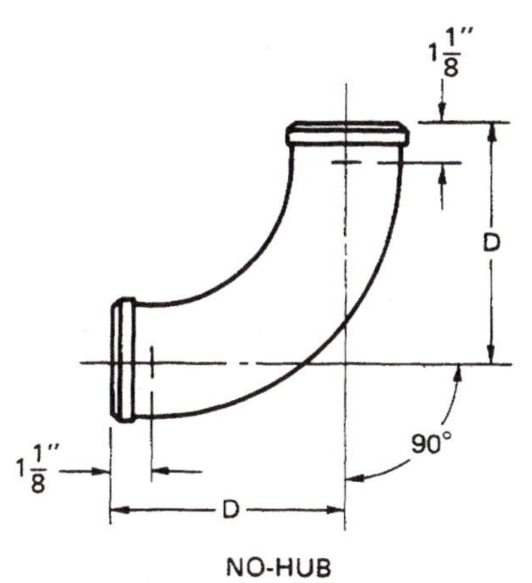

1/4 Bend

Size (inches)	Hub D (inches)	Hub C (inches)	Hub X (inches)	No-Hub D (inches)
2	8	5 1/4	7 3/4	6 1/2
3	9	6	8 3/4	7
4	10	6 1/2	9 1/2	7 1/2

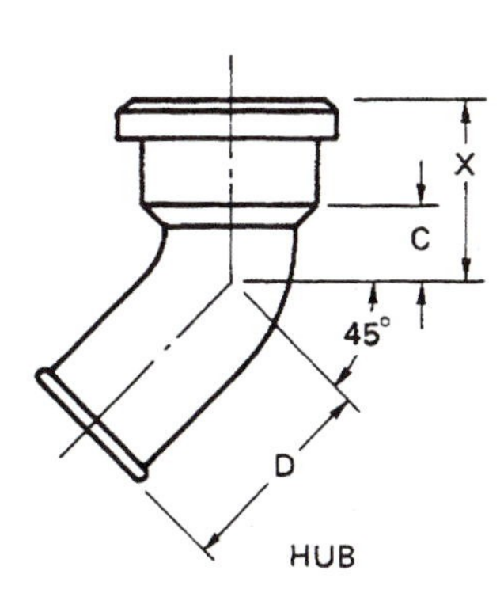

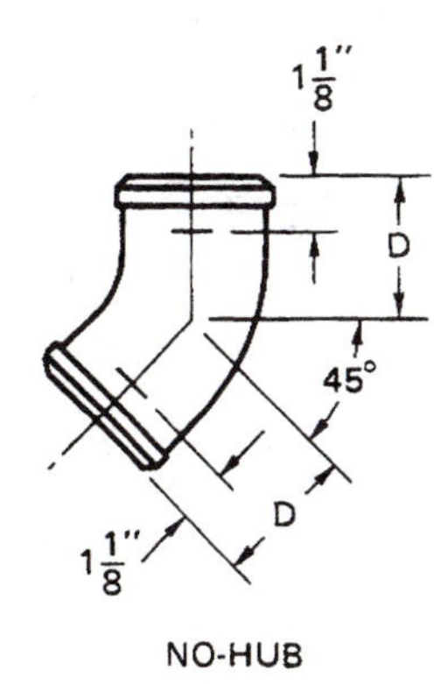

1/8 Bend

Size (inches)	Hub D (inches)	Hub C (inches)	Hub X (inches)	No-Hub D (inches)
2	4 1/4	1 1/2	4	2 3/4
3	5	2	4 3/4	3
4	5 3/4	2 1/4	5 1/4	3 1/8

DATA 13
BENDS – 1/5 and 1/6 – HUB AND NO-HUB

D is Fitting-Allowance Measure (long run)
C is Fitting-Allowance Measure (short run)
X is Center-to-Face Measure

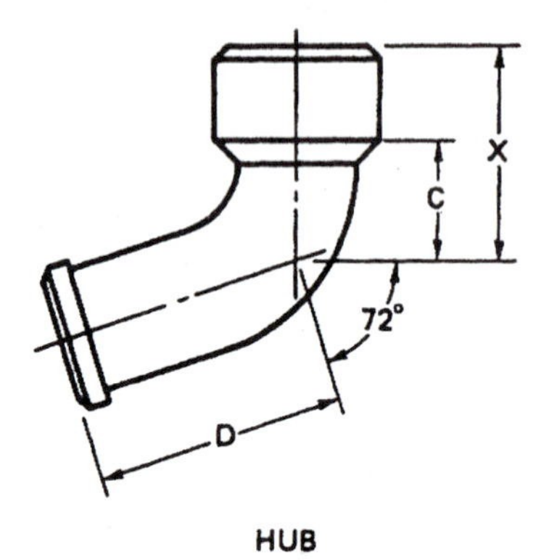

HUB

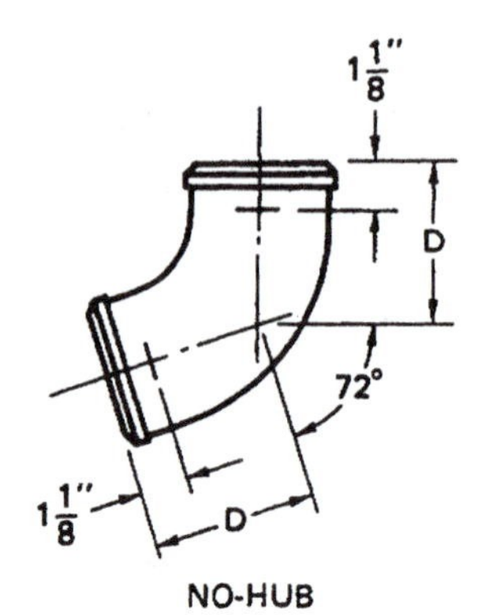

NO-HUB

1/5 Bend

Size (inches)	Hub			No-Hub
	D (inches)	C (inches)	X (inches)	D (inches)
2	5 1/4	2 1/2	5	3 3/4
3	6	3	5 3/4	4
4	7	3 1/2	6 1/2	4 1/2

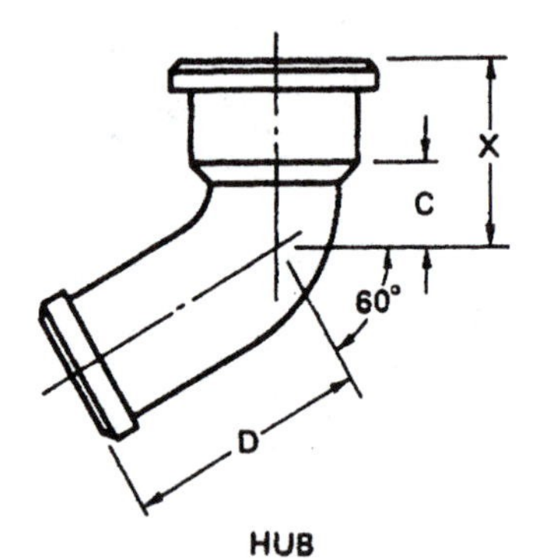

HUB

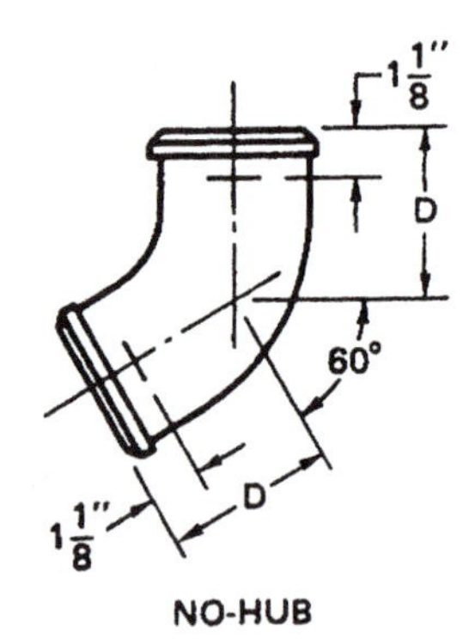

NO-HUB

1/6 Bend

Size (inches)	Hub			No-Hub
	D (inches)	C (inches)	X (inches)	D (inches)
2	4 3/4	2	4 1/2	3 1/4
3	5 1/2	2 1/2	5 1/4	3 1/2
4	6 1/4	2 3/4	5 3/4	3 7/8

DATA 14
BENDS – 1/16 – HUB AND NO-HUB

D is Fitting-Allowance Measure (long run)
C is Fitting-Allowance Measure (short run)
X is Center-to-Face Measure

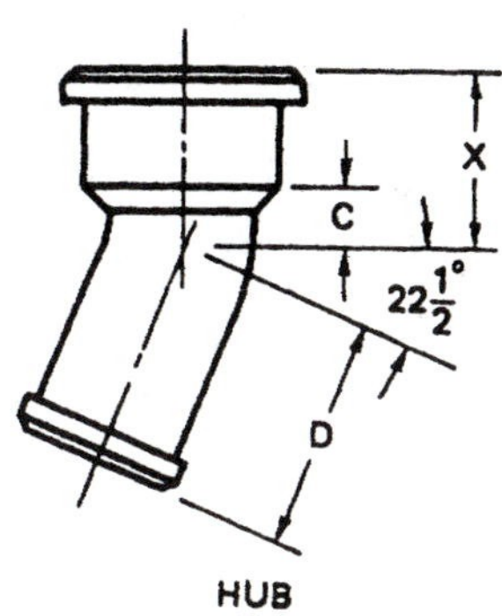

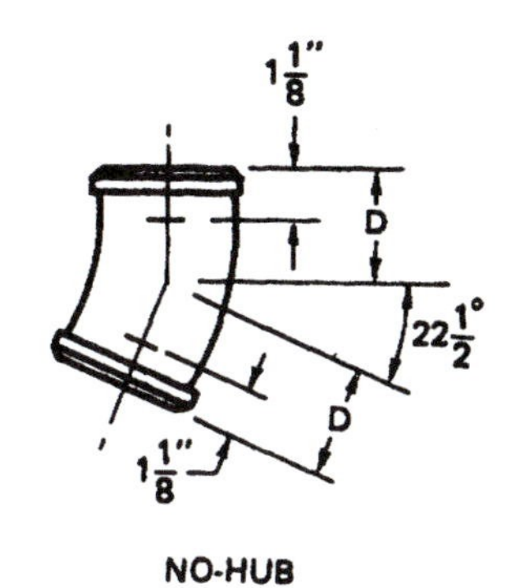

1/16 Bend

Size (inches)	Hub			No-Hub
	D (inches)	C (inches)	X (inches)	D (inches)
2	3 5/8	7/8	3 3/8	2 1/8
3	4 1/4	1 1/4	4	2 1/4
4	4 3/4	1 1/4	4 1/4	2 3/8

DATA 15
WYE-BRANCHES – HUB AND NO-HUB

D is Fitting-Allowance Measure (long run)
C is Fitting-Allowance Measure (short run)
E is Fitting-Allowance Measure (branch)
X is Center-to-Face Measure

Size (inches)	Hub			
	D (inches)	C (inches)	E (inches)	X (inches)
2	4	4	4	6 1/2
3	5 1/2	5	5 1/2	8 1/4
4	6 3/4	5 1/4	6 3/4	9 3/4
3 x 2	4 3/4	4 1/4	5	7 5/8
4 x 2	5 3/8	3 5/8	5 3/4	8 3/8
4 x 3	6	4 1/2	6 1/4	9

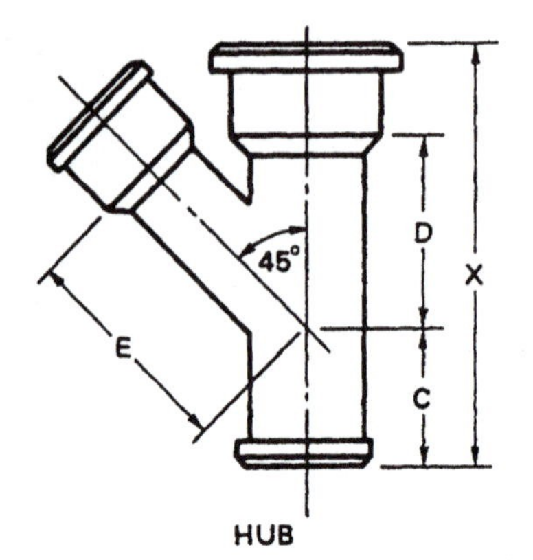

Size (inches)	No-Hub		
	D (inches)	C (inches)	E (inches)
2	4 5/8	2	4 5/8
3	5 3/4	2 1/4	5 3/4
4	7	2 1/2	7
3 x 2	5 1/8	1 1/2	5 1/4
4 x 2	5 5/8	1	6
4 x 3	6 1/4	1 3/4	6 1/2

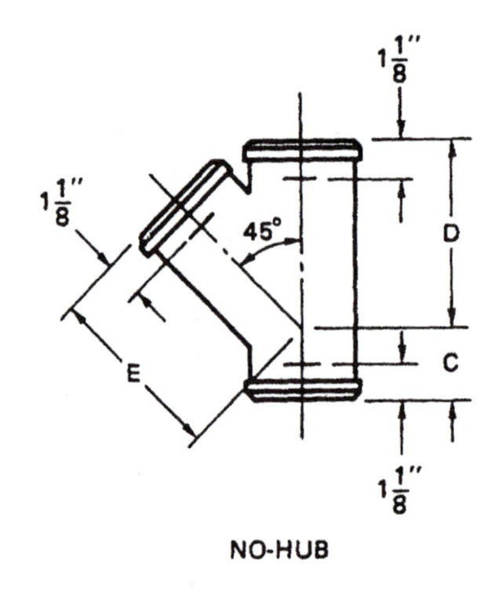

DATA 16
TEE WYE-BRANCHES – HUB AND NO-HUB

D is Fitting-Allowance (long run)
C is Fitting-Allowance (short run)
E is Fitting-Allowance (branch)
X is Center-to-Face Measure (short run)

Size (inches)	Hub			
	D (inches)	C (inches)	E (inches)	X (inches)
2	8	1 3/4	2 3/4	4 1/4
3	10	2 1/2	4	5 1/4
4	11	3	4 1/2	6
3 x 2	9	2	4	4 3/4
4 x 2	9	2	4 1/2	5
4 x 3	10	2 1/2	4 1/2	5 1/2

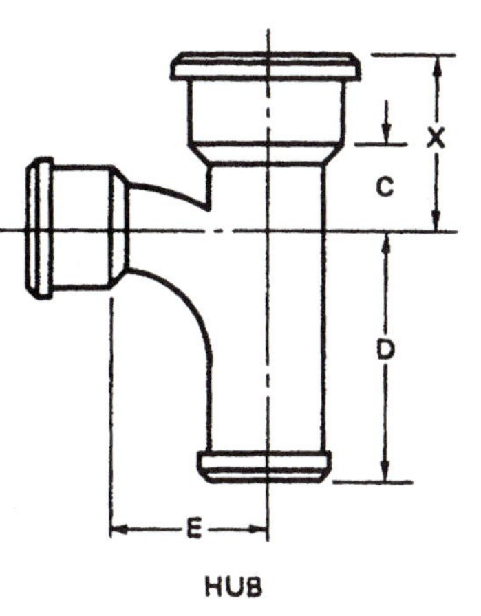

Size (inches)	No-Hub		
	D (inches)	C (inches)	E (inches)
2	4 1/2	2 3/8	4 1/2
3	5	3	5
4	5 1/2	3 5/8	5 1/2
3 x 2	4 1/2	2 3/8	5
4 x 2	4 1/2	2 3/8	5 1/2
4 x 3	5	3	5 1/2

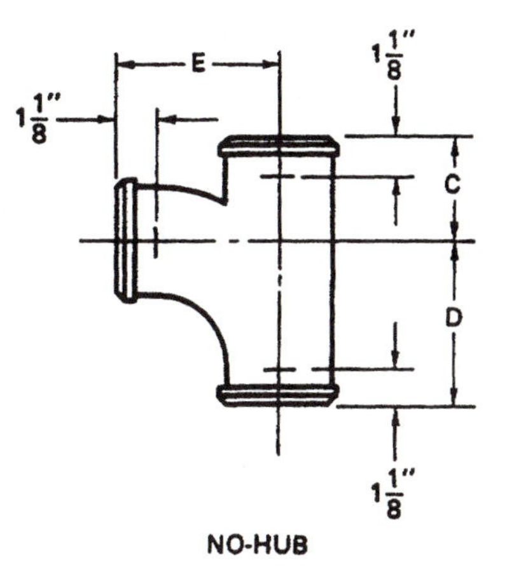

DATA 17
TAPPED TEES, HUB AND TAPPED TEE WYES, NO-HUB

D is Fitting-Allowance Measure (long run)
C is Fitting-Allowance Measure (short run)
E is Center-to-Face Measure (branch)
B is Thread-in Measure

Size (inches)	Tapped Tees Hub				
	D (inches)	C (inches)	E (inches)	B (inches)	I.P.S. Tapping (inches)
2 x 2	6 1/4	1 3/4	2	1/2	1 1/4, 1 1/2, 2
3 x 1 1/2 3 x 2	7	2	2 1/2	1/2	1 1/4, 1 1/2, 2
4 x 1 1/2 4 x 2	7	2	3	1/2	1 1/4, 1 1/2, 2

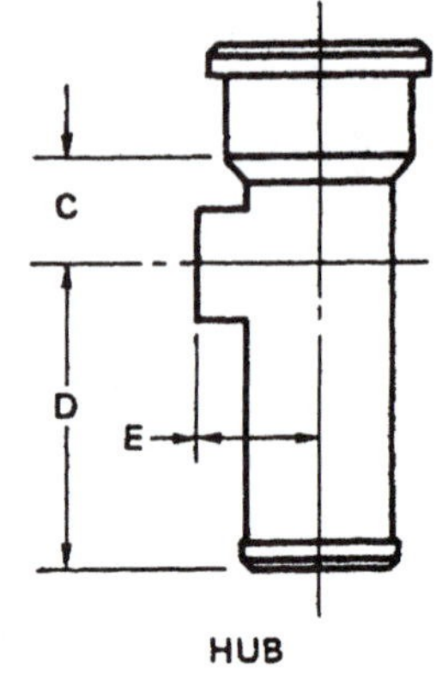

Size (inches)	Tapped Tee Wyes No-Hub			
	D (inches)	C (inches)	E (inches)	B (inches)
2 x 1 1/2	3 1/4	2 1/2	2 3/4	1/2
2 x 2	3 3/4	2 5/8	3 1/8	1/2
3 x 1 1/2	3 1/4	2 1/2	3 3/8	1/2
3 x 2	3 3/4	2 5/8	3 5/8	1/2
4 x 1 1/2	3 1/4	2 1/2	3 7/8	1/2
4 x 2	3 3/4	2 5/8	4 1/8	1/2

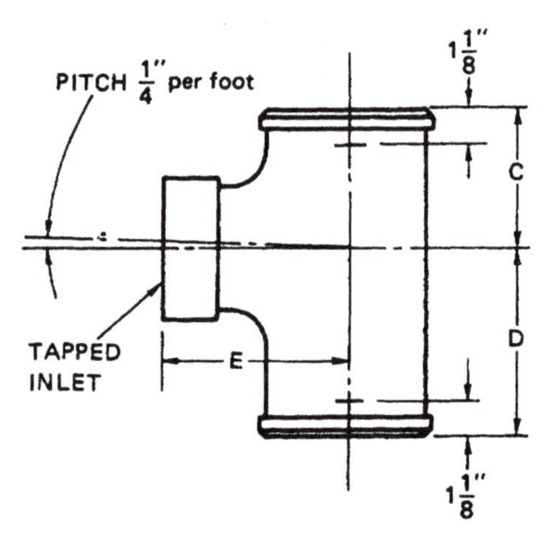

Note: A tee wye used as a vent connection is installed with the long run in the up position. The fitting is installed upside down compared to a drain when used as a vent fitting.

DATA 18
SLIP- AND CAULK-TYPE DRAINAGE FITTINGS

C is Center-to-Face Measure (long run and branch)
D is Fitting-Allowance Measure (short run) (includes 2 thread-in allowances)
B is Thread-in Measure

Pipe Size (inches)	D (inches)	C (inches)	B (inches)
1 1/2	3/4	2 3/4	1/2
2	3/4	3	1/2

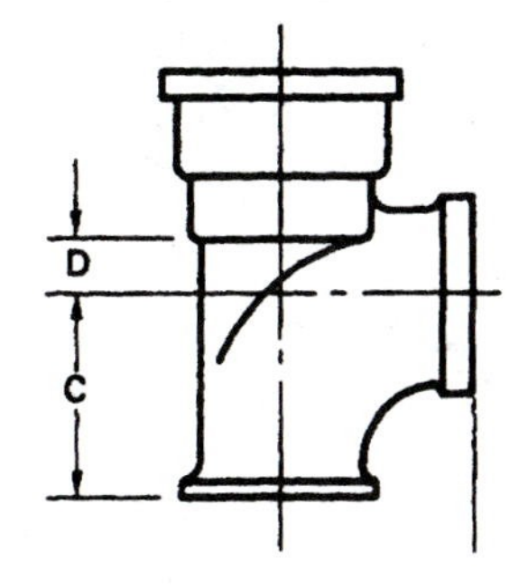

DATA 19
FLANGED FITTINGS

A is Center-to-Face Measure
C is Center-to-Face Measure
D is Face-to-Face Measure
E is Center-to-Face (long run and branch)
F is Center-to-Face (short run)
B is Thread-in Measure for Screwed Flanges

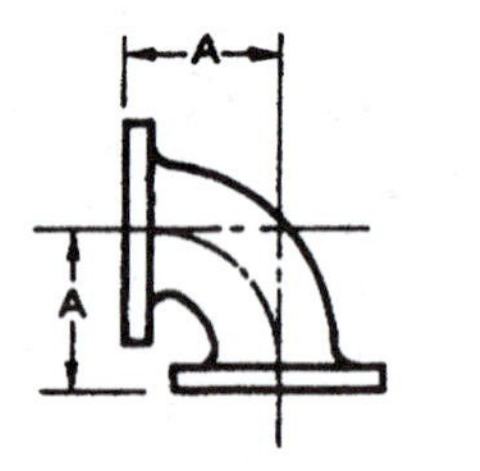

90° Elbow

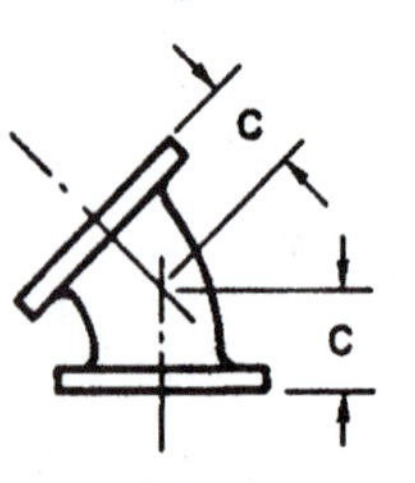

45° Elbow

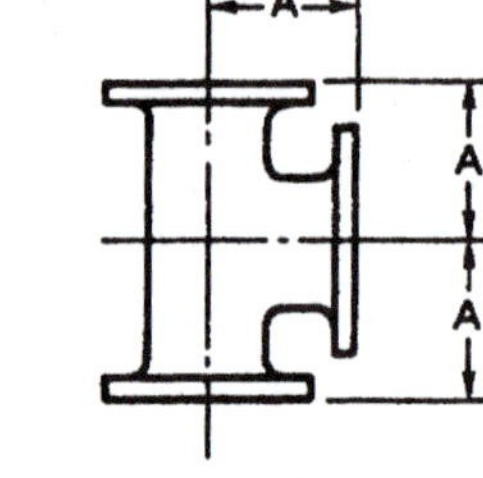

Tee

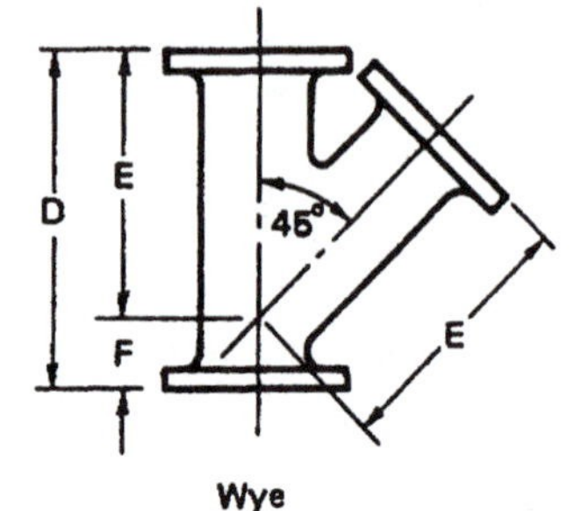

Wye

Size (inches)	A (inches)	C (inches)	D (inches)	E (inches)	F (inches)	B (inches)
1 1/2	4	2 1/4				1/2
2	4 1/2	2 1/2	10 1/2	8	2 1/2	1/2
2 1/2	5	3	12	9 1/2	2 1/2	3/4
3	5 1/2	3	13	10	3	1
3 1/2	6	3 1/2				
4	6 1/2	4	15	12	3	1
5	7 1/2	4 1/2	17	13 1/2	3 1/2	
6	8	5	18	14 1/2	3 1/2	
8	9	5 1/2	22	17 1/2	4 1/2	
10	11	6 1/2	25 1/2	20 1/2	5	
12	12	7 1/2	30	24 1/2	5 1/2	
14	14	7 1/2				
16	15	8				

Screwed Flanges

Y is Length through Hub
O is Diameter
Q is Thickness
B is Thread-in

Size (inches)	B (inches)	O (inches)	Q (inches)	Y (inches)	Number of Bolts
1	½	4¼	7⁄16	11⁄16	4
1¼	½	4⅝	½	13⁄16	4
1½	½	5	9⁄16	⅞	4
2	½	6	⅝	1	4
2½	¾	7	11⁄16	1⅛	4
3	1	7½	¾	1 3⁄16	4
3½	1	8½	13⁄16	1¼	8
4	1	9	15⁄16	1 5⁄16	8

DATA 20
CONSTANTS FOR 45° FITTINGS

VERTICAL

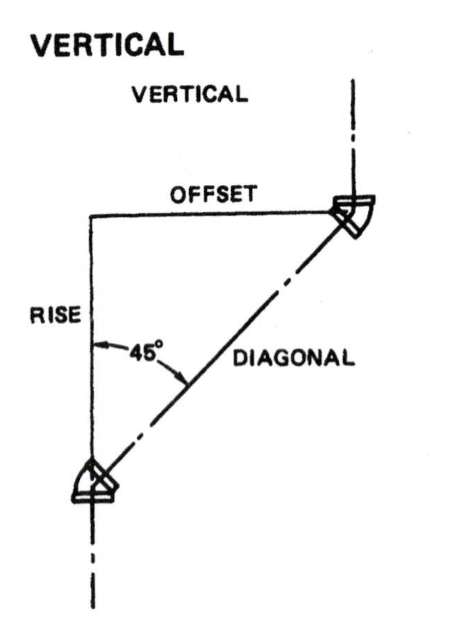

Offset Measured
Rise = offset
Diagonal = 1.414 X offset

Rise Measured
Offset = rise
Diagonal = 1.414 X rise

Diagonal Measured
Rise = 0.707 X diagonal
Offset = 0.707 X diagonal

HORIZONTAL

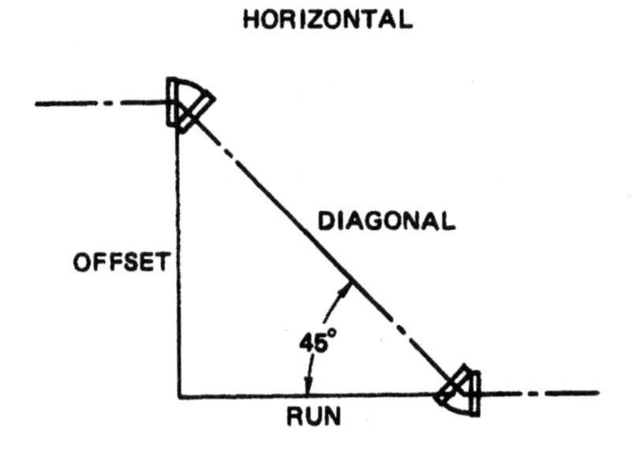

Offset Measured
Run = offset
Diagonal = 1.414 X offset

Run Measured
Offset = run
Diagonal = 1.414 X run

Diagonal Measured
Run = 0.707 X diagonal
Offset = 0.707 X diagonal

DATA 21
CONSTANTS FOR 60° FITTINGS

VERTICAL

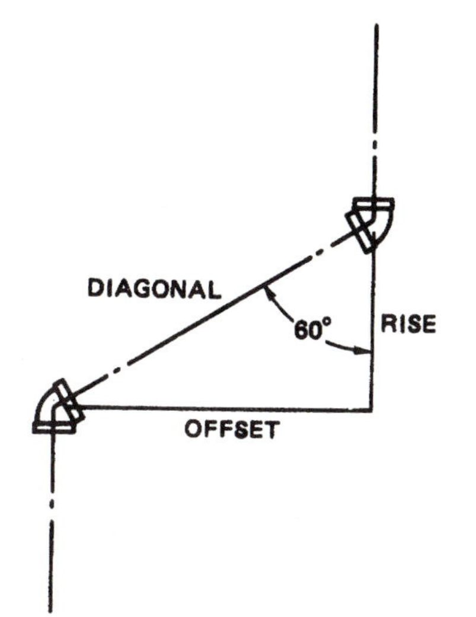

Offset Measured
Rise = 0.577 X offset
Diagonal = 1.155 X offset

Rise Measured
Diagonal = 2 X rise
Offset = 1.732 X rise

Diagonal Measured
Rise = 0.5 X diagonal
Offset = 0.866 X diagonal

HORIZONTAL

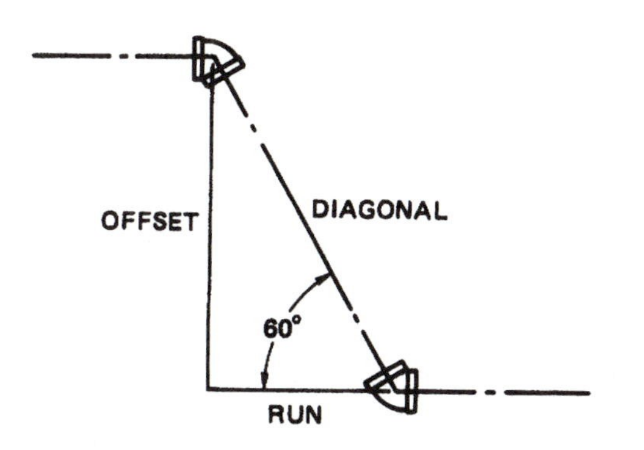

Offset Measured
Run = 0.577 X offset
Diagonal = 1.155 X offset

Run Measured
Diagonal = 2 X run
Offset = 1.732 X run

Diagonal Measured
Run = 0.5 X diagonal
Offset = 0.866 X diagonal

DATA 22
CONSTANTS FOR 22 1/2° FITTINGS

VERTICAL

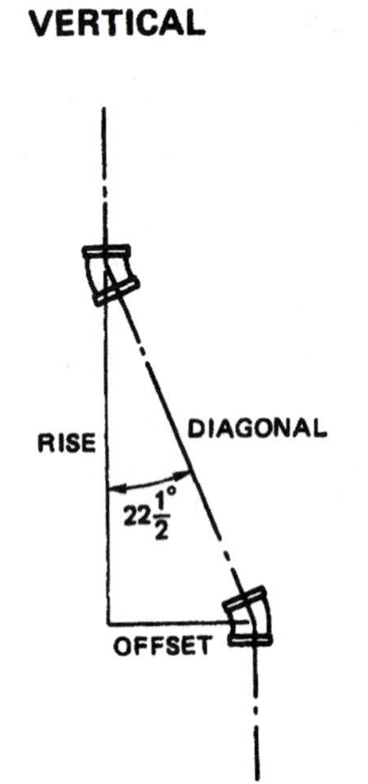

Offset Measured
Rise = 2.414 X offset
Diagonal = 2.613 X offset

Rise Measured
Offset = 0.414 X rise
Diagonal = 1.082 X rise

Diagonal Measured
Rise = 0.924 X diagonal
Offset = 0.383 X diagonal

HORIZONTAL

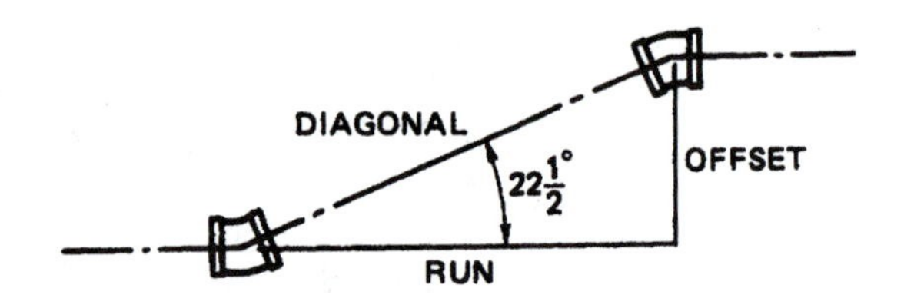

Offset Measured
Run = 2.414 X offset
Diagonal = 2.613 X offset

Run Measured
Offset = 0.414 X run
Diagonal = 1.082 X run

Diagonal Measured
Run = 0.924 X diagonal
Offset = 0.383 X diagonal

DATA 23
CONSTANTS FOR 11 1/4° FITTINGS

VERTICAL

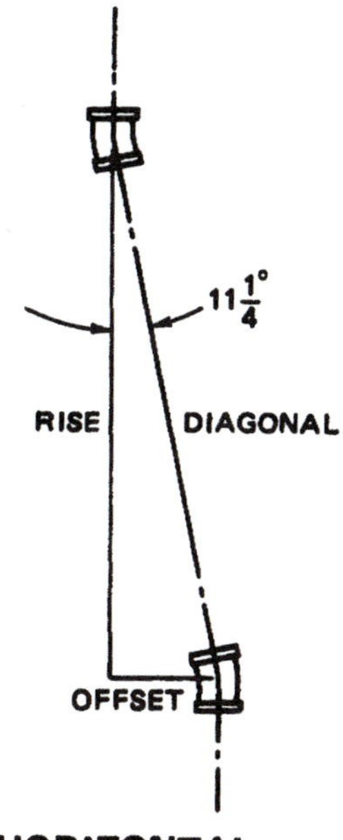

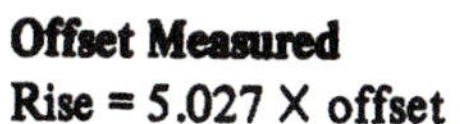

Offset Measured
Rise = 5.027 X offset
Diagonal = 5.126 X offset

Rise Measured
Offset = 0.199 X rise
Diagonal = 1.02 X rise

Diagonal Measured
Rise = 0.981 X diagonal
Offset = 0.195 X diagonal

HORIZONTAL

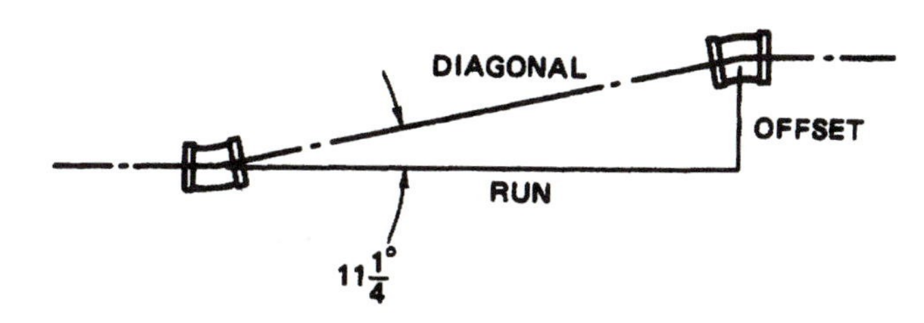

Offset Measured
Run = 5.027 X offset
Diagonal = 5.126 X offset

Run Measured
Offset = 0.199 X run
Diagonal = 1.02 X run

Diagonal Measured
Run = 0.981 X diagonal
Offset = 0.195 X diagonal

DATA 24
CONSTANTS FOR 1/5 BENDS

VERTICAL

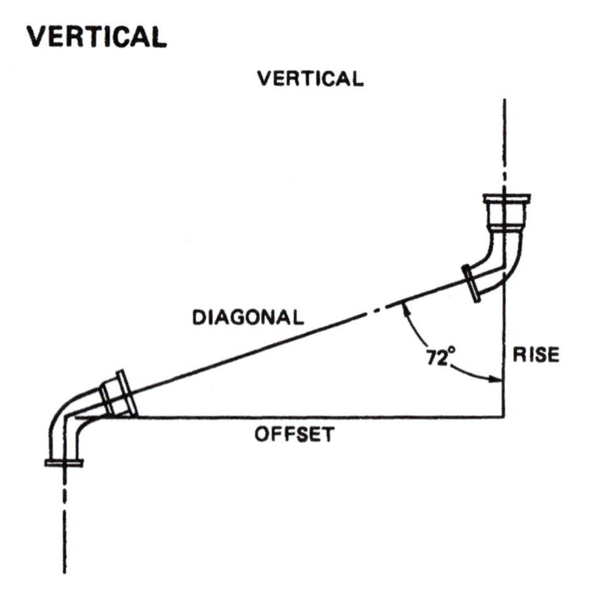

Offset Measured
Rise = 0.325 X offset
Diagonal = 1.052 X offset

Rise Measured
Offset = 3.078 X rise
Diagonal = 3.236 X rise

Diagonal Measured
Rise = 0.309 X diagonal
Offset = 0.951 X diagonal

HORIZONTAL

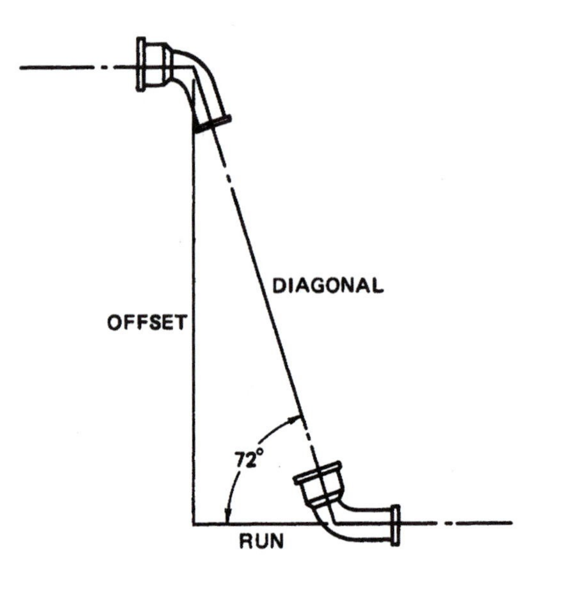

Offset Measured
Run = 0.325 X offset
Diagonal = 1.052 X offset

Run Measured
Offset = 3.078 X run
Diagonal = 3.236 X run

Diagonal Measured
Run = 0.309 X diagonal
Offset = 0.951 X diagonal

DATA 25
CONSTANTS FOR PARALLEL OFFSETS

Fitting Angle	**90°**	**72°**	**60°**	**45°**	**22 1/2°**	**11 1/4°**
Diagonal = Offset X	–	1.052	1.155	1.414	2.613	5.126
Rise (Run) = Offset X	–	0.325	0.577	1.000	2.414	5.027
Parallel Angle	45°	36°	30°	22 1/2°	11 1/4°	5 5/8°
Difference in Length = Spread X	1.000	0.727	0.577	0.414	0.199	0.098

DATA 26
CONSTANTS FOR ROLLING OFFSETS

Fitting Angle	90°	72°	60°	45°	22 1/2°	11 1/4°
Diagonal = true offset X	1.000	1.052	1.155	1.414	2.613	5.126
Setback = true offset X	0	0.325	0.577	1.000	2.414	5.027

DATA 27
TRIGONOMETRIC FUNCTIONS

Angle	Sine	Cosine	Tangent	Cotangent	Secant	Cosecant
1	0.0175	0.9998	0.0175	57.2900	1.0002	57.2987
2	0.0349	0.9994	0.0349	28.6363	1.0006	28.6537
3	0.0523	0.9986	0.0524	19.0811	1.0014	19.1073
4	0.0698	0.9976	0.0699	14.3007	1.0024	14.3356
5	0.0872	0.9962	0.0875	11.4301	1.0038	11.4737
5 5/8	0.0980	0.9952	0.0985	10.1532	1.0048	10.2023
6	0.1045	0.9945	0.1051	9.5144	1.0055	9.5668
7	0.1219	0.9925	0.1228	8.1443	1.0075	8.2055
8	0.1392	0.9903	0.1405	7.1154	1.0098	7.1853
9	0.1564	0.9877	0.1584	6.3138	1.0125	6.3925
10	0.1736	0.9848	0.1763	5.6713	1.0154	5.7588
11	0.1908	0.9816	0.1944	5.1446	1.0187	5.2408
11 1/4	0.1951	0.9808	0.1989	5.0273	1.0196	5.1258
12	0.2079	0.9781	0.2126	4.7046	1.0223	4.8097
13	0.2250	0.9744	0.2309	4.3315	1.0263	4.4454
14	0.2419	0.9703	0.2493	4.0108	1.0306	4.1336
15	0.2588	0.9659	0.2679	3.7321	1,0353	3.8637
16	0.2756	0.9613	0.2867	3.4874	1.0403	3.6280
17	0.2924	0.9563	0.3057	3.2709	1.0457	3.4203
18	0.3090	0.9511	0.3249	3.0777	1.0515	3.2361
19	0.3256	0.9455	0.3443	2.9042	1.0576	3.0716
20	0.3420	0.9397	0.3640	2.7475	1.0642	2.9238
21	0.3584	0.9336	0.3839	2.6051	1.0711	2.7904
22	0.3746	0.9272	0.4040	2.4751	1.0785	2.6695
22 1/2	0.3827	0.9239	0.4142	2.4142	1.0824	2.6131
23	0.3907	0.9205	0.4245	2.3559	1.0864	2.5593
24	0.4067	0.9135	0.4452	2.2460	1.0946	2.4586
25	0.4226	0.9063	0.4663	2.1445	1.1034	2.3662
26	0.4384	0.8988	0.4877	2.0503	1.1126	2.2812
27	0.4540	0.8910	0.5095	1.9626	1.1223	2.2027
28	0.4695	0.8829	0.5317	1.8807	1.1326	2.1301
29	0.4848	0.8746	0.5543	1.8040	1.1434	2.0627
30	0.5000	0.8660	0.5774	1.7321	1.1547	2.0000
31	0.5150	0.8572	0.6009	1.6643	1.1666	1.9416
32	0.5299	0.8480	0.6249	1.6003	1.1792	1.8871
33	0.5446	0.8737	0.6494	1.5399	1.1924	1.8361

(*Continued*)

DATA 27 (Continued)

Angle	Sine	Cosine	Tangent	Cotangent	Secant	Cosecant
34	0.5592	0.8290	0.6745	1.4826	1.2062	1.7883
35	0.5736	0.8192	0.7002	1.4281	1.2208	1.7434
36	0.5878	0.8090	0.7265	1.3764	1.2361	1.7013
37	0.6018	0.7986	0.7536	1.3270	1.2521	1.6616
38	0.6157	0.7880	0.7813	1.2799	1.2690	1.6243
39	0.6293	0.7771	0.9098	1.2349	2.2868	1.5890
40	0.6428	0.7660	0.8391	1.1918	1.3054	1.5557
41	0.6561	0.7547	0.8693	1.1504	1.3250	1.5243
42	0.6691	0.7431	0.9004	1.1106	1.3456	1.4945
43	0.6820	0.7314	0.9325	1.0724	1.3673	1.4663
44	0.6947	0.7193	0.9657	1.0355	1.3902	1.4396
45	0.7071	0.7071	1.0000	1.0000	1.4142	1.4142
46	0.7193	0.6947	1.0355	0.9657	1.4396	1.3902
47	0.7314	0.6820	1.0724	0.9325	1.4663	1.3673
48	0.7431	0.6691	1.1106	0.9004	1.4945	1.3456
49	0.7547	0.6561	1.1504	0.8693	1.5243	1.3250
50	0.7660	0.6428	1.1918	0.8391	1.5557	1.3054
51	0.7771	0.6293	1.2349	0.9098	1.5890	1,2868
52	0.7880	0.6157	1.2799	0.7813	1.6243	1.2690
53	0.7986	0.6018	1.3270	0.7536	1.6616	1.2521
54	0.8090	0.5878	1.3764	0.7265	1.7013	1.2361
55	0.8192	0.5736	1,4281	0.7002	1.7434	1.2208
56	0.8290	0.5592	1.4826	0.6745	1.7883	1.2062
57	0.8387	0.5446	1.5399	0.6494	1.8361	1.1924
58	0.8480	0.5299	1.6003	0.6249	1.8871	1.1792
59	0.8572	0.5150	1.6643	0.6009	1.9416	1.1666
60	0.8660	0.5000	1.7321	0.5774	2.0000	1.1547
61	0.8746	0.4848	1.8040	0.5543	2.0627	1.1434
62	0.8829	0.4695	1.8807	0.5317	2.1301	1.1326
63	0.8910	0.4540	1.9626	0.5095	2.2027	1.1223
64	0.8988	0.4384	2.0503	0.4877	2.2812	1.1126
65	0.9063	0.4226	2.1445	0.4663	2.3662	1.1034
66	0.9135	0.4067	2.2460	0.4452	2.4586	1.0946
67	0.9205	0.3907	2.3559	0.4245	2.5593	1.0864
68	0.9272	0.3746	2.4751	0.4040	2.6695	1.0785
69	0.9336	0.3584	1.6051	0.3839	2.7904	1.0711
70	0.9397	0.3420	2.7475	0.3640	2.9238	1.0642
71	0.9455	0.3256	2.9042	0.3443	3.1916	1.0576

DATA 27 (Continued)

Angle	Sine	Cosine	Tangent	Cotangent	Secant	Cosecant
72	0.9511	0.3090	3.0777	0.3249	3.2361	1.0515
73	0.9563	0.2924	3.2709	0.3057	3.4203	1.0457
74	0.9613	0.2756	3.4874	0.2867	3.6280	1.0403
75	0.9659	0.2588	3.7321	0.2679	3.8637	1.0353
76	0.9703	0.2419	4.0108	0.2493	4.1336	1.0306
77	0.9744	0.2250	4.3315	0.2309	4.4454	1.0263
78	0.9781	0.2079	4.7046	0.2126	4.8097	1.0223
79	0.9816	0.1908	5.1446	0.1944	5.2408	1.0187
80	0.9848	0.1736	5.6713	0.1763	5.7588	1.0154
81	0.9877	0.1564	6.3138	0.1584	6.3925	1.0125
82	0.9903	0.1392	7.1154	0.1405	7.1853	1.0098
83	0.9925	0.1219	8.1443	0.1228	8.2055	1.0075
84	0.9945	0.1045	9.5144	0.1051	9.5668	1.0055
85	0.9962	0.0872	11.4301	0.0875	11.4737	1.0038
86	0.9976	0.0698	14.3007	0.0699	14.3356	1.0024
87	0.9986	0.0523	19.0811	0.0524	19.1073	1.0014
88	0.9994	0.0349	28.6363	0.0349	28.6537	1.0006
89	0.9998	0.0175	57.2900	0.0175	57.2987	1.0002

APPENDIX B

Using the Calculator

The calculator is a device for doing arithmetic. It is handy as a time saver and increases accuracy. The groups of problems presented in the following table provide some practice in using a calculator. The first problem in each group shows the correct key sequence to use. Use the same key sequence for the other problems given in the group and check your answer against the one shown in the **Display** column. The usual key sequence is the same as the problem statement in words or in formula shorthand.

Problem	Calculator Entry	Display
A. Clear Calculator	[ON/C]	0
B. Add		
1. 28 + 4.5 + 127	28 [+] 4 [.] 5 [+] 127 [=]	159.5
2. 57 + 32.5 + 29.25		118.75
3. $16.87 + $156.75		173.62
4. 2319 + 37625		39944
C. Subtract		
1. 129.67 less 18.7	129 [.] 67 [−] 18 [.] 7 =	110.97
2. 83.5 – 21.75 – 11		50.75
3. $133.50 – $16.62		116.88
D. Multiply		
1. 13 x 15.5 x 3	13 [X] 15 [.] 5 [X] 3 =	604.5
2. 17.75 x 17.75 x 3.1416		989.80035
3. 13 X 1.414		18.382
4. $26.79 x 7		187.53

E. Divide		
1. 53.5 ÷ 4	53 [.] 5 [÷] 4 [=]	13.375
2. $67.50 ÷ 3		22.5
3. 83" ÷ 3.1416		26.419658
F. Square		
1. 8.75^2	8 [.] 75 [×] 8 [.] 75 [=]	76.5625
2. 19.5^2		380.25
G. Square Root		
1. $\sqrt{30.625}$	30 [.] 625 [√]	5.5339859
2. Sq. Rt. of 123.6		11.117553
H. Percent		
1. $9 is what % of $125?	9 [÷] 125 [%]	7.2
2. 53.5 is what % of 460?		11.63043
3. 1234 is what % of 950?		129.89473
I. Chain Problems		
1. (19.875 + 12) ÷ 2 x 9	19 [.] 875 [+] 12 [÷] 2 [×] 9 [=]	143.4375
2. 26 x 7 - 89.75 + 26		118.25
J. Series of Problems All Using the Same Number *Note: Calculators vary in keying and function. Read the manufacturer's instructions if there are difficulties with answers.*		
1a. .067 x $87.50	[.] 067 [×] 87 [.] 5 [=]	5.86
1b. .067 x $128.20	128 [.] 2 [=]	8.59
1c. .067 x $215.98	215 [.] 98 [=]	14.47
2a. $875 ÷ .72	875 [÷] [.] 72 [=]	1215.28

2b. $1365.50 ÷ .72	1365 ⊡ 5 ⊟	1896.53
2c. $75.80 ÷ .72	75 ⊡ 8 ⊟	105.28
K. Step Problems		
1. $\sqrt{9^2 + 12^2}$	Step 1: 9 ⊠ 9 [M+]	M 81
	Step 2: 12 ⊠ 12	M 144
	Step 3: [M^R_C]	M 225
	Step 4: [√]	M 15
Clear Memory and Calculator	[M^R_C] [M^R_C] [ON/C]	0.
2. $\sqrt{19.875^2 + 27.5^2}$	Step 1:	M 395.01562
	Step 2:	M 756.25
	Step 3:	M 1151.2656
	Step 4:	M 33.930305
3. 26.5 x 26.5–(3.14 x 11 x 11)	Step 1: 26 ⊡ 5	
	⊠ 26 ⊡ 5 [M+]	M 702.25
	Step 2: 3 ⊡ 14	
	⊠ 11 ⊠ 11 [M–]	M 379.94
	Step 3: [M^R_C]	M 322.31
4. 9" x 8"	Step 1:	M 72
– 4 (1.5" x 1.25")	Step 2:	M 7.5
– 5 (3.14 x .75" x .75")	Step 3:	M 8.83
	Step 4:	M 55.67